# Machine Learning

Rodrigo Fernandes de Mello
Moacir Antonelli Ponti

# Machine Learning

## A Practical Approach on the Statistical Learning Theory

 Springer

Rodrigo Fernandes de Mello
Institute of Mathematics
and Computer Science
University of São Paulo
São Carlos, São Paulo, Brazil

Moacir Antonelli Ponti
Institute of Mathematics
and Computer Science
University of São Paulo
São Carlos, São Paulo, Brazil

ISBN 978-3-030-06949-0        ISBN 978-3-319-94989-5   (eBook)
https://doi.org/10.1007/978-3-319-94989-5

This Springer imprint is published by the registered company Springer Nature Switzerland AG
The registered company address is: Gewerbestrasse 11, 6330 Cham, Switzerland

*"To my lovely wife Claudia for her unconditional support along all these great years, my mother and my father for all background, and specially to every person who has dedicated time, effort, and love to study."*

— *Rodrigo Fernandes de Mello*

*"The dedication of this book is split five ways: to Paula, love of all past times and book; to my mother Clara who always believed that I could do anything; to you (the reader) the theory and practice I wanted someone had taught me 15 years ago; to caffeine for making it happen; and for the most beautiful of the goddesses."*

— *Moacir Antonelli Ponti*

# Foreword

Although Machine Learning is currently a hot topic, with several research break-throughs and innovative applications, the area is missing a literature that provides, in a practical and accessible way, theoretical issues that provided the foundation of the area. Statistical learning theory plays an important role in many research breakthroughs in machine learning. There is a classical and excellent literature in the area. However, most of them assume a strong mathematical and statistical knowledge. The book, written by Rodrigo Fernandes de Mello and Moacir Antonelli Ponti, both with much experience in the area, provides an innovative and clear approach to present the main concepts in statistical learning theory. The authors were very competent in the choice of the issues to be covered and on how to explain them in an easy and didactic way. The idea of providing elegant algorithms for the implementation of several important issues in statistical learning theory, the book motivates the reader and demystifies the impression that this is a dry and difficult issue. Using the R Statistical Software, the examples of source code are easy to follow and give deep insights to the reader. This book will be a very important source of knowledge for those interested in learning and pursuing a career in machine learning.

The authors present concepts, methods, and algorithms in a simple and intuitive way. The textbook could be a relevant tool for master's and PhD students interested in understanding and exploring the theoretical aspects of ML.

LTCI, Télécom ParisTech, Paris, France            Albert Bifet
University of São Paulo, São Paulo, Brazil    André Carlos P. de L. F. de Carvalho
University of Porto, Porto, Portugal                 João Gama

# Acknowledgments

This book was only possible due to the direct and indirect cooperation of several friends, colleagues, and students throughout the last 7 years, after deciding to propose and to teach the course on Statistical Learning Theory to postgraduate students at the Institute of Mathematics and Computer Science at the University of São Paulo, Brazil.

While studying several books and articles on theoretical aspects of machine learning, linear algebra, linear and convex optimization, and support vector machines, we decided to offer a first version of such postgraduate course, including and detailing all necessary subjects so any student could picture the theoretical foundation of supervised learning and implement its most relevant algorithm, i.e., the support vector machine. Several course amendments were made based on the questions from students and further discussions with friends and colleagues, including Tiago Santana de Nazaré who motivated us to develop the theoretical formulation for the distance-weighted nearest neighbors, Felipe Simões Lage Gomes Duarte and Gabriel de Barros Paranhos da Costa who helped us to improve slides and pay attention to additional material which was included later on, Thiago Bianchi for the discussions, and Victor Hugo Cândido de Oliveira and Daniel Moreira Cestari for helping with slide improvements as well as theoretical discussions. We thank Carlos Henrique Grossi Ferreira for supporting the discussions and helping us to prove the general-purpose shattering coefficient.

We also thank our students and collaborators who somehow supported us during this work: Martha Dais Ferreira, Débora Cristina Corrêa, Lucas de Carvalho Pagliosa, Yule Vaz, Ricardo Araújo Rios, Cássio Martini Martins Pereira, Rosane Maria Maffei Vallim, André Mantini Eberle, Eduardo Alves Ferreira, Vinícius de Freitas Reis, José Augusto Andrade Filho, Matheus Lorenzo dos Santos, Fausto Guzzo da Costa, Marcelo Keese Albertini, Paulo Henrique Ribeiro Gabriel, Renato Porfirio Ishii, Evgueni Dodonov (in memoriam), Anderson Caio Santos Silva, Fernando Pereira dos Santos, Luiz Fernando Coletta, Camila Tatiana Picon, Isadora Rossi, Mateus Riva, Leonardo Sampaio Ferraz Ribeiro, Gabriela Thumé, and Fábio Rodrigues Jorge.

After four consecutive years of this course on statistical learning theory, we now share this textbook to bring those concepts and practical aspects to the whole community. We expect this content to help students and researchers to understand and open new paths for the machine learning area.

# Contents

# Acronyms

# Chapter 1
# A Brief Review on Machine Learning

## 1.1 Machine Learning Definition

The area of Machine Learning (ML) is interested in answering how a computer can "learn" specific tasks such as recognize characters, support the diagnosis of people under severe diseases, classify wine types, separate some material according to its quality (e.g. wood could be separated according to its weakness, so it could be later used to build either pencils or houses). Those and many other applications make evident the usefulness of ML to tackle daily problems and support specialists to take decisions, attracting researchers from different areas of knowledge [1, 17].

In this situation you may ask: but how can a computer learn? Or how can one ensure an algorithm is in fact learning? Before answering such questions we should start by addressing human learning about some subject. Consider a child looking at a chair for the first time ever. How could a child associate the physical object to the concept of a chair? Parents or relatives would first tell him/her that the object being observed is a chair, but (s)he must find a way to extract relevant features to characterize it as a chair. For example: is color an important variable in order to define the concept chair? Well, probably not because one chair could be black, but there are chairs in several other colors. As an alternative, someone would say the shape provides an important feature. Yes, the shape would be indeed a more adequate feature to associate this object to the class or label "chair". By analyzing this simple example, we notice human beings rely on features (also called attributes or even variables) to build up some type of classification function (or simply classifier) $f$ to map an object $x$ into a label $y$, i.e. $f(x) = y$.

As humans, we are constantly observing different scenarios and attempting to extract relevant features to take conclusions on different tasks. For example, based on the variables of temperature and humidity, we could conclude about the feasibility of playing soccer this evening [8]. We refer to those variables as **input variables** because they will be taken as input to learn some concept (e.g. viability of playing soccer). In this example, the input variables are only associated to weather

© Springer International Publishing AG, part of Springer Nature 2018
R. Fernandes de Mello, M. Antonelli Ponti, *Machine Learning*,
https://doi.org/10.1007/978-3-319-94989-5_1

**Fig. 1.1** Letter "A" in a
bitmap representation

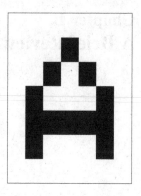

conditions (temperature and humidity), but we could also include health conditions (e.g. past injuries, pain). As in this situation we simply want to predict whether some person will play soccer or not, the **output variable** (a.k.a. class or label) is either positive (yes) or negative (no). Therefore, based on our past experiences and given the current input variables, we must decide whether the output is either yes or no.

There are more complex situations in which the output variable can assume more than two labels. For example, consider a bitmap representing a letter from the alphabet (for instance the letter "A" in Fig. 1.1). Now consider the output variable as an integer value, which assumes 0 for "A", 1 for "B", and so on. This is the sort of situation we have more than two classes. But how could we learn those bitmap representations for the whole alphabet?

To make the computer learn the alphabet representation, we should first build up a **training set** containing $n$ examples. This term "example" is used to refer to the features provided to represent a given letter so it can be learned. For now, let us consider the features as the image of such a letter. We could have letter "A" graphed in different manners, every one constituting one different example (Fig. 1.2 illustrates one example of each letter to be learned). In order to make a simple representation, we will consider letter "A" organized as a binary row vector, as listed in Table 1.1, having 1 as a black and 0 as a white pixel. This vector (a.k.a. **feature vector**) organization is typically used in ML to represent an example. We can see every element of this vector as associated to one input variable, as illustrated in Table 1.2. In this case, each input variable is the pixel value at a given space location (0 or 1). After defining such input variables for our first training example, we add an extra column to our table to represent the output variable, which corresponds to the class (see the last variable in Table 1.2).

Before moving on, we recall a summary of the most important terminology used until now:

1. an **example** (also known as instance or observation) is an observed object of interest (e.g. a letter from the alphabet);
2. a **feature** (also known as attribute or characteristic) is an input variable that is used to characterize the example (e.g. black and white pixels for images);

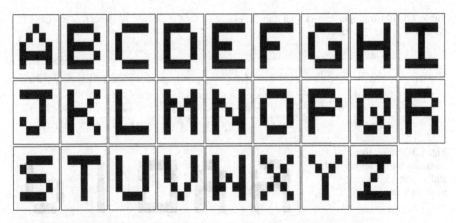

**Fig. 1.2** Alphabet in a bitmap representation

**Table 1.1** Letter "A" organized as a binary row vector

| A: | 1 | 1 | 1 | 1 | 1 | 1 | 1 | 1 | 1 | 1 | 0 | 1 | 1 | 1 | 1 | 1 | 0 | 1 | 0 | 1 | 1 | 1 | 0 | 1 | 1 | 1 | 0 | 1 | 1 | 0 | 0 | 0 | 0 | 0 | 1 | 1 | 0 | 1 | 1 | 1 | 0 | 1 | 1 | 1 | 1 | 1 | 1 | 1 | 1 |
|---|---|---|---|---|---|---|---|---|---|---|---|---|---|---|---|---|---|---|---|---|---|---|---|---|---|---|---|---|---|---|---|---|---|---|---|---|---|---|---|---|---|---|---|---|---|---|---|---|---|

**Table 1.2** Illustrating the variables composing the binary row vector for letter "A", in which $P_i$ refers to the $i$-th pixel of the bitmap

| Letter | $P_1$ | $P_2$ | $P_3$ | $P_4$ | $P_5$ | $P_6$ | $P_7$ | $P_8$ | $P_9$ | $P_{10}$ | $P_{11}$ | $P_{12}$ | $P_{13}$ | $P_{14}$ | $P_{15}$ | $P_{16}$ | $P_{17}$ | |
|---|---|---|---|---|---|---|---|---|---|---|---|---|---|---|---|---|---|---|
| A: | 1 | 1 | 1 | 1 | 1 | 1 | 1 | 1 | 1 | 1 | 0 | 1 | 1 | 1 | 1 | 1 | 0 | |
| | $P_{18}$ | $P_{19}$ | $P_{20}$ | $P_{21}$ | $P_{22}$ | $P_{23}$ | $P_{24}$ | $P_{25}$ | $P_{26}$ | $P_{27}$ | $P_{28}$ | $P_{29}$ | $P_{30}$ | $P_{31}$ | $P_{32}$ | $P_{33}$ | $P_{34}$ | |
| | 1 | 0 | 1 | 1 | 1 | 0 | 1 | 1 | 1 | 0 | 1 | 1 | 0 | 0 | 0 | 0 | 0 | |
| | $P_{35}$ | $P_{36}$ | $P_{37}$ | $P_{38}$ | $P_{39}$ | $P_{40}$ | $P_{41}$ | $P_{42}$ | $P_{43}$ | $P_{44}$ | $P_{45}$ | $P_{46}$ | $P_{47}$ | $P_{48}$ | $P_{49}$ | | | class |
| | 1 | 1 | 0 | 1 | 1 | 1 | 0 | 1 | 1 | 1 | 1 | 1 | 1 | 1 | 1 | | | 0 |

3. the **class** of an example is its label typically defined by some specialist of the application domain (e.g. class "A" has label 0);
4. a **classifier** is a function $f$ that maps the set of features $x$ into some label $y$, therefore it is expected to be capable of classifying examples which were never seen before (typically referred to as unseen examples).

By adding up more rows to Table 1.2, we obtain something similar to Table 1.3, which contains several examples to be learned (one per row). This is the notation considered throughout this book. This data table is commonly referred to as **dataset**, which contains both the input variables (features) and the **supervision**, i.e., the class or label, which was (most probably) defined by a specialist. Datasets like this are provided to a classification algorithm which must infer a classifier $f : X \rightarrow Y$, having $x_i \in X$ as the features of the example at row $i$ and $y_i \in Y$ as the corresponding class to be learned (last column). We expect this algorithm to find the best possible classifier for such a problem. In summary, this is how we typically approach ML tasks involving supervision (i.e., when we have classes) (Fig. 1.3).

**Table 1.3** Illustrating a dataset with training examples for the classes 'A' (label 0), 'B' (label 1), 'C' (label 2) and 'D' (label 3)

| Features | | | | | | | | | | | | | | | | | | | | | | | | | | | | | | | | | | | | | | | | | | | | | | | Label |
|---|---|---|---|---|---|---|---|---|---|---|---|---|---|---|---|---|---|---|---|---|---|---|---|---|---|---|---|---|---|---|---|---|---|---|---|---|---|---|---|---|---|---|---|---|---|---|---|---|
| 1 | 1 | 1 | 1 | 1 | 1 | 1 | 1 | 1 | 0 | 1 | 1 | 1 | 1 | 1 | 0 | 1 | 0 | 1 | 1 | 1 | 0 | 1 | 1 | 1 | 0 | 1 | 1 | 0 | 0 | 0 | 0 | 0 | 1 | 1 | 0 | 1 | 1 | 1 | 0 | 1 | 1 | 1 | 1 | 1 | 1 | 1 | | 0 |
| 1 | 1 | 1 | 1 | 1 | 1 | 1 | 1 | 1 | 0 | 0 | 1 | 1 | 1 | 1 | 0 | 1 | 1 | 0 | 1 | 1 | 1 | 0 | 0 | 0 | 0 | 1 | 1 | 1 | 0 | 1 | 1 | 0 | 1 | 1 | 0 | 1 | 1 | 1 | 0 | 1 | 1 | 0 | 1 | 1 | 1 | 1 | | 0 |
| 1 | 1 | 1 | 1 | 1 | 1 | 1 | 1 | 0 | 0 | 0 | 0 | 1 | 1 | 1 | 0 | 1 | 1 | 1 | 1 | 0 | 1 | 1 | 0 | 0 | 0 | 0 | 1 | 1 | 1 | 0 | 1 | 1 | 1 | 0 | 1 | 1 | 0 | 0 | 0 | 0 | 1 | 1 | 1 | 1 | 1 | 1 | | 1 |
| 1 | 1 | 1 | 1 | 1 | 1 | 1 | 1 | 0 | 0 | 0 | 0 | 1 | 1 | 0 | 1 | 0 | 1 | 1 | 1 | 1 | 1 | 1 | 0 | 1 | 1 | 1 | 1 | 1 | 1 | 0 | 1 | 1 | 1 | 1 | 1 | 1 | 0 | 0 | 0 | 0 | 1 | 1 | 1 | 1 | 1 | 1 | | 2 |
| 1 | 1 | 1 | 1 | 1 | 1 | 1 | 1 | 0 | 0 | 0 | 0 | 1 | 1 | 1 | 0 | 1 | 1 | 1 | 1 | 0 | 1 | 1 | 0 | 1 | 1 | 0 | 1 | 1 | 0 | 1 | 1 | 0 | 1 | 1 | 0 | 0 | 0 | 0 | 1 | 1 | 1 | 1 | 1 | 1 | 1 | 1 | | 3 |

**Fig. 1.3** Alternative instances for the letters of the alphabet in a bitmap representation ('A','A','C','I', and 'J')

There are other types of learning as briefly discussed in the next section, however the type we approach throughout this book is the one we have just introduced which is referred to as **supervised learning**. Just to motivate association, the reader is invited to think about real-world problems and how to obtain relevant features and define classes. This may even motivate philosophical thinking about how humans often define hard labels (sometimes even binary labels) for every day problems and how that could influence in prejudging and stereotyping things. Also observe that classes or labels are often dependent on the bias imposed by our relatives, friends and other peers, as well as the environment we live in.

## 1.2 Main Types of Learning

The area of Machine Learning is typically organized in two main branches: supervised learning; and non-supervised learning. **Supervised learning** was introduced in the previous section. In that type of learning, the ML algorithm receives pre-labeled input examples and intends to converge to the best as possible classifier $f : X \rightarrow Y$, so one can predict labels for unseen examples with high accuracy. **Non-supervised learning** is associated to the process of building up models after analyzing the similarities among input data [4]. For example, the clustering algorithm k-Means attempts to find $k$ representative groups according to the relative distance of points in $\mathbb{R}^m$ [1]. The main characteristic of this second type of learning is that algorithms do not have access to labels, therefore the problem is no longer to find a map $f$, but instead analyze how points are organized in the input space. Later on, a specialist may assess the results and conclude whether the groups are relevant or not. Such analysis may not be even possible in some scenarios due to the large amount of data [9].

Supervised learning has its foundations on the Statistical Learning Theory (SLT), proposed by Vapnik [15, 16], which defines the conditions that learning is ensured. Although the importance of non-supervised learning, it still requires studies to provide stronger theoretical guarantees [3, 14]. This book intends to tackle only supervised learning.

## 1.3   Supervised Learning

As previously introduced, supervised learning consists in finding the best as possible classifier $f : X \rightarrow Y$ for a given problem. The algorithm responsible for finding this mapping is referred to as **classification algorithm**, which infers a model from every input example $x \in X$ and its respective class $y \in Y$. As we will see in the next chapter, this model is an approximation for the **joint probability distribution** (JPD) of variables $X$ and $Y$. For now, we are much more interested in discussing about the issues involved when searching for the best classifier $f$.

As a classifier can be seen as a function $f : X \rightarrow Y$, the easiest way to understand how to obtain the best as possible mapping is probably by considering a regression problem as provided in [13, 17]. In a regression, $Y$ is not a set of discrete labels, but often a range of real values, and, therefore, it will be easier to depict an example for $f$ and understand how well (or how bad) a classifier can be. For instance, consider we want to find a function to map the input variable atmospheric pressure to the output class rain probability (which is seen as a continuous variable in range $[0, 1]$), as shown in Table 1.4. Thus, our problem is to answer whether it will rain next, according to the current atmospheric pressure measurement. In order to proceed, take a training set of previous observations from which we learn the best function, given the current atmospheric pressure, to produce the probability of raining as output. Listing 1.1 illustrates the program used to generate Table 1.4. Throughout this book, our programs are written to run on the R Statistical Software, provided by the R Foundation for Statistical Computing.[1]

---

[1]Complete source-codes are available at the repository https://github.com/maponti/ml_statistical_learning. For more information, we suggest the reader to study the R manuals available at https://www.r-project.org.

**Table 1.4** Training examples
of the regression problem

| x | y |
|---|---|
| 63.16 | 0.28 |
| 64.54 | 0.29 |
| 67.62 | 0.39 |
| 57.81 | 0.14 |
| 71.43 | 0.50 |
| 71.17 | 0.44 |
| 62.73 | 0.27 |
| 65.44 | 0.35 |
| 75.71 | 0.54 |
| 64.47 | 0.32 |
| 62.02 | 0.23 |
| 70.92 | 0.46 |

**Listing 1.1** Read and show regression data

```
# Code to produce the examples of regression problem
# Data columns are located in different files
x <- read.table("atmospheric_pressure.dat")
y <- read.table("rain_probability.dat")

x <- pressure$pressure[7:19]
x <- log(x)
plot(x,y)
```

Figure 1.4 illustrates both dimensions, i.e., the input variable and the output class, for this toy problem. We observe there is some sort of linear relationship between those variables, therefore we could consider a set of linear functions to estimate classifier $f$. But, by using a linear $f$, some points of this training set will not coincide with the function, i.e., we will not avoid errors on the training sample, as distances of points from $f$ confirm. So, how about using a polynomial function $g$ as shown in Fig. 1.5. Function $g$ would fit perfectly on this sample. This question motivates the **Bias-Variance Dilemma** from Statistics [17], which is here basically translated to: is it the linear or the polynomial function more appropriate to fit this training sample? To answer that question, we need to quantify appropriateness.

**Loss Function**   This measure of appropriateness requires what is referred to as **loss function**, which is responsible for quantifying error or risk [16, 17]. Two of the most commonly used loss functions are: (1) the 0–1-loss function, and (2) the squared-error loss function. The first basically consists in counting how many times the output class is misclassified by $f$, as defined in Eq. (1.1). The squared-error loss function quantifies errors in a continuous way such as shown in Eq. (1.2), thus it is adequate when the classifier output is in $\mathbb{R}$, such as in regression problems.

**Fig. 1.4** Dataset of pressure
$x$ and probability of rain $y$

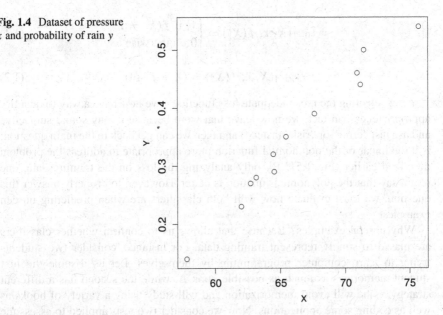

**Fig. 1.5** Regression of
pressure $x$ and probability of
rain $y$ using linear and
polynomial functions

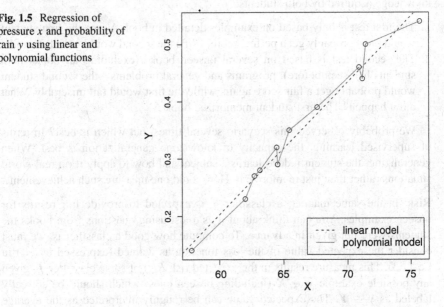

$$\ell_{0-1}(X, Y, f(X)) = \begin{cases} 1, & \text{if } f(X) \neq Y \\ 0, & \text{otherwise.} \end{cases} \tag{1.1}$$

$$\ell_{\text{squared}}(X, Y, f(X)) = (Y - f(X))^2 \tag{1.2}$$

After selecting the most adequate loss function,[2] we now have a way to quantify appropriateness (in fact, we now leave that word because it may sound subjective and use just "error" or "risk" instead) and then we can get back to the main question: is it the linear or the polynomial function more appropriate to address the problem described earlier (Fig. 1.5)? By only analyzing the loss on the training data, one might say that the polynomial function is better. However, to properly answer this question, we must evaluate how well both classifiers are when predicting unseen examples.

Why unseen examples? Because that allows us to confirm whether classifiers are biased to simply represent training data. For instance, consider two students trying to learn computer programming by themselves. Let us assume the first student memorizes as much as possible book A, while the second has a different strategy: (s)he will avoid memorization and will study using a variety of books as well as coding some applications. Now we consider two tests applied to assess the knowledge acquired by both students:

1. The first test is only based on examples detailed in book A—so the first student would most probably get a perfect score, while the second would get a fair grade;
2. The second test is based on several unseen books (excluding all books both students had seen before), programs and general problems—the second student would probably get a fair score again, while the first would fail miserably. What did it happen? The first student memorized book A!

We probably observed this scenario several times, but which is best? In terms of supervised learning, the capacity of knowledge generalization is best. When generalizing, the student indeed learns a subject and how to apply it on real-world situations rather than just memorize it. How could one measure such achievement?

**Risk** In the same manner, a classifier $f$ is expected to provide fair results for unseen examples. You can think about it as a test using questions from books the students were not given in advance. To compute how good a classifier is, we must consider the expected value of the loss function as defined (expressed by $:=$) in Eq. (1.3). This measure results in the expected risk $R(.)$ of some classifier $f$, given any possible example $x \in X$ (including unseen ones) which should be correctly labeled as $y \in Y$. The expected value can be roughly interpreted as the average value of the loss, given every possible input example. Notice this is not feasible for real-world problems which hardly ever provide access to all examples covering the whole space of possibilities.

$$R(f) := E(\ell(X, Y, f(X))). \tag{1.3}$$

---

[2]Observe one can still define another function instead of using those.

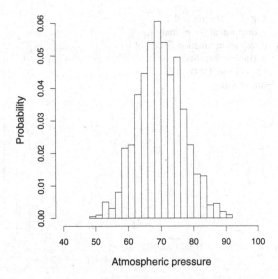

**Fig. 1.6** Discretized estimation of the probability distribution function (PDF) of $X$ (atmospheric pressure) considering 1000 measurements

To compute the expected value, we need the **joint probability distribution** (JPD) of variables $X$ and $Y$ so that there is full access to every combination of $x \in X$ versus their corresponding classes $y \in Y$. To introduce the concept of this JPD, consider 1000 measurements of the atmospheric pressure as input variable $X$, which were collected along years. Figure 1.6 illustrates a discretized estimation of the probability distribution function (PDF) of $X$ considering those 1000 measurements. Let us consider this PDF of $X$ as a good estimation for its real PDF function, as if we had access to the universe of all measurements for such a World region. Then, observe pressure values around 70 kPa are much more probable, thus we will have many more training examples assuming values close to it while inferring classifier $f$, just because values around 70 kPa are the most common.

Now consider the classes or labels (i.e. rain probability) associated to those 1000 measurements of atmospheric pressure, whose PDF estimative is shown in Fig. 1.7. Observe that both PDFs for $X$ and $Y$ are very close to the Normal distribution (they were in fact synthetically produced using that distribution). The joint probability distribution (JPD) $P(X \times Y)$ is responsible for associating the input examples $x \in X$ to their corresponding output classes $y \in Y$, as illustrated for those 1000 measurements in Fig. 1.8. This is just an example in which the complete JPD is not available, because such distribution would require **every possible** pair (atmospheric pressure, rain probability) and not just 1000 measurements. The actual JPD would require infinite data observations once each variable is in $\mathbb{R}$. In fact, ML supervised algorithms attempt to find the best classifier $f$ fitting $P(X \times Y)$ with the lowest possible error. Now we can infer what most probably happens when a given atmospheric pressure is observed. For instance, consider the current atmospheric pressure is 65 kPa, so we analyze range [64.5, 65.5] (centered at 65 kPa) to check for the most probable value for $Y$ (see Fig. 1.9).

**Fig. 1.7** Discretized
estimation of the probability
distribution function (PDF) of
$Y$ (rain probability)
considering 1000
measurements

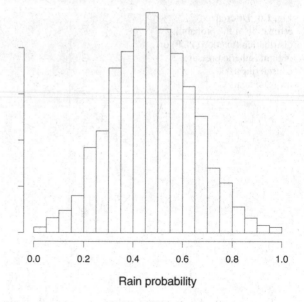

**Fig. 1.8** Joint probability
distribution (JPD) $P(X \times Y)$
associating the input
examples $x \in X$ (atmospheric
pressure) to the output classes
$y \in Y$ (rain probability) for
the 1000 measurements

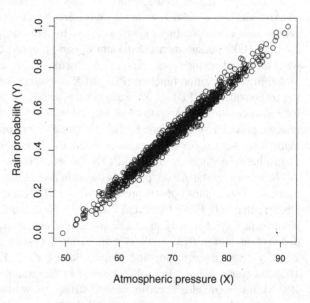

**Fig. 1.9** Defining range [64.5, 65.5] for variable $X$ (centered at 65 kPa)

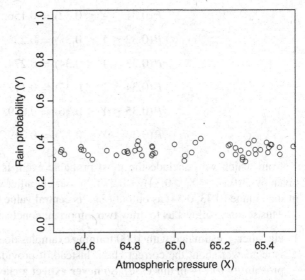

**Fig. 1.10** Histogram illustrating the possible values for $Y$ (rain probability), given the range of interest for $X \in [64.5, 65.5]$ (centered at 65 kPa)

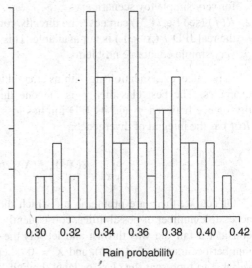

This is exactly what the mapping $f$ is expected to produce. Afterwards, we analyze the PDF for $Y$ but only when $64.5 \le X \le 65.5$, as illustrated in Fig. 1.10, providing the following probabilities:

$$P(0.27 \le Y \le 0.28) = 0.019,$$

$$P(0.28 \le Y \le 0.29) = 0.058,$$

$$P(0.29 \le Y \le 0.30) = 0.019,$$

$$P(0.30 \le Y \le 0.31) = 0.058,$$

$$P(0.31 \leq Y \leq 0.32) = 0.156,$$

$$P(0.32 \leq Y \leq 0.33) = 0.235,$$

$$P(0.33 \leq Y \leq 0.34) = 0.274,$$

$$P(0.34 \leq Y \leq 0.35) = 0.117,$$

$$P(0.35 \leq Y \leq 0.36) = 0.039,$$

$$P(0.36 \leq Y \leq 0.37) = 0.019,$$

from which we conclude the most probable event is in range $[0.33, 0.34]$, as given by $P(0.33 \leq Y \leq 0.34) = 0.274$. So our classifier would provide some value in such range $[0.33, 0.34]$ as output (e.g. its central value $0.335$).

This scenario allows us to draw two important conclusions:

1. classifier $f$ estimated with 1000 training examples does not provide a deterministic answer about the correct class. Instead, it provides an answer under some probability (0.274 in this case). So **never expect a classifier to be perfect**, only for very simple toy scenarios;
2. $R(f)$ (see Eq. (1.3)) can never be directly computed for real-world problems, as the real JPD $P(X \times Y)$ is not available. This is only possible when dealing with very simple countable problems.

In practice, classification algorithms attempt to estimate the JPD relying on input examples. The best classifier $f$ is the one that better fits $P(X \times Y)$, and any divergence between $f$ and its JPD implies loss. Therefore, we can formulate risk $R(f)$ as the integral of divergence:

$$\int_{X \times Y} \ell(X, Y, f(X)) \, d(X \times Y). \tag{1.4}$$

Of course, there are problems for which the JPD $P(X \times Y)$ has a finite and accessible number of possibilities, consequently $R(f)$ is computable. For example, consider a fair die was thrown and let: (1) the input variable $X = 1$, if an even number occurs (i.e., 2, 4 or 6), and $X = 0$ otherwise; (2) in addition, let a second variable to represent the class or label, having $Y = 1$ if the number is prime (i.e., 2, 3 or 5) and $Y = 0$, otherwise.

In this problem, a single thrown will produce the following joint probabilities:

$$P(X = 0 \times Y = 0) = P(1) = \frac{1}{6}, \text{ probability of observing } 1,$$

$$P(X = 0 \times Y = 1) = P(3, 5) = \frac{2}{6}, \text{ probability of observing } 3 \text{ or } 5,$$

$$P(X = 1 \times Y = 0) = P(4, 6) = \frac{2}{6}, \text{ probability of observing } 4 \text{ or } 6,$$

**Fig. 1.11** Throwing a die: joint probability distribution for variables $X$ and $Y$

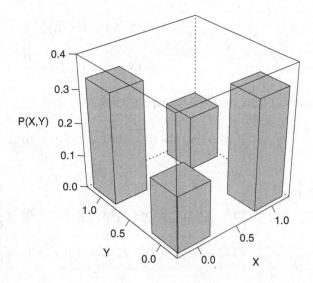

$$P(X = 1 \times Y = 1) = P(2) = \frac{1}{6}, \text{ probability of observing } 2,$$

which are used to compose the JPD for variables $X$ and $Y$ as illustrated in Fig. 1.11.

In this particular situation, we can compute all necessary probabilities to have the JPD and, because of that, $R(f)$ is computable. This simple situation allows us to assess what is the best possible classifier: any function $f$ that makes the perfect regression over the JPD, as shown in Fig. 1.11. As result, if this classifier receives $X = 0$, it will have two options to select from:

$$P(X = 0 \times Y = 0) = P(1) = \frac{1}{6}, \text{ probability of having value } 1$$

$$P(X = 0 \times Y = 1) = P(3, 5) = \frac{2}{6}, \text{ probability of having either value } 3 \text{ or } 5,$$

and given the most probable is $P(X = 0 \times Y = 1) = \frac{2}{6}$, $f$ should provide $Y = 1$ as output. However, it outputs the incorrect answer for $\frac{1}{6}$ of examples. That allows us to conclude that even such perfect classifier is prone to make mistakes, with an average error of $\frac{1}{3}$ for every input.

For the sake of comparison, consider that any other algorithm converged to another classifier $g$ with the following representation for the same JPD:

$$P_g(X = 0 \times Y = 0) = \frac{0}{6},$$

$$P_g(X = 0 \times Y = 1) = \frac{3}{6},$$

$$P_g(X = 1 \times Y = 0) = \frac{3}{6},$$

$$P_g(X = 1 \times Y = 1) = \frac{0}{6},$$

in this latter situation, the expected risk $R(g)$ is:

$$P(X = 0 \times Y = 0) - P_g(X = 0 \times Y = 0) = \frac{1}{6} - \frac{0}{6} = \frac{1}{6},$$

$$P(X = 0 \times Y = 1) - P_g(X = 0 \times Y = 1) = \frac{2}{6} - \frac{3}{6} = -\frac{1}{6},$$

$$P(X = 1 \times Y = 0) - P_g(X = 1 \times Y = 0) = \frac{2}{6} - \frac{3}{6} = -\frac{1}{6},$$

$$P(X = 1 \times Y = 1) - P_g(X = 1 \times Y = 1) = \frac{1}{6} - \frac{0}{6} = \frac{1}{6},$$

whose sum provides an incorrect value of $R(g) = 0$, due to negative values cancel out positive ones. Thus, we should use the squared-error loss function instead:

$$R(g) = \left(\frac{1}{6}\right)^2 + \left(-\frac{1}{6}\right)^2 + \left(-\frac{1}{6}\right)^2 + \left(\frac{1}{6}\right)^2 = \frac{4}{36} = \frac{1}{9},$$

making evident the error of $g$ given unseen examples.

If the JPD is available though, we could compute $R(.)$ for any classifier learned during the training stage. Note that most real-world problems do not fit in this case, that is why **Machine Learning has been widely studied and employed: to estimate good enough classifiers from limited data, without having access to the real JPD**.

**Empirical Risk**   At this point, we are prepared to take a next step. There is no way of computing $R(.)$ assuming $P(X \times Y)$ is unknown at the time of learning. So how could one measure the loss of classifier $f$ and conclude about its usefulness for some problem? We start by estimating the risk based only on the training examples, given they are sampled from the same JPD, as defined in Eq. (1.5). This estimator is referred to as the **empirical risk** of classifier $f$.

$$R_{emp}(f) := \frac{1}{n} \sum_{i=1}^{n} \ell(X_i, Y_i, f(X_i)) \tag{1.5}$$

Can one use this risk estimative to select the best classifier? This question is central for the SLT and it is not that simple to answer. To investigate whether the empirical risk $R_{emp}(.)$ can be used as a good estimator for risk $R(.)$, Vapnik considered the **Law of Large Numbers**, one of the most important theorems from Statistics [5]. However, he faced many challenges to employ this law, requiring a set of assumptions about input examples. To begin with, the Law of Large Numbers (Eq. (1.6)) states the mean of some random variable $\Xi$ converges to the expected value of the underlying distribution $P$ as the sample size $n$ goes to infinity, provided $\Xi_i$ is drawn in an independent and identical manner from $P$, i.e.:

$$\frac{1}{n}\sum_{i=1}^{n}\Xi_i \rightarrow E(\Xi), \text{ for } n \rightarrow \infty. \tag{1.6}$$

Observe this theorem is very close to what Vapnik needed. So, by showing that the empirical risk $R_{\text{emp}}(.)$ converges to risk $R(.)$ for any classifier $f$, then one could select the best candidate among them (see Eq. (1.7)).

$$\frac{1}{n}\sum_{i=1}^{n}\ell(X_i, Y_i, f(X_i)) \rightarrow E(\ell(X_i, Y_i, f(X_i))), \text{ for } n \rightarrow \infty \tag{1.7}$$

Recall that the best classifier is the one providing minimal risk $R(.)$. However, there are several **assumptions to be held**:

1. The joint probability distribution $P(X \times Y)$ is unknown at the time of training;
2. No assumption is made about the joint probability distribution $P(X \times Y)$;
3. Labels may assume nondeterministic values;
4. The joint probability distribution $P(X \times Y)$ is static/fixed;
5. Examples must be sampled in an independent and identical manner;

The first assumption states the joint probability distribution $P(X \times Y)$ is unknown at the time of training, as a consequence the supervised algorithm must estimate it without any previous knowledge but only using samples drawn from this JPD. Observe $P(X \times Y)$ is estimated using an indirect access to itself, once training examples provide a pointwise perspective of this distribution. If the joint probability distribution is known, one could simply estimate its parameters by using some fitting process. By assuming the JPD is unknown, learning would be possible given any data probability distribution.

The second assumption states the joint probability distribution $P(X \times Y)$ may come from any family of distributions. It can either be Normal, Poisson, Binomial or any other [12]. This assumption makes impossible the selection of a given family to later estimate its parameters. As a consequence, the Statistical Learning Theory works in an agnostic way about $P(X \times Y)$.

In the third assumption, labels may have nondeterministic values. The first reason is that data collection may impose noise to labels. For instance, suppose a person is classifying customers from a Web commerce application and misses the correct label in some situations. It should not be impossible to learn even in the presence of such eventual errors. The second reason for this assumption is that classes may overlap, e.g. when throwing a die, as previously seen. For example, the classification of people according to sex, given their height as input variable $X$ [17]. Such a variable is not sufficient to completely separate male and female classes apart,

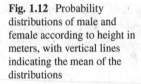

**Fig. 1.12** Probability
distributions of male and
female according to height in
meters, with vertical lines
indicating the mean of the
distributions

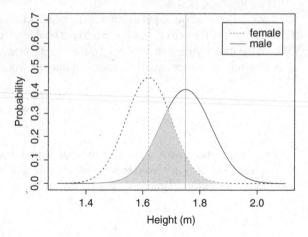

especially due to some height intervals as illustrated in Fig. 1.12. It is difficult to
find a good classifier for this problem once there is a significant overlap of class
distributions, as depicted in the grey area.

This third assumption is very important because data may contain incorrect labels
or a mixture of them. To exemplify this mixture, consider Fig. 1.13 which shows a
$P(X \times Y)$ for this problem involving heights. A classification would bisect the
heights into two regions, but there is still a significant overlap. Compare it with
Fig. 1.12, which shows heights versus densities. To improve classification results,
we should consider more meaningful variables to compose the input set $X$.

The fourth assumption states the joint probability distribution $P(X \times Y)$ is
static, i.e., it never changes along time. This is necessary because the empirical
risk (Eq. (1.5)) is required to approximate $R(.)$ as the sample size $n$ tends to
infinity, otherwise variations in the mapping from $X$ to $Y$ would cause conflicts to
ensure such approximation. It is obvious that if the JPD starts representing a given
association from $X$ to $Y$ and then changes it, all previous training is useless. Time
series and data stream applications suffer in such a context, as data behavior may
change and therefore impact in the joint probability distribution.

As last assumption, examples must be statistically independent from each other
and be sampled in an identical manner. To illustrate, see Fig. 1.14 in which a bag
contains three balls enumerated as 1, 2 and 3. Now consider the probability of
drawing either ball 1, 2 or 3 which is equal to $P(\text{ball} = 1) = P(\text{ball} = 2) =
P(\text{ball} = 3) = 1/3$. After drawing ball 1 **without replacement**, the probabilities
change to $P(\text{ball} = 1) = 0$, $P(\text{ball} = 2) = 1/2$ and $P(\text{ball} = 3) = 1/2$. This
happens because the event of drawing ball 1 influences the probability of future
events, what is assumed not to occur in the context of the Statistical Learning
Theory. Therefore, the SLT considers the sampling **with replacement** from the JPD.
As part of this last assumption, examples must be sampled in an identical manner,
meaning each one of the three balls must have the same probability of being chosen,
consequently there is no bias to define one of them as preferred.

All those assumptions are necessary to employ the Law of Large Numbers, which
is used in the context of the SLT to ensure the empirical risk $R_{\text{emp}}(.)$ is a good

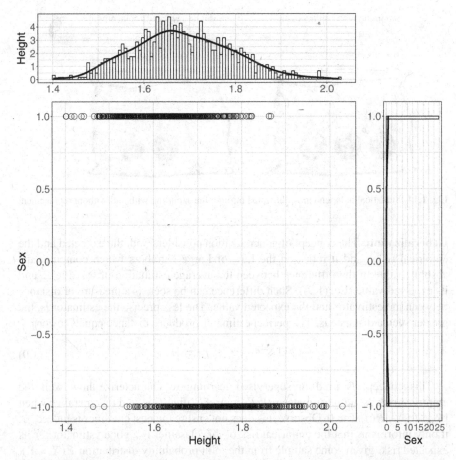

**Fig. 1.13** Classifying sex according to heights. Top: a histogram and density of observed heights; right: the distribution of classes Male ($-1$) and Female (1); center: joint probability distribution (observe the need for plotting variables $X$, $Y$ and the densities separately). A classifier would bisect the height axis into two classes

estimator for the expected risk $R(.)$. Equation (1.8) defines this concept in a reduced form when compared to Eq. (1.7).

$$R_{\text{emp}}(f) \rightarrow R(f), \text{ for } n \rightarrow \infty \qquad (1.8)$$

In fact, the best classifier $f_i$ is the one whose risk $R(.)$ is the lowest, i.e. $R(f_i) \leq R(f_j)$ for all $j$. However, there is no approach to compute $R(.)$ for real-world problems, so $R_{\text{emp}}(.)$ is computed instead for all classifiers $f_1, \ldots, f_k$, so that the one with the smallest empirical risk can be selected. As seen in the next chapter, this strategy is not that simple, once several other conditions rely on the described assumptions to make the empirical risk a good estimator for the (expected) risk.

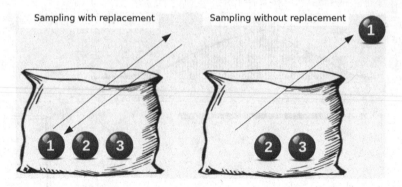

Sampling with replacement          Sampling without replacement

**Fig. 1.14** Statistical independence illustrated through the sampling with and without replacement

**Generalization** The concept of generalization considers both the expected and the empirical risks, and it relies on the Law of Large Numbers which computes the absolute value of the difference between the average estimation of variable $\varXi$ and its expected value (Eq. (1.9)). Such difference can be seen as a measure of distance between the estimator and the expected value. The less precise the estimator is, the greater such a distance is. The perfect estimator produces distance equals to zero.

$$\left| \frac{1}{n} \sum_{i=1}^{n} \varXi_i - E(\varXi) \right| \tag{1.9}$$

This concept is used in supervised learning to characterize how well the empirical risk $R_{\mathrm{emp}}(.)$ estimates risk $R(.)$. A classifier $f_i$ is said to generalize when $|R_{\mathrm{emp}}(f_i) - R(f_i)| \approx 0$. Observe this does not ensure the best classifier is indeed $f_i$, it only informs us that the empirical risk of this classifier is a good estimator for its expected risk, given some sample from the joint probability distribution $P(X \times Y)$. Being the empirical risk a good estimator of $R(.)$, we can use $R_{\mathrm{emp}}(.)$ as a proxy measure to select the best classifier.

Now, suppose all empirical risks of classifiers $f_1, \ldots, f_k$ provide a good enough generalization. Those risks can be used to sort classifiers out according to their losses, in form $R_{\mathrm{emp}}(f_i) \leq R_{\mathrm{emp}}(f_j) \leq \ldots$ in order to select the best one. From this perspective, we see how important is to build up some theoretical foundation to ensure good estimators.

In a practical point of view, people usually have a dataset with several examples which are divided into two sets: a training and a test set. The **training set** (seen examples) is used to infer classifiers $f_1, \ldots, f_k$ from. The **test set** contains unseen examples and is used to verify if the inferred classifiers are indeed good enough and which is the best. This simple evaluation technique relies on this concept of generalization and attempts to verify the risk of every classifier $f_i$, given unseen examples.

We can use this simple scenario to evaluate the generalization of any classifier $f$. Let a supervised algorithm infer $f$ over the training set and consider the empirical risk $R_{\mathrm{emp}}(f) = 0.01$. Then, $f$ is used to classify examples from the test set, producing a loss equals to 0.11. The absolute difference $|R_{emp}(f) - 0.11| = |0.01 - 0.11| = 0.1$ provides a measure of how good the risk estimation is. Note that, instead of computing $|R_{emp}(f) - R(f)|$, we calculated $|R_{emp}(f) - R'_{emp}(f)|$, given $R'_{emp}(f)$ is the empirical risk computed for an unseen finite sample (test set). As discussed in the next chapter, this last generalization in terms of two samples (one for training and another for testing) is an approximation for $|R_{emp}(f) - R(f)|$ after the Symmetrization lemma [17].

In order to assess the generalization of both classifiers illustrated in Fig. 1.4, suppose the linear classifier $f$ has the following empirical risk $R_{emp}(f) = 0.12$ while the expected is $R(f) = 0.10$, and the polynomial classifier $g$ has $R_{emp}(g) = 0$ but $R(g) = 0.9$. In this situation, the classifier providing the best generalization is $f$, once $|R_{emp}(f) - R(f)| = 0.02$ is minimal so that it is the best estimative for risk $R(.)$. By having a set of classifiers with a good enough generalization capacity, we can choose the one with the smallest empirical risk.

## 1.4 How a Supervised Algorithm Learns?

A supervised learning algorithm attempts to induce the best classifier $f$ as possible so that $f : X \rightarrow Y$ provides the minimal risk $R(f)$. In this context, observe classifier $f$ can also work as an **approximation or regression function** for the joint probability distribution $P(X \times Y)$, as illustrated in Fig. 1.15. So any supervised learning algorithm (e.g. Naive Bayes, SVM, Multilayer Perceptron, etc.) builds up some regression function, which is the same as inferring a classifier from a given training sample. Of course, if the sample collected from $P(X \times Y)$ is not sufficient, such regression will produce unacceptable losses.

Classifier $f$ comes from a space of admissible functions $\mathscr{F}$. This space defines the **bias of the supervised algorithm**, that is, functions in $\mathscr{F}$ are often restricted to some types or families (e.g. linears, 2-order polynomials, etc.). As a consequence, if $P(X \times Y)$ has nonlinear characteristics but $f$ has a linear bias, the supervised algorithm will produce an insufficient approximation that misses output classes more often than we wish. Although a classifier is in fact a regression function for the JPD, we can see it in another way which is by using hyperplanes or functions to "shatter" (or split) the input space into different regions. To better understand this approach, let us analyze a two-class dataset.

Let dataset $\mathscr{D} = \{(x_1, y_1), (x_2, y_2), \ldots, (x_n, y_n)\}$ in which $x_i \in \mathbb{R}^2$ corresponds to the input variables (or input space) of example $i$, and $y_i \in \{+1, -1\}$ is its corresponding class. Consider the two input variables are temperature and humidity measurements, while the class defines whether a person plays ($+1$) or does not play ($-1$) soccer under such conditions. Figure 1.16 illustrates this dataset, having $n = 100$ examples in which the circles represents label $-1$ (no), while the plus sign is associated with $+1$ (yes). By analyzing this plot, we have a good idea about how points are distributed in this bidimensional space.

**Fig. 1.15** Regression function produced to approximate the joint probability distribution illustrated in Fig. 1.13

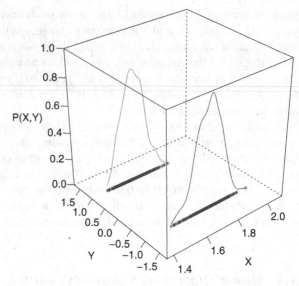

**Fig. 1.16** Input space of examples containing two variables (temperature in Celsius and humidity in percentages) and two classes ($\{-1, +1\}$) for the problem of playing soccer

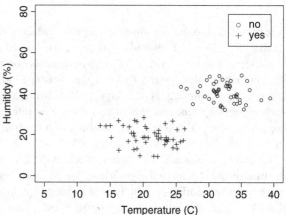

A supervised learning algorithm attempts to estimate $P(X \times Y)$ according to the available data sample $\mathcal{D}$ drawn from $P(X \times Y)$ in an independent and identical manner. As observed, there is a concentration of positive instances around a given mean point, and a concentration of negative instances around another mean. Let us assume that $P(X \times Y)$ is modeled by two Normal (or Gaussian) distributions as illustrated in Fig. 1.17, as the sample size $n \to \infty$.

Consider the JPD is known, so that one is able to employ a supervised learning algorithm whose space of admissible functions (or bias) $\mathcal{F}$ only contains Gaussian functions. Now consider our algorithm found a classifier $f$ which is a composition of two Gaussian functions, each one centered at a different mean to represent each class. By assigning an identifier to each Gaussian (either $-1$ or $+1$), we can define

**Fig. 1.17** Joint probability distribution $P(X \times Y)$ for the problem of playing soccer

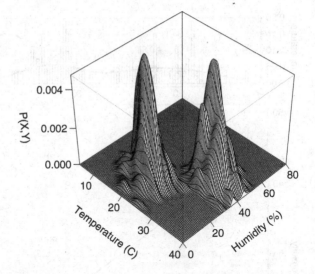

the output labels for examples. All those steps make sense, but most well-known supervised learning algorithms are typically based on something simpler than that.

As an alternative to Gaussian functions, most of the supervised learning algorithms build **linear hyperplanes** to "shatter" the input space into regions, as shown in Fig. 1.18. Thus, every point on one side of the hyperplane will be classified as $+1$, while points on the other will be labeled as $-1$. Points lying on the hyperplane represent a tie, and, in principle, have no class assigned. Observe that this simpler approach was as effective (for this problem) as the Gaussians. In this case, we say the supervised learning algorithm has a linear bias instead.

The space containing all possible functions considered by a supervised algorithm, that is, the algorithm bias, is illustrated using a box as shown in Fig. 1.19. Inside such a box, the algorithm bias is circumscribed according to its restrictions. For example, the supervised algorithm called Perceptron [6] considers a single linear hyperplane to separate examples of binary problems, while the Multilayer Perceptron (MLP) may employ more than a single hyperplane to separate two or more classes. Therefore, the Perceptron bias is more restricted than the MLP.[3] As a consequence, the region illustrating the Perceptron bias is smaller than the one for MLP. We suggest the reader to associate a smaller region as a space containing less admissible functions to infer classifiers from.

We now illustrate the binary problem AND (examples defined in Table 1.5) in Fig. 1.20. Let $\mathbb{B}$ be the set of binary numbers. The input space contains examples $x_i \in \mathbb{B}^2$ and the corresponding output classes $y_i = \{0, 1\}$. This problem is used to discuss the Perceptron and MLP biases. By analyzing Fig. 1.20, we see a single

---

[3] When this algorithm employs more than a single hyperplane, otherwise it is the same as the Perceptron.

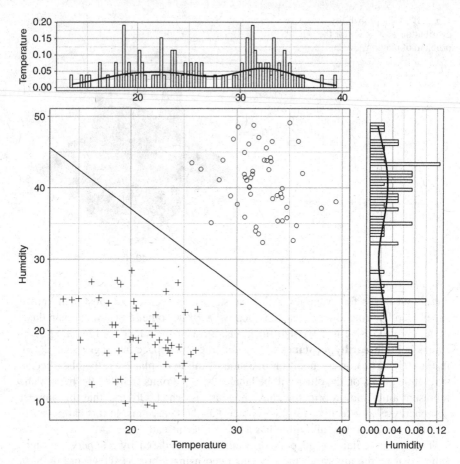

**Fig. 1.18** Shattering of the input space of examples containing two variables (temperature and humidity) and two classes ($\{-1, +1\}$) for the problem of playing soccer

linear hyperplane is enough to provide the correct labels for AND, therefore the Perceptron algorithm satisfies such a task. However, what happens if we decide to employ the Perceptron to solve the problem XOR (Fig. 1.21)? Observe this new task requires a space with more admissible functions than a single hyperplane. This could be solved by using a Gaussian function as shown in Fig. 1.22, so it would provide an activation to characterize whether the class is 0 or 1. Another possible solution, considered by the Multilayer Perceptron (MLP), is to use more than a single linear hyperplane as shown in Fig. 1.23.

By using more than one hyperplane, MLP is capable of representing nonlinear behaviors. Comparing the Perceptron against the Multilayer Perceptron, we conclude the first has a linear while the second has a nonlinear bias if more hyperplanes are combined, as illustrated in Fig. 1.19. The Perceptron bias is more restricted (also referred to as a stronger bias), as it contains less functions (only linear ones) inside

**Fig. 1.19** Box illustrating the universe of admissible functions $\mathcal{F}_{all}$ for any supervised learning algorithm. Subspace $F_{Perceptron}$ corresponds to the Perceptron bias while $F_{MLP}$ to the MLP bias (for MLP using two or more hyperplanes)

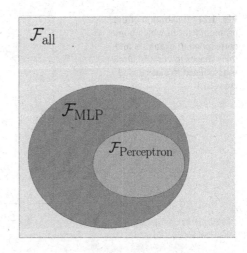

**Table 1.5** Problem AND: Input space of examples and the corresponding classes

| $x_1$ | $x_2$ | $y$ |
|---|---|---|
| 0 | 0 | 0 |
| 0 | 1 | 0 |
| 1 | 0 | 0 |
| 1 | 1 | 1 |

**Fig. 1.20** Problem AND whose points correspond to examples and their shapes to classes: circle is associated to class 0, and cross to 1. Hyperplane illustrates a possible classifier

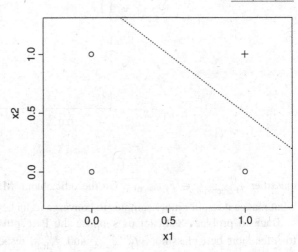

such space region, while the MLP bias is less restricted (also referred to as weaker), containing nonlinear functions. This is the case if and only if MLP considers two or more linear hyperplanes, otherwise it will have the same bias as the Perceptron.

All this analysis supports the study of the **Bias-Variance Dilemma** (discussed in Sects. 1.3 and 1.4). By having a smaller subspace (a.k.a. stronger bias) of admissible functions, the Perceptron has less variance associated to the convergence to its best

**Fig. 1.21** Illustration of the problem XOR in which points correspond to examples and their shapes to classes: circle is associated to class 0, and cross to 1

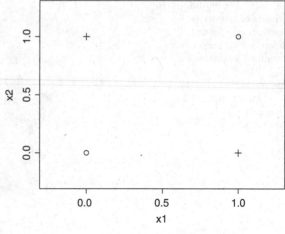

**Fig. 1.22** Problem XOR approached with a Gaussian function

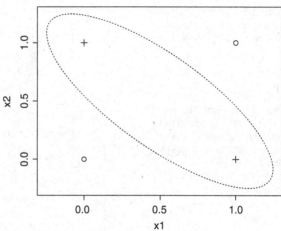

classifier $f_{\text{Perceptron}}^{(b)} \in F_{\text{Perceptron}}$. On the other hand, MLP has a greater variance, what makes it require more samples to converge to its best classifier $f_{\text{MLP}}^{(b)} \in F_{\text{MLP}}$.

Back to problem XOR, let us suppose the Perceptron and MLP were trained to infer their best classifiers $f_{\text{Perceptron}}^{(b)}$ and $f_{\text{MLP}}^{(b)}$, respectively. In such scenario, $f_{\text{Perceptron}}^{(b)}$ is not enough to solve this task, producing a greater risk. On the other hand, MLP would provide a solution using two hyperplanes as shown in Fig. 1.23. How does that affect the biases of both algorithms? Figure 1.24 illustrates a possible location for those best classifiers. Observe the Perceptron would never converge to $f_{\text{MLP}}^{(b)}$, because it is outside of its space of admissible functions.

For the problem AND, a single hyperplane is enough (Fig. 1.20), consequently the two-hyperplane MLP and the Perceptron represent the same solution as illustrated in Fig. 1.25. Note the biases of both algorithms and the location of the best classifier $f^{(b)}$, which ends up to be the same for both classification algorithms.

**Fig. 1.23** Problem XOR
approached with the
Multilayer Perceptron (MLP)
using two linear hyperplanes

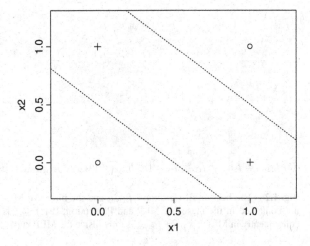

**Fig. 1.24** The universe of
functions $\mathscr{F}_{all}$ which contains
the Perceptron and MLP
biases. The best Perceptron
classifier is different from the
best for MLP

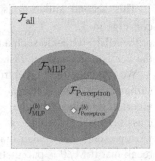

**Fig. 1.25** The universe of
functions $\mathscr{F}_{all}$ contains all
biases, including a more
restricted subspace of
functions related to the
Perceptron and MLP biases.
For the problem AND, the best
Perceptron classifier is the
same as for MLP

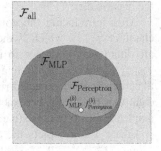

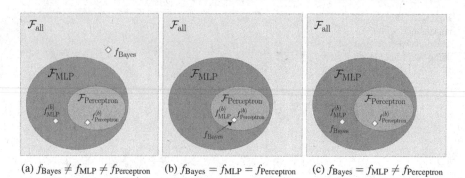

(a) $f_{\text{Bayes}} \neq f_{\text{MLP}} \neq f_{\text{Perceptron}}$    (b) $f_{\text{Bayes}} = f_{\text{MLP}} = f_{\text{Perceptron}}$    (c) $f_{\text{Bayes}} = f_{\text{MLP}} \neq f_{\text{Perceptron}}$

**Fig. 1.26** The best of all classifiers $f_{\text{Bayes}}$ and the biases of MLP and Perceptron. (**a**) $f_{\text{Bayes}}$ is not contained in the biases of MLP and Perceptron; (**b**) $f_{\text{Bayes}}$ is contained in Perceptron, and consequently in MLP bias; (**c**) $f_{\text{Bayes}}$ is only inside the MLP bias

This second situation is very interesting. Both algorithms will attempt to converge to $f^{(b)} = f_{\text{Perceptron}}^{(b)} = f_{\text{MLP}}^{(b)}$. Which one should we use (the Perceptron or the MLP) to tackle such a task? Both can be used, but the smaller variance helps the Perceptron converge faster to $f^{(b)}$, so it would be undoubtedly a better choice.

We conclude the Perceptron bias is enough for the problem AND, while it is not sufficient for XOR. The reader is invited to analyze the biases of other supervised algorithms. Whenever the algorithm bias is not sufficient to induce a suitable classifier for a given task, we should allow more admissible functions. That is why we represent such weaker bias with a greater region as shown in Fig. 1.19. Now we recall that if MLP is parametrized to use a single hyperplane, it will have the same bias as the Perceptron. By using more hyperplanes, we make the space larger, i.e. the bias weaker.

After introducing the concept of algorithm bias, let us introduce the Bayes classifier $f_{\text{Bayes}}$, which is the best function in $\mathscr{F}$. As illustrated in Fig. 1.26, an algorithm may converge to it only if such function is inside its bias. Classifier $f_{Bayes}$ is the best approximation for the joint probability distribution $P(X \times Y)$, thus reducing the expected risk $R(f_{\text{Bayes}})$ as much as possible. Depending on the data, that might not be exactly zero, because classes may be overlapped due to nondeterminism or noise (e.g. the problem involving a die).

Figure 1.26a depicts the situation in which the Perceptron and MLP do not contain $f_{\text{Bayes}}$, so they would converge to $f_{\text{Perceptron}}^{(b)}$ and $f_{\text{MLP}}^{(b)}$, respectively. In Fig. 1.26b, both algorithms can converge to $f_{\text{Bayes}}$, so $f_{Bayes} = f_{\text{Perceptron}}^{(b)} = f_{\text{MLP}}^{(b)}$; and finally, in Fig. 1.26c, the MLP bias contains $f_{\text{Bayes}}$, so $f_{\text{Bayes}} = f_{\text{MLP}}^{(b)}$ but $f_{\text{Perceptron}}^{(b)} \neq f_{\text{Bayes}}$, thus MLP may converge to such best function as more training examples are provided.

It is also worth mentioning that a given algorithm bias may contain more than one classifier providing the same solution, i.e. more than a single $f_{\text{Bayes}}$, $f_{\text{Perceptron}}^{(b)}$ and $f_{\text{MLP}}^{(b)}$ in $\mathscr{F}_{all}$. Observe we could slightly move the hyperplanes from Figs. 1.20 and 1.23 and still have the same classification results.

**Fig. 1.27** When the space of admissible functions $\mathscr{F}_{\text{algorithm}}$ grows, its probability of containing the memory-based classifier increases. This is just an illustration: this probability function is not necessarily a sigmoid such as depicted

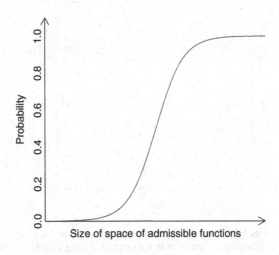

**Algorithm Bias and the Memory-Based Classifier** The algorithm bias is a central concept for the Statistical Learning Theory. For example, consider an algorithm has a bias containing a **memory-based classifier** $f_m$ capable of producing an empirical risk $R_{\text{emp}}(f_m) = 0$ by memorizing all training examples, but failing for unseen examples. In detail, in a binary classification problem, for any unseen example received as input, this algorithm produces a random output: let 50% of examples lying in class $+1$ while the other half in $-1$. Obviously, in average, it will output $+1$ correctly for half of unseen examples, misclassifying the other half. Assessing $f_m$ with the 0−1-loss function, we observe $R(f_m) = 0.5$, as half of unseen examples will be associated to the wrong class. As a consequence, $f_m$ does not provide a good generalization, i.e., $|R_{\text{emp}}(f_m) - R(f_m)| = 0.5$, once it will never approach zero as the sample size $n$ tends to infinity. Therefore, any algorithm capable of admitting the memory-based classifier may eventually converge to $f_m$ and, thus, never generalize knowledge.

This is one of the most important issues studied by the SLT: an algorithm should never have the memory-based classifier inside its bias $\mathscr{F}_{\text{algorithm}}$, otherwise no learning is guaranteed, but solely the memorization of training examples. That represents a challenge: by setting a bias with more admissible functions to address a given problem, space $\mathscr{F}_{\text{algorithm}}$ will contain more classifiers to choose from, increasing the probability of containing the memory-based classifier (as shown in Fig. 1.27).

From Fig. 1.27, we conclude some bias is necessary to ensure learning, once a larger space of admissible functions may contain the memory-based classifier. For example, consider the data depicted in Fig. 1.28 and assume the supervised algorithm is capable of inferring any polynomial function. Therefore, it could converge to a perfect nonlinear function with $R_{\text{emp}}(f) = 0$, i.e., it does not make mistakes in the training sample, however it could fail miserably on unseen examples, an unacceptable result.

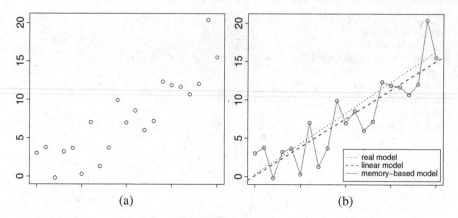

**Fig. 1.28** The dotted line shown is the real model, from which noisy sample data (circles) were obtained. A regression is approached by two algorithms with different biases: the first includes only linear functions, while the second includes a high-order polynomial function, resulting in a memory-based model. (**a**) Noisy sample of points, (**b**) different models to fit the points

The concepts introduced so far are detailed, in the point of view of SLT, in Chap. 2. However, before that, it is important to have tools to instantiate such concepts in practical scenarios. The Perceptron and the Multilayer Perceptron were selected since those algorithms provide excellent case studies for supervised learning.

## 1.5    Illustrating the Supervised Learning

This section introduces the Perceptron and the Multilayer Perceptron, including their formulations and implementations using the R language [10].

### 1.5.1    The Perceptron

As discussed in Sect. 1.4, the Perceptron bias is linear and therefore it contains less functions inside space $\mathscr{F}_{\text{Perceptron}}$ than any MLP using more than a single hyperplane. As a consequence, this algorithm is adequate to model linearly separable problems such as AND and OR. On the other hand, the Perceptron does not suit more complex problems (even toy problems such as XOR).

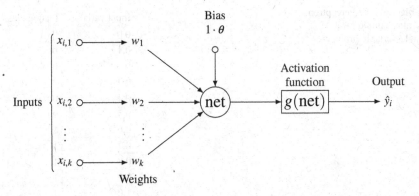

**Fig. 1.29** The Perceptron: input $x_i \in \mathbb{R}^k$, $k$ connection weights $w_j$ and the bias term $\theta$. This neuron computes net as a sum of inputs multiplied by weights plus $\theta$. The real-valued output is given by $\hat{y}_i = g(\text{net})$

The Perceptron algorithm was proposed by Rosenblatt [11] in 1957. It considers an artificial neuron or unit[4] that receives an input example $x_i$ to produce an output class $y_i$, as illustrated in Fig. 1.29. Such input example $x_i \in \mathbb{R}^k$, i.e. it contains a list of indexed variables $j = 1, \ldots, k$.

The neuron multiplies every input value $x_{i,j}$ by a weight $w_j$ as well as constant 1 by the bias term[5] denoted by $\theta$. As a next step, this algorithm computes a term net $= \sum_{j=1}^{k} x_{i,j} w_j + \theta$. Note this represents a linear combination of $x_{i,j}$, for which $\theta$ is the interception in relation to the axes defined by $x_{i,j} \forall i$. Next, the Perceptron applies a heaviside (or step) function $g(\text{net})$ (Eq. (1.10)), in which $\epsilon$ is a user-defined parameter, producing the output class $\hat{y}_i$ for example $x_i$.

$$g(\text{net}) = \begin{cases} 1, & \text{iff net} > \epsilon \\ 0, & \text{otherwise} \end{cases} \tag{1.10}$$

In order to better understand the Perceptron algorithm, we approach a problem with a single variable $x_{i,1}, \forall i$ provided as input. Let every real input value $0 \leq x_{i,1} \leq 0.5$ be assigned to class 0, while $0.5 < x_{i,1} \leq 1.0$ is assigned to class 1, such as the training sample provided in Table 1.6. Figure 1.30 shows the expected classes $y_i$ for every possible input value in range $[0, 1]$.

In this case, the Perceptron adapts parameters $w_1$ (weight) and $\theta$ (bias) to build up a linear function net $= x_{i,1} w_1 + \theta$, in form of $ax + b$, having $x = x_{i,1}$ and $b = \theta$, as shown in Fig. 1.31. If we set $\epsilon = 0$ and apply every value of net into Eq. (1.10), we obtain the expected outputs as shown in the same figure. The Perceptron learning process consists in searching for adequate weights $w_j$ and bias $\theta$ such that $ax + b$ has the correct linear shape.

---

[4]The term neuron is used due to the biological neuron motivation.
[5]The reader may not confuse this bias term with the Bias-Variance Dilemma. The bias term $\theta$ is simply a space interception value.

**Table 1.6** The Perceptron: training sample for a single-variable problem

| Input variable | Output class |
|---|---|
| 0.0 | 0 |
| 0.1 | 0 |
| 0.2 | 0 |
| 0.3 | 0 |
| 0.4 | 0 |
| 0.5 | 0 |
| 0.6 | 1 |
| 0.7 | 1 |
| 0.8 | 1 |
| 0.9 | 1 |
| 1.0 | 1 |

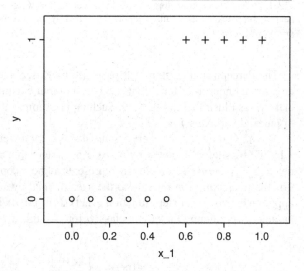

**Fig. 1.30** The Perceptron on a simple problem: the expected classes for the range of the input variable

To adapt weights and the bias term, the Perceptron considers the empirical risk $R_{emp}(f_{candidate})$, i.e., the error considering the training set for every candidate classifier. The squared-error is used as the loss function to compute the empirical risk (as defined in Eq. (1.11)), because it provides a convex function when the classes involved in the problem are linearly separable, i.e., separable using a linear hyperplane. We discuss about this issue later on. For now, let us employ this function to find the solution.

$$\ell_{squared}(x_i, y_i, g(x_i)) = (y_i - g(x_i))^2 \qquad (1.11)$$

At this point, it may be clear that we want to minimize error (or risk) considering the training sample. In that sense, $g(.)$ should produce the smallest error when analyzing an input example $x_i$, as defined in Eq. (1.12). Figure 1.32 illustrates this squared-error function in terms of the free variables, i.e., variables adapted by

**Fig. 1.31** The Perceptron on a simple problem: linear approximation given net

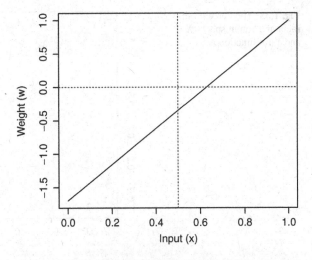

**Fig. 1.32** Squared-error loss function computed for some training sample with $n$ observations. The $x$-axis corresponds to parameter $w_1$, while the $y$-axis to bias term $\theta$

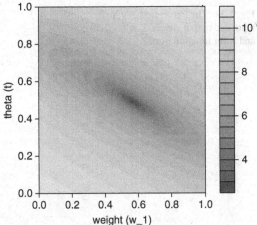

Perceptron while finding a good linear approximation function to solve this problem. The Perceptron aims to converge to the minimum of function $E^2$.

$$E^2 = \sum_{i=1}^{n} \ell_{\text{squared}}(x_i, y_i, g(x_i)) = \sum_{i=1}^{n} (y_i - g(x_i))^2 \qquad (1.12)$$

Let us assume this squared-error function provides a perfect paraboloid as shown in Fig. 1.32, what might not occur once it depends on the available training examples and also on the loss function. Consider also initial values $w_1 = -1$ and $\theta = 1$. We would have a linear function net as shown in Fig. 1.33, when analyzing the input and output spaces $X$ and $Y$, respectively. After applying the heaviside function $g(\text{net})$ (Eq. (1.10)) with $\epsilon = 0.5$, we would obtain a plot as shown in Fig. 1.34.

**Fig. 1.33** The Perceptron:
input and output spaces $X$
and $Y$ for function net

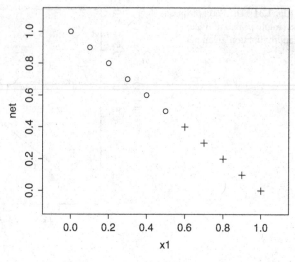

**Fig. 1.34** The Perceptron:
input and output spaces $X$
and $Y$ for function $g$(net)

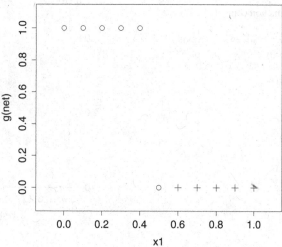

For the sake of simplicity, observe $w_1 = -1$ as a projected point touching the squared error function illustrated in Fig. 1.32. Now, let the Perceptron adapt $w_1$ so it converges to a good enough function $g$(net). The gradient information on the projected point indicates the direction, i.e. how to modify $w_1$, in order to approach the solution with the smallest risk as possible. Note the same happens for variable $\theta$.

As there is a negative tendency at point $w_1 = -1$, given by the tangent slope, the Perceptron should increase the value of $w_1$. This is exactly what is defined by the Gradient Descent (GD) method as formalized in Eq. (1.13), in which $\eta \in (0, \infty)$ is a user-defined parameter, a.k.a. gradient step.

In the case of a negative derivative, $\frac{\partial E^2}{\partial w_1} < 0$, its multiplication by $-\eta$ results in an increment of the current $w_1(t)$. Of course, if $\eta$ is too small, this increment

**Fig. 1.35** Squared-error
function $E^2$ in terms of the
free variable $w_1$

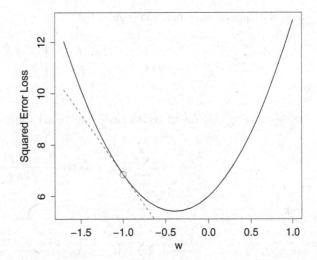

will be also small, taking too long to converge to the minimal squared error. On the
other hand, if $\eta$ is too great, then $w_1(t+1)$ may cross the minimum and reach the
other side, thus the derivative $\frac{\partial E^2}{\partial w_1}$ will be positive in the next step (Fig. 1.35). Note
$\eta$ influences the convergence rate and, in some cases, can even cause divergence.
Equation (1.14) defines the GD method for the second free variable $\theta$.

$$w_1(t+1) = w_1(t) - \eta \frac{\partial E^2}{\partial w_1} \tag{1.13}$$

$$\theta(t+1) = \theta(t) - \eta \frac{\partial E^2}{\partial \theta} \tag{1.14}$$

Now, the formulation to adapt the Perceptron variables is presented. As first
requirement, terms $\frac{\partial E^2}{\partial w_1}$ and $\frac{\partial E^2}{\partial \theta}$ must be computed, given $E^2 = \sum_{i=1}^{n}(y_i - f(x_i))^2$,
and $f(x_i)$ written as:

$$f(x_i) = g(\text{net}_i) = g\left(\sum_{j=1}^{k} x_{i,j} w_j + \theta\right).$$

However, as $g(.)$ is the heaviside function, we could not directly differentiate it.
As an approximation, Rosenblatt relaxed such differentiation by considering:

$$f(x_i) = \text{net}_i = \sum_{j=1}^{k} x_{i,j} w_j + \theta,$$

then, the squared-error function is given by:

$$E^2 = \sum_{i=1}^{n}(y_i - f(x_i))^2 = \sum_{i=1}^{n}\left(y_i - \left[\sum_{j=1}^{k}x_{i,j}w_j + \theta\right]\right)^2,$$

so, $\frac{\partial E^2}{\partial w_1}$ and $\frac{\partial E^2}{\partial \theta}$ are found via the chain rule for derivatives as follows:

$$\frac{\partial E^2}{\partial w_1} = 2\sum_{i=1}^{n}(y_i f - f(x_i))\frac{\partial [y_i - f(x_i)]}{\partial w_1},$$

and:

$$\frac{\partial E^2}{\partial \theta} = 2\sum_{i=1}^{n}(y_i - f(x_i))\frac{\partial [y_i - f(x_i)]}{\partial \theta}.$$

Recall $y_i$ is the expected output class for example $i$ (i.e., a label defined by some specialist), thus when differentiating it in terms of either $w_1$ or $\theta$, $y_i$ is disconsidered. Differentiating $\frac{\partial - f(x_i)}{\partial w_1}$ and $\frac{\partial - f(x_i)}{\partial \theta}$:

$$\frac{\partial - f(x_i)}{\partial w_1} = -\frac{\partial [x_{i,1}w_1 + \theta]}{\partial w_1} = -x_{i,1},$$

and:

$$\frac{\partial - f(x_i)}{\partial \theta} = -\frac{\partial [x_{i,1}w_1 + \theta]}{\partial \theta} = -1.$$

The Gradient Descent method is formulated as follows:

$$w_1(t+1) = w_1(t) - \eta\frac{\partial E^2}{\partial w_1}$$

$$= w_1(t) - \eta\, 2\sum_{i=1}^{n}(y_i - f(x_i))(-x_{i,1}),$$

and:

$$\theta(t+1) = \theta(t) - \eta\frac{\partial E^2}{\partial \theta}$$

$$= \theta(t) - \eta\, 2\sum_{i=1}^{n}(y_i - f(x_i))(-1),$$

which, in open-form, is written as:

$$w_1(t+1) = w_1(t) - \eta \frac{\partial E^2}{\partial w_1}$$

$$= w_1(t) - \eta\, 2 \sum_{i=1}^{n} (y_i - [x_{i,1} w_1 + \theta])(-x_{i,1}), \quad (1.15)$$

and, finally:

$$\theta(t+1) = \theta(t) - \eta \frac{\partial E^2}{\partial \theta}$$

$$= \theta(t) - \eta\, 2 \sum_{i=1}^{n} (y_i - [x_{i,1} w_1 + \theta])(-1). \quad (1.16)$$

Observe constant 2 is irrelevant due to another constant $\eta$ defining the gradient step. Those two last equations correspond to the Perceptron learning process, as implemented in Listing 1.2. Instead of adapting $w_1$ and $\theta$ after analyzing all input examples, as the summation used in Eqs. (1.15) and (1.16), our code performs the GD method on an example basis.

The R functions in Listing 1.2 implement a simple classification task in which a single variable represents the input, i.e. $x_i \in \mathbb{R}$, producing output classes in set $\{0, 1\}$. This listing contains: (1) the heaviside function (lines 4-9); (2) the Perceptron training function, which returns the model, i.e. the learned parameters $w_1$ and $\theta$ (lines 14-87); (3) a function responsible to perform the classification of new instances, given the trained model (lines 92-115); and, finally, (4) a demo function with training and test sets (lines 127-169).

**Listing 1.2** The Perceptron—implementation of the simplest classification task ("perceptron.r")

```
1   # Source code file: "perceptron.r"
2
3   # Heaviside function with a default epsilon
4   g <- function(net, epsilon =0.5) {
5       if (net > epsilon) {
6           return (1)
7       } else {
8           return (0)
9       }
10  }
11
12  # This is the function to train the Perceptron
13  # Observe eta and threshold assume default values
14  perceptron.train <- function(train.table, eta =0.1,
15                               threshold=1e-2) {
16
17      # Number of input variables
```

```
18    nVars = ncol(train.table)-1
19
20    cat("Randomizing weights and theta in range [-0.5, 0.5]\n
         ")
21
22    # Randomizing weights
23    weights = runif(min=-0.5, max=0.5, n=nVars)
24
25    # Randomizing theta
26    theta = runif(min=-0.5, max=0.5, n=1)
27
28    # This sum of squared errors will accumulate all errors
29    # occurring along training iterations. When this error is
30    # below a given threshold, learning stops.
31    sumSquaredError = 2*threshold
32
33    # Learning iterations
34    while (sumSquaredError > threshold) {
35
36       # Initializing the sum of squared errors as zero
37       # to start counting and later evaluate the total
38       # loss for this dataset in train.table
39       sumSquaredError = 0
40
41       # Iterate along all rows (examples) contained in
42       # train.table
43       for (i in 1:nrow(train.table)) {
44
45          # Example x_i
46          x_i = train.table[i, 1:nVars]
47
48          # Expected output class
49          # Observe the last column of this table
50          # contains the output class
51          y_i = train.table[i, ncol(train.table)]
52
53          # Now the Perceptron produces the output
54          # class using the current values for
55          # weights and theta, then it applies the
56          # heaviside function
57          hat_y_i = g(x_i %*% weights + theta)
58
59          # This is the error, referred to as (y_i - g(x_i))
60          # in the Perceptron formulation
61          Error = y_i - hat_y_i
62
63          # As part of the Gradient Descent method, we here
64          # compute the partial derivative of the Squared
                Error
65          # for the current example i in terms of weights and
66          # theta. Observe constant 2 is not necessary, once
                we
67          # can set eta using the value we desire
68          dE2_dw1 = 2 * Error * -x_i
```

```
69          dE2_dtheta = 2 * Error * −1
70
71          # This is the Gradient Descent method to adapt
72          # weights and theta as defined in the formulation
73          weights = weights − eta * dE2_dw1
74          theta = theta − eta * dE2_dtheta
75
76          # Accumulating the squared error to define
77          # the stop criterion
78          sumSquaredError = sumSquaredError + Error^2
79        }
80
81        cat("Sum_of_squared_errors_=_", sumSquaredError , "\n")
82      }
83
84    # Returning weights and theta , once they represent
85    # the solution
86    ret = list ()
87    ret$weights = weights
88    ret$theta = theta
89
90    return (ret)
91  }
92
93  # This is the function to execute the Perceptron
94  # over unseen data (new examples)
95  perceptron.test <− function(test.table , weights , theta) {
96
97      # Here we print out the expected class (yi) followed by
               the
98      # obtained one (hat_yi) when considering weights and
               theta.
99      # Of course , function perceptron.train should be called
100     # previously to find the values for weights and theta
101     cat("#yi\that_yi\n")
102
103     # Number of input variables
104     nVars = ncol(test.table )−1
105
106     # For every row in the test.table
107     for (i in 1:nrow(test.table)) {
108
109         # Example i
110         x_i = test.table[i, 1:nVars]
111
112         # Expected class for example i
113         y_i = test.table[i, ncol(test.table)]
114
115         # Output class produced by the Perceptron
116         hat_y_i = g(x_i %*% weights + theta)
117
118         cat(y_i, "\t", hat_y_i, "\n")
119     }
120  }
```

```
121
122   # This is an example of learning the simplest problem
123   # To run this example:
124   #    1) Open the R Statistical Software
125   #    2) source("perceptron.r")
126   #    3) perceptron.run.simple()
127   #
128   perceptron.run.simple <- function() {
129
130       # This is a table with training examples
131       train.table = matrix(c(0.0, 0,
132                              0.1, 0,
133                              0.2, 0,
134                              0.3, 0,
135                              0.4, 0,
136                              0.5, 0,
137                              0.6, 1,
138                              0.7, 1,
139                              0.8, 1,
140                      .       0.9, 1,
141                              1.0, 1),
142                      nrow=11,
143                      ncol=2,
144                      byrow=TRUE)
145
146       # This is a table with test examples.
147       # The last column only shows the expected
148       # output and it is not used in the testing stage
149       test.table = matrix(c(0.05, 0,
150                             0.15, 0,
151                             0.25, 0,
152                             0.35, 0,
153                             0.45, 0,
154                             0.55, 1,
155                             0.65, 1,
156                             0.75, 1,
157                             0.85, 1,
158                             0.95, 1),
159                      nrow=10,
160                      ncol=2,
161                      byrow=TRUE)
162
163       # Training the Perceptron to find weights and theta
164       training.result = perceptron.train(train.table)
165
166       # Testing the Perceptron with the weights and theta found
167       perceptron.test(test.table, training.result$weights,
168                       training.result$theta)
169
170       return (training.result)
171   }
```

To run this example, the reader must load the source code in the R Statistical Software. Afterwards, (s)he may execute function `perceptron.run.simple()`.[6] Notice the expected and obtained output classes may contain errors when the input variable $x_{i,1}$ is nearby the transition value 0.5, as it is the point splitting up labels 0 and 1.

By running `perceptron.run.simple()`, the textual output should be similar to the one illustrated in Listing 1.3. At first, $w_1$ and $\theta$ are randomized, then the Perceptron starts iterating on training examples until the sum of squared errors (for the entire set) converges to zero. The expected $y_i$ and the obtained $\hat{y}_i$ classes for every test example are listed. A careful reader may note that one of the output classes $\hat{y}_i$ is not correct in Listing 1.3 because: (1) the hyperplane was fitted according to the training set; and (2) the test set contains unseen examples whose variations were not seen during training.

**Listing 1.3** Text output produced by function perceptron.run.simple()

```
Randomizing weights and theta in range [-0.5, 0.5]...
Accumulated sum of squared errors = 2
Accumulated sum of squared errors = 2
Accumulated sum of squared errors = 2
Accumulated sum of squared errors = 2
Accumulated sum of squared errors = 1
Accumulated sum of squared errors = 0
#yi       hat_yi
0         0
0         0
0         0
0         0
0         0
1         0
1         1
1         1
1         1
1         1
$weights
[1] 0.3979496

$theta
[1] 0.2652314
```

In order to visualize how the Perceptron separates the input space, see function `perceptron.simple.hyperplane.plot(w,t)` in Listing 1.4 which plots the resulting hyperplane. It requires weight $w_1$ and theta $\theta$ obtained after running function `perceptron.run.simple()` from Listing 1.4. An example of hyperplane is illustrated in Fig. 1.36 (note the effect of the heaviside function, which discretizes output values as either 0 or 1).

---

[6]We suggest to execute this function several times, in order to see the effects of using different starting random values for weight $w_1$ and $\theta$.

**Listing 1.4** The Perceptron—function for plotting the hyperplane found for the simplest classification task

```
source("perceptron.r")

# This function plots the hyperplane found for this simplest
# problem which considers a single input variable.
# Variables range.start and range.end define the interval of
# values for the single input variable composing the problem

# This simple problem has a single variable composing every
# example i, which is x_i,1
perceptron.simple.hyperplane.plot <- function(weight, theta,
                                                     range.start=0,
                                                     range.end=1) {

    # Number of variables is 1
    nVars = 1

    # We will now define the same range for the input
        variable.
    # This range will contain 100 discretized values
    range_of_every_input_variable =
                seq(range.start, range.end, length=100)
    x_1 = range_of_every_input_variable

    # Computing net for every input value of variable x_i,1
    all_nets = cbind(x_1, 1) %*% c(weight, theta)

    # This variable all_nets contains all net values for all
    # values assumed by variable x_1. Variable hat_y will
    # contain the Perceptron outputs after applying the
    # heaviside function
    hat_y = rep(0, length(all_nets))
    for (i in 1:length(all_nets)) {
        hat_y[i] = g(all_nets[i])
    }

    # Variable hyperplane will contain two columns, the first
    # corresponds to the input value of x_i,1 and the second
    # to the class produced by the Perceptron
    hyperplane = cbind(x_1, hat_y)

    # Plotting the hyperplane found by the Perceptron
    plot(hyperplane)

    return (hyperplane)
}
```

In order to analyze the obtained hyperplane, the heaviside function is omitted so that the output is only net(.) in function `perceptron.simple.hyperplane.plot.without.g()`. The Perceptron was executed five times using function `perceptron.run.simple()`, recording the corresponding weights and thetas

**Fig. 1.36** Illustration of the hyperplane (dashed line) produced by the Perceptron, with outputs $\hat{y}_i$ along different input values of $x_{i,1}$

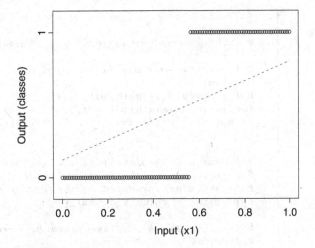

(see Listing 1.5). Then, each pair weight and theta were used to compute five different functions net(.) using function `perceptron.simple.hyperplane.plot.without.g()`.

**Listing 1.5** The Perceptron—plotting the hyperplane in terms of function net

```
source("perceptron.r")

# This function plots the hyperplane found for this simplest
# problem which considers a single input variable and
# function net only.
# Variables range.start and range.end define the interval of
# values for the single input variable composing the problem

# This simple problem has a single variable composing every
# example i, which is x_i,1
perceptron.simple.hyperplane.plot.without.g <-
                function(weight, theta, range.start=0,
                                        range.end=1) {

    # Number of variables is 1
    nVars = 1

    # We will now define the same range for the input
        variable.
    # This range will contain 100 discretized values
    range_of_every_input_variable =
            seq(range.start, range.end, length=100)
    x_1 = range_of_every_input_variable

    # Computing net for every input value of variable x_i,1
    all_nets = cbind(x_1, 1) %*% c(weight, theta)
```

```
# This variable all_nets contains all net values for
    every
# value assumed by variable x_1. Variable hat_y will
    contain
# Perceptron outputs before applying the heaviside
    function
hat_y = rep(0, length(all_nets))
for (i in 1:length(all_nets)) {
    hat_y[i] = all_nets[i] # No heaviside function g(net)
}

# Variable hyperplane will contain two columns, the first
# corresponds to the input value of x_i,1 and the second
# to the class produced by the Perceptron
hyperplane = cbind(x_1, hat_y)

# Plotting the hyperplane found by Perceptron in terms of
# function net
plot(hyperplane)

return (hyperplane)
}
```

Listing 1.6 implements the repeated training executions. Figure 1.37 exemplifies five net(.) functions. Each training stage is executed after calling perceptron.run.simple(), which invokes perceptron.train() and, consequently, the heaviside function $g$(net), using the threshold $\epsilon = 0.5$. By omitting the heaviside function in Fig. 1.37, one can see the effects produced by parameter $\epsilon$, as defined in Eq. (1.10).

Figure 1.37 makes evident that parameter $\epsilon$ is responsible for defining the same central point for hyperplanes in terms of the $y$-axis, so their slopes change but they preserve the same pivot. Even for different positive slopes, the classification results are the same after applying $g$(net). Also observe that training with different values of $\epsilon$ make hyperplanes assume different pivots as shown in Fig. 1.38 for $\epsilon = \{0.25, 0.50, 0.75\}$. In fact, any other value of $\epsilon$ would not affect the final result, once Eq. (1.10) simply uses it as a heaviside threshold, to separate label 0 from 1.

**Listing 1.6** Details about the several training executions

```
source("perceptron-hyperplane-without-g.r")

# This function is used to run the training stage for the
# simplest classification task. Each training will produce
# a different pair of weight and theta, which are then used
# to plot function net
run.several.times <- function(times=5) {

    # Saving the results for each function net
    net.functions = list()

    # For each execution
```

```
for (i in 1:times) {
    # Call training
    training.execution = perceptron.run.simple()

    # Obtaining function net
    net.functions[[i]] =
            perceptron.simple.hyperplane.plot.without.g(
                training.execution$weight,
                training.execution$theta)
}

# Plotting
plot(net.functions[[1]], col=1)
for (i in 2:times) {
    points(net.functions[[i]], col=i)results
}
}
```

We have already advanced in several aspects of the Perceptron algorithm, but illustrations are still needed to understand the error and the squared-error functions. In Listing 1.7, function `perceptron.simple.error()` produces the error, i.e., the difference between the expected and obtained classes $y_i - \hat{y}_i$. It also plots such an error in terms of weight and theta, as shown in Fig. 1.39. Observe weight and theta variations imply in a linear surface of errors.

By using the gradient to modify the free variables so that they minimize the error, the solution would tend to minus infinity because the function has an undefined minimal point, therefore there is no stop condition. One might even think about applying the Newton-Raphson method [18] to find successive approximations for the roots of this real-valued error function, but positive and negative errors might cancel out each other. Consequently, the roots may not represent the zero error as expected.

**Fig. 1.37** Results provided by function net under five different training stages with parameter $\epsilon = 0.5$ (Eq. (1.10))

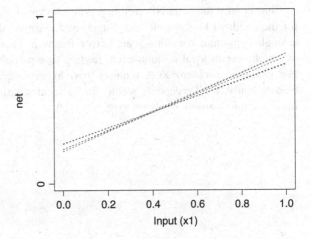

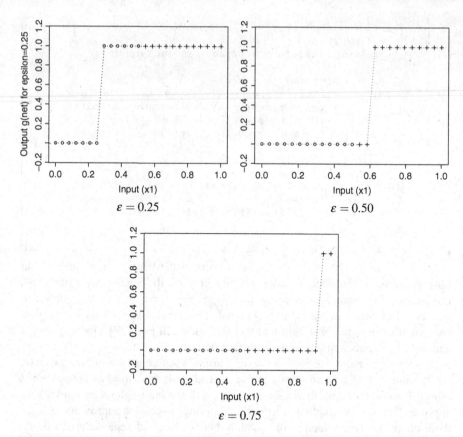

**Fig. 1.38** Result provided by function net and the use of different values for parameter epsilon (Eq. (1.10)): $\epsilon = \{0.25, 0.50, 0.75\}$

That is why the squared-error function (shown in Fig. 1.40) is more adequate for the Gradient Descent method. Supervised learning algorithms assume there is a single minimum for this squared-error function. However, there are scenarios in which several local minima exist, leading to a partial and insufficient solution. For example, problem XOR requires two hyperplanes, as shown in Fig. 1.23. Consequently, the Perceptron would find a local solution, confirming its bias is insufficient to learn the problem XOR.

**Listing 1.7** Functions to study how the Perceptron error behaves

```
source("perceptron.r")

perceptron.simple.error <- function(range.start=-1,
                                     range.end=1, mu=1e-10) {

    # Defining the table with examples
    # Negatives
    table = cbind(seq(0, 0.5, length=100), rep(0,100))
    # Positives
    table = rbind(table, cbind(seq(0.5+mu, 1, length=100),
                                                 rep(1,100)))

    # We will now define the same range for the free
        variables
    # weight and theta. Range contains 100 discretized values
    range_for_free_variables =
                seq(range.start, range.end, length=100)
    weight = range_for_free_variables
    theta = range_for_free_variables

    # Sum of errors while varying weight and theta
    error_function = matrix(0, nrow=length(weight),
                               ncol=length(theta))

    # For each weight
    for (w in 1:length(weight)) {
        # For each theta
        for (t in 1:length(theta)) {
            # Compute all net values
            net = cbind(table[,1], rep(1, nrow(table))) %*%
                                    c(weight[w], theta[t])
            # Defining a vector to save the Perceptron outputs
            hat_y = rep(0, length(net))
            # Producing the output classes
            for (n in 1:length(net)) {
                # g(net) was removed to improve illustration
                hat_y[n] = net[n]
            }

            # These are the expected classes
            y = table[,2]

            # Computing the error
            error = y - hat_y

            # Saving the total error in the matrix
            error_function[w, t] = sum(error)
            # This last instruction makes positive and negative
            # terms cancel each other
        }
    }
}
```

```r
   # Plotting the error
   filled.contour(error_function)
}

perceptron.simple.squared.error <- function(range.start=-10,
                                             range.end=10,
                                             mu=1e-10) {

   # Defining the table with examples
   table = cbind(seq(0, 0.5, length=100), rep(0, 100))
   table = rbind(table, cbind(seq(0.5+mu, 1, length=100),
                                             rep(1, 100)))

   # We will now define the same range for the free
      variables
   # weight and theta. Range contains 100 discretized values
   range_for_free_variables =
         seq(range.start, range.end, length=50)
   weight = range_for_free_variables
   theta = range_for_free_variables

   # Sum of squared errors while varying weight and theta
   error_function = matrix(0, nrow=length(weight),
                              ncol=length(theta))

   # For each weight
   for (w in 1:length(weight)) {
     # For each theta
     for (t in 1:length(theta)) {
       # Compute all net values
       net = cbind(table[,1], rep(1, nrow(table))) %*%
                              c(weight[w], theta[t])
       # Defining a vector to save the Perceptron outputs
       hat_y = rep(0, length(net))
       # Producing the output classes
       for (n in 1:length(net)) {
         # g(net) was removed to improve illustration
         hat_y[n] = net[n]
       }

       # These are the expected classes
       y = table[,2]

       # This is squared to avoid negative and positive
       # values to amortize each other
       squared.error = (y - hat_y)^2

       # Saving the total squared error in the matrix
       error_function[w, t] = sum(squared.error)
     }
   }

   # We apply a log on the squared error function to improve
   # illustration, otherwise we do not see the paraboloid as
```

```
        # clear as in this form.
        filled.contour(log(error_function))
}
```

The concept of convexity is what supports the use of the squared-error function to guide learning. In a simple way, any line segment connecting two function points must lie above or on the curve to make it convex, as shown in Fig. 1.41. The most important reference for the reader is that every convex function has minima. Also notice the set of points above the function (a.k.a. epigraph) must form a convex set, as illustrated in Fig. 1.42.[7] If the supervised learning algorithm considers a convex error function, then the adaptation of free variables will eventually converge to a minimum. As cases of study, this chapter covers the Perceptron and Multilayer Perceptron (MLP) algorithms.

$$\alpha f(a) + (1 - \alpha)f(b) \geq f(\alpha a + (1 - \alpha)b), \quad \forall \alpha \in [0, 1] \qquad (1.17)$$

**Fig. 1.39** The Perceptron: error function $y_i - \hat{y}_i$

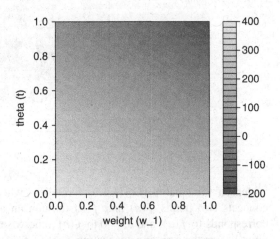

**Fig. 1.40** The Perceptron: squared-error function $(y_i - \hat{y}_i)^2$

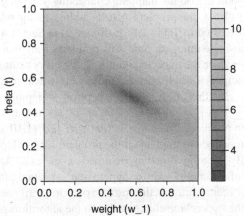

---

[7]The book Convex Optimization [2] is suggested.

**Fig. 1.41** Example of a
convex function

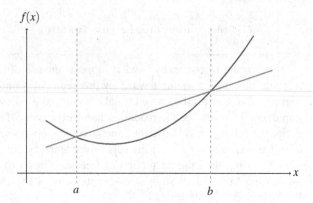

**Fig. 1.42** The epigraph of a
convex function must form a
convex set of points

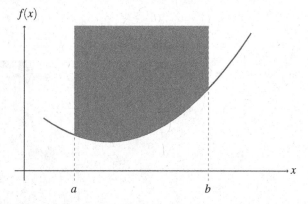

Jensen's inequality (Eq. (1.17)) formalizes when a function $f(.)$ is convex. It connects two points $f(a)$ and $f(b)$ through an affine function, so that $\alpha = 1$ corresponds to $f(a)$, and $\alpha = 0$ to $f(b)$, otherwise, provided $\alpha \in [0, 1]$, the affine point lies on the mapping connecting $f(a)$ to $f(b)$ (see Fig. 1.41).

Next, the problem AND is addressed for which the input and the output spaces are listed in Table 1.5, as discussed in Sect. 1.4 (see Fig. 1.20). Instead of receiving a single input variable, this problem deals with two, i.e. $x_{i,1}$ and $x_{i,2}$, in attempt to produce the expected output $y_i$, for every example $i$. Observe the Perceptron always outputs a single variable for any classification task.

In this task, the Perceptron computes the following function net $=$ $\sum_{j=1}^{2} x_{i,j} w_j + \theta = x_{i,1} w_1 + x_{i,2} w_2 + \theta$ and then applies the activation function to produce classes $\hat{y}_i = g(\text{net})$, $\forall i$ (Eq. (1.10)). Listing 1.8 provides two additional functions to implement the problem AND, which also considers our previous codes. By running function `perceptron.run.AND()`, the Perceptron is trained to infer a model, whose hyperplane is plotted afterwards (Fig. 1.43). We suggest the reader to run function `perceptron.run.AND()` several times and observe how the hyperplane changes, due to the algorithm converges to different solutions given weights and theta are randomly initialized.

**Listing 1.8**  Additional functions to implement the problem AND

```r
source("perceptron.r")

# This is an example of learning the problem AND
#
# To run this example:
#
#       1) Open the R Statistical Software
#       2) source("perceptron-AND.r")
#       3) perceptron.run.AND()
#
perceptron.run.AND <- function() {

    # This is a table with all possible examples
    # for the problem AND. In this case we will
    # use the same table for training and testing,
    # just because we know all possible binary
    # combinations
    table = matrix(c(0, 0, 0,    # 0 AND 0 = 0
                     0, 1, 0,    # 0 AND 1 = 0
                     1, 0, 0,    # 1 AND 0 = 0
                     1, 1, 1),   # 1 AND 1 = 1
                  nrow=4,
                  ncol=3,
                  byrow=TRUE)

    # Training the Perceptron to find weights and theta
    training.result = perceptron.train(table)

    # Testing the Perceptron with the weights and theta found
    perceptron.test(table, training.result$weights,
                            training.result$theta)

    # Plotting the hyperplane found
    perceptron.hyperplane.plot(training.result$weights,
                               training.result$theta)
}

# This function plots the hyperplane found for a given
# classification task with two input variables only.
# Variables range.start and range.end define the interval
# for variables composing the problem. The problem AND
# has two variables composing each example i, which are
# x_i,1 and x_i,2
perceptron.hyperplane.plot <- function(weights, theta,
                                        range.start=0,
                                        range.end=1) {

    # Variable weights define the number of input variables
    # we have, so we can use this information to create
    # axes in a multidimensional input space in order to
    # see how inputs modify the output class provided by
    # the Perceptron
```

```
nVars = length(weights)

# We will now define the same range for every input
# variable. This range will contain 100 discretized
# values
range_of_every_input_variable =
                seq(range.start, range.end, length=100)

x_1 = range_of_every_input_variable
x_2 = range_of_every_input_variable

# Function outer combines every possible value for
# variable x_1 against every possible value for x_2.
# Observe they are continuous values which were never
# seen (we expect either 0 or 1) by this Perceptron
# during the training stage. Also observe value 1
# inside the cbind, which refers to the 1 * theta while
# computing function net. Operation %*% corresponds
# to the dot product.
all_nets = outer(x_1,x_2, function(x,y) {
                cbind(x,y,1) %*% c(weights,theta) } )

# This variable all_nets contains all net values for
# every combination between variables x_1 and x_2.
# Variable y will contain the Perceptron outputs after
# applying the heaviside function
y = matrix(0, nrow=nrow(all_nets), ncol=ncol(all_nets))
for (row in 1:nrow(all_nets)) {
    for (col in 1:ncol(all_nets)) {
        y[row, col] = g(all_nets[row, col])
    }
}

# Plotting the hyperplane found by the Perceptron
filled.contour(x_1, x_2, y)
}
```

The reader should also apply the Perceptron on the problem XOR to notice that it never finds a suitable hyperplane. It will fail for at least 25% of examples because when the hyperplane answers correctly for one of the classes, it will misclassify half of examples associated to the other class, as seen in Fig. 1.44. Since the error does not converge to zero, the threshold must be significantly increased so that the hyperplane can be plotted. As already discussed, the hyperplane will change along every run, providing different classification answers.

Problems such as XOR motivated the design of supervised learning algorithms using multiple hyperplanes. The Multilayer Perceptron (MLP) is a remarkable example among all other algorithms, what motivated its discussion along the next section.

**Fig. 1.43** Hyperplane found
by the Perceptron trained on
the problem AND

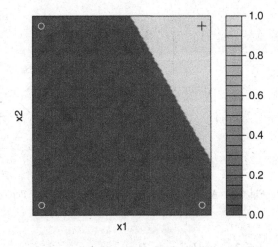

**Fig. 1.44** Different attempts to use a single hyperplane to solve the problem XOR

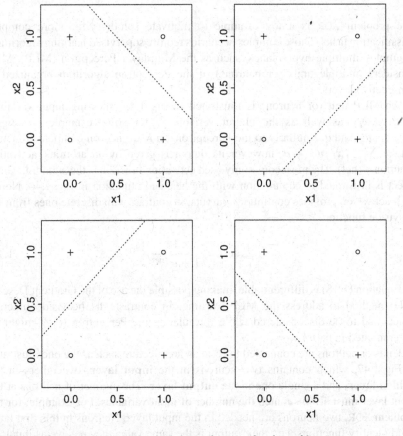

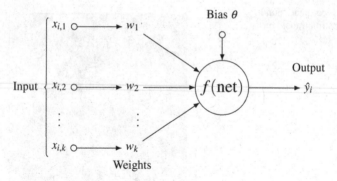

**Fig. 1.45** A single unit or neuron of the Multilayer Perceptron, in which $f(.)$ is the activation function

## 1.5.2  Multilayer Perceptron

The problem XOR is a toy example to motivate solutions for more complex classification tasks. Those complex scenarios require supervised learning algorithms to employ multiple hyperplanes, such as the Multilayer Perceptron (MLP). MLP considers multiple units (or neurons) of the Perceptron algorithm organized in consecutive layers.

An MLP unit (or neuron) is illustrated in Fig. 1.45, showing input variables $x_{i,1}, \ldots, x_{i,k}$ as well as the output variable $\hat{y}_i$ for some example $i$, weights $w_1, \ldots, w_k$ and $\theta$. Similarly to the Perceptron, an MLP neuron computes function net $= \sum_{j=1}^{k} x_{i,j} w_j + \theta$, however its output is given by an arbitrary activation function $f(\text{net})$. The most commonly used activation function is the sigmoid, whose effect is illustrated in conjunction with the heaviside function in Fig. 1.46. Notice such activation provides continuous outputs, in contrast with discrete ones from the heaviside function.

$$f(\text{net}) = \frac{1}{1 + e^{-\text{net}}} \tag{1.18}$$

Equation (1.18) is differentiable, making possible the use of the Gradient Descent (GD) method to address the MLP learning. In contrast, the heaviside function $g(\text{net})$ had to be disconsidered while formulating the Perceptron (or it might be differentiated in parts).

Units or neurons are connected to form an architecture such as the one illustrated in Fig. 1.47, which contains two neurons in the **input layer**, two others in the **hidden layer**, and a single one in the **output layer**. The number of neurons at the input layer must always match the number of input variables. For example, for the problem XOR, two neurons are needed in the input layer. Neurons in this first layer build identity functions, i.e., their output is the same value they receive as input, so weights and theta are not necessary.

**Fig. 1.46** The outputs
produced by the heaviside
versus the sigmoid function
according to the input value
net

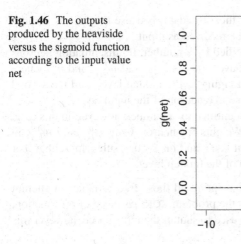

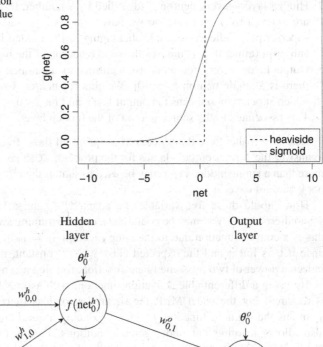

**Fig. 1.47** An example of an MLP architecture with two neurons at the input layer, two others at
the hidden layer, and, finally, a single neuron at the output layer

The hidden layer defines the number of hyperplanes to shatter (or divide) the
input space. Again, the problem XOR requires two hyperplanes, consequently this
layer must be set with two neurons. There are several problems that cannot be
plotted due to the input space dimensionality, thus classification results must be
considered in order to set an adequate number of neurons. At last, the output layer
must contain sufficient neurons to encode all output classes. As XOR is a two-class
problem, a single output neuron is enough, finally concluding the MLP architecture
to tackle this specific classification task.

For such architecture, the following free variables are set:

1. Input layer—there is no weight nor theta to adapt, because they act as identity functions. They output the same value received as input;
2. Hidden layer—every neuron is identified by a number, in this situation neurons are indexed by $j = \{0, 1\}$, so we have weights $w^h_{j,0}$ and $w^h_{j,1}$, and $\theta^h_j$, where superscript $h$ indicates variables belonging to the hidden layer, and the second subscript (either 0 or 1) indexes the source neuron of the input layer;
3. Output layer—every neuron is also identified by a number, however, in this case, there is a single neuron $k = \{0\}$. Weights for neuron $k$ are $w^o_{k,0}$ and $w^o_{k,1}$, in which superscript $o$ means the output layer, and the second subscript (either 0 or 1) is associated to the source neuron of the hidden layer.

Learning is therefore the process of adapting all those free variables in attempt to answer the correct output classes for the problem XOR. As piece of information, more than a single hidden layer may be used to shatter the input space, however this book will not cover it.

How should those free variables be adapted? Again, a loss function such as squared-error $(y_i - \hat{y}_i)^2$ may be computed along the training set. For the Perceptron, this is a convex function due to the simplification $\hat{y}_i = \text{net}_i$, which is easily seen once $\text{net}_i$ is linear and the expected class $y_i$ is a constant, therefore the result is indeed a power of two for some linear function, forming a convex function.

By using a differentiable activation function, such as in MLP, no simplification is required. For the usual MLP, the squared-error loss function takes a constant $y_i$ minus the sigmoid function $f(\text{net}_i)$, producing a quasi-convex function which also allows learning but under some conditions. Figure 1.48 illustrates quasi-convex[8] functions typically associated with the Multilayer Perceptron. Jensen's Inequality 1.17 connects a pair of points $f(a)$ to $f(b)$ so the region between them can be evaluated in terms of convexity. For example, Fig. 1.49 shows the convex interval as a shaded area and the non-convex intervals are outside.

The reader should plot function $(c - f(\text{net}))^2$ for values of net $\in \mathbb{R}$ varying constant $c$.[9] Observe how net influences in the squared-error function, as depicted in Fig. 1.48. If net is too great or too small, observe we may be out of the convex region and any neighborhood around provides no direction for the GD method, i.e., the derivative of this error function is numerically equal to zero.

Figure 1.49 shows a more interesting situation, in which besides interval $[a, b]$ is not convex, the GD method can be applied once the derivative is not null. This scenario can indeed happen to MLP and it is not difficult to understand why. Consider one has two variables $x_{i,1}$ and $x_{i,2}$ for some input example $i$. Let $x_{i,1}$ be in range $[10^5, 10^6]$ and $x_{i,2}$ be in $[-1, 1]$, and weights and thetas be randomly initialized in interval $[-1, 1]$. Consequently, function $\text{net} = x_{i,1}w_1 + x_{i,2}w_2 + \theta$ will produce very large or very small values, impacting the convergence to the minimal squared error. To solve this issue, either $w_1$ could be initialized in a smaller interval

---

[8]We suggest the book Convex Optimization [2] for further references and to complement studies.
[9]Constant $c$ is the expected class and it is in range $[0, 1]$.

**Fig. 1.48** Illustration of quasi-convex functions produced when using the squared-error loss function for the Multilayer Perceptron using different values for $c = \{0, 0.25, 0.5, 0.75, 1\}$ in a function $c - f(\text{net})$

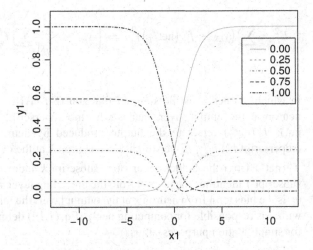

**Fig. 1.49** The effect of net in the quasi-convex function produced by using the squared-error loss function for the Multilayer Perceptron, given constant $c = 0.5$. The region below the line is quasi-convex in terms of net

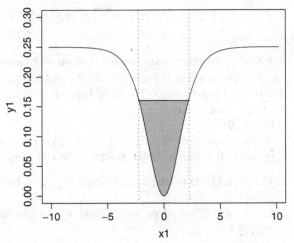

to force $x_{i,1}$ to fit in the same scale as $x_{i,2} w_2$ or all inputs could be normalized in range $[-1, 1]$, so one input will not dominate net. That is why normalization is so used while tackling classification problems with neural networks. By ensuring that, we can proceed with the GD method to find the squared-error minimum.

We formalize MLP with the squared-error function:

$$E_i^2 = (y_i - f(\text{net}_i))^2, \qquad (1.19)$$

in which $f(\text{net}_i)$ corresponds to the output MLP produces for example $i$. Rewriting term $f(\text{net}_i)$:

$$E_i^2 = \sum_k \left(y_{i,k} - f_k^o\left(\text{net}_{i,k}^o\right)\right)^2 \;\; = \left(y_{i,k} - f_k^o\left[\sum_j f_j^h(\text{net}_j^h)w_{k,j}^o + \theta_k^o\right]\right)^2,$$

$$(1.20)$$

in which $f_k^o(\text{net}_{i,k}^o)$ is the sigmoid function (Eq. (1.18)) applied by the $k$th MLP neuron at the output layer, that is why the superscript $o$ is used (meaning output); $f_j^h(\text{net}_j^h)$ refers to the output produced by neuron $j$ at the hidden layer (superscript $h$); $w_{k,j}^o$ is the weight from neuron $k$ at the output layer receiving input $f_j^h(\text{net}_j^h)$ from the hidden layer (first subscript $k$ indexes the owner neuron, and subscript $j$ identifies the neuron from the preceding layer which provides the input); $\theta_k^o$ is the bias term from neuron $k$ at the output layer; the internal summation indexed with $j$ is responsible for computing $\text{net}_{i,k}^o$ (Eq. (1.19) defined this same term as $\text{net}_i$ for simplification purposes), thus:

$$\text{net}_{i,k}^o = \sum_j f_j^h(\text{net}_j^h)w_{k,j}^o + \theta_k^o,$$

and, finally, the external summation in terms of $k$ computes the overall divergences for all output neurons. In particular, for the problem XOR, term $k = 0$ meaning there is only one output neuron. The same happens with the expected output $y_{i,k}$, which is a single value for every example $i$, that is why we used the simplified version $y_i$ in Eq. (1.19).

This squared-error function (Eq. (1.20)) allows us to proceed with the GD method to train all MLP free variables. For XOR, the GD method will consider:

1. Equation (1.21) to adapt every weight $w_{j,l}^h$ connecting input neuron $l$ to hidden neuron $j$;
2. Equation (1.22) to adapt every weight $w_{k,j}^o$ connecting hidden neuron $j$ to output neuron $k$.

$$w_{j,l}^h(t+1) = w_{j,l}^h(t) - \eta \frac{\partial E_i^2}{\partial w_{j,l}^h} \qquad (1.21)$$

$$w_{k,j}^o(t+1) = w_{k,j}^o(t) - \eta \frac{\partial E_i^2}{\partial w_{k,j}^o} \qquad (1.22)$$

To complement, Eqs. (1.23) and (1.24) provide the GD method to adapt the free variable $\theta$ for every neuron at either the hidden or the output layer along iterations:

$$\theta_j^h(t+1) = \theta_j^h(t) - \eta \frac{\partial E_i^2}{\partial \theta_j^h}, \tag{1.23}$$

$$\theta_k^o(t+1) = \theta_k^o(t) - \eta \frac{\partial E_i^2}{\partial \theta_k^o}. \tag{1.24}$$

Despite the Gradient Descent equations are the same for weights in both layers as well as for thetas, the partial derivatives:

$$\frac{\partial E_i^2}{\partial w_{j,l}^h}, \frac{\partial E_i^2}{\partial w_{k,j}^o}, \frac{\partial E_i^2}{\partial \theta_j^h}, \text{ and } \frac{\partial E_i^2}{\partial \theta_k^o}$$

change. Consequently, we must compute the partial derivatives for this problem XOR to exemplify the MLP formulation, however this solution still supports a general-purpose algorithm.

The partial derivatives for the output layer are found in advance, once they are simpler to obtain:

$$\frac{\partial E_i^2}{\partial w_{k,j}^o} = \frac{\partial \sum_k (y_{i,k} - \hat{y}_{i,k})^2}{\partial w_{k,j}^o}$$

$$= \frac{\sum_k \left( y_{i,k} - f_k^o \left( \sum_j f_j^h(\text{net}_j^h) w_{k,j}^o + \theta_k^o \right) \right)^2}{\partial w_{k,j}^o}, \tag{1.25}$$

given a training example $i$, neuron $k$ at the output layer, and neuron $j$ at the hidden layer.

The differentiation employs the chain rule as follows:

$$\frac{\partial E_i^2}{\partial w_{k,j}^o} = \sum_k 2 \left( y_{i,k} - \hat{y}_{i,k} \right) \frac{\partial \left( y_{i,k} - \hat{y}_{i,k} \right)}{\partial w_{k,j}^o},$$

as $y_{i,k}$ is a constant defining the expected class to be produced by output neuron $k$, assuming zero while deriving in terms of $w_{k,j}^o$:

$$\frac{\partial E_i^2}{\partial w_{k,j}^o} = \sum_k 2(y_{i,k} - \hat{y}_{i,k}) \frac{\partial - \hat{y}_{i,k}}{\partial w_{k,j}^o},$$

thus, we still need to solve the derivative $\frac{\partial - \hat{y}_{i,k}}{\partial w_{k,j}^o}$. As the reader may recall, $\hat{y}_{i,k} = f_k^o(\text{net}_{i,k}^o)$, i.e., the output value produced by output neuron $k$ is the result of the sigmoid function, defined in Eq. (1.18):

$$\frac{\partial - \hat{y}_{i,k}}{\partial w_{k,j}^o} = \frac{\partial - f_k^o(\text{net}_{i,k}^o)}{\partial w_{k,j}^o}$$

$$= \frac{\partial - (1 + e^{-\text{net}_{i,k}^o})^{-1}}{\partial w_{k,j}^o}$$

$$= \frac{\partial - \left(1 + e^{-\left[\sum_j f_j^h(\text{net}_j^h) w_{k,j}^o + \theta_k^o\right]}\right)^{-1}}{\partial w_{k,j}^o}.$$

We know the sigmoid function (Eq. (1.18)) has the following derivative in terms of net:

$$\frac{\partial f(\text{net})}{\partial \text{net}} = f(\text{net})(1 - f(\text{net})),$$

what simplifies our formulation:

$$\frac{\partial - f_k^o(\text{net}_{i,k}^o)}{\partial w_{k,j}^o} = -\left[ f_k^o(\text{net}_{i,k}^o)(1 - f_k^o(\text{net}_{i,k}^o)) \frac{\partial \text{net}_{i,k}^o}{\partial w_{k,j}^o} \right],$$

so, we still need to find the following partial derivative:

$$\frac{\partial \text{net}_{i,k}^o}{\partial w_{k,j}^o} = \frac{\partial \sum_j f_j^h(\text{net}_{i,j}^h) w_{k,j}^o + \theta_k^o}{\partial w_{k,j}^o} = f_j^h(\text{net}_{i,j}^h).$$

Connecting all terms, the update rule for weights at the output layer is:

$$w_{k,j}^o(t + 1) = w_{k,j}^o(t) - \eta \frac{\partial E_i^2}{\partial w_{k,j}^o}$$

$$w_{k,j}^o(t + 1) = w_{k,j}^o(t) - \eta \, 2(y_i - \hat{y}_{i,k}) \frac{\partial - \left[ f_k^o(\text{net}_{i,k}^o) \right]}{\partial w_{k,j}^o}$$

$$w_{k,j}^o(t + 1) = w_{k,j}^o(t)$$
$$- \eta \, 2(y_i - f_k^o(\text{net}_{i,k}^o)) \left( -\left[ f_k^o(\text{net}_{i,k}^o)(1 - f_k^o(\text{net}_{i,k}^o)) \right] \right) \frac{\partial \text{net}_{i,k}^o}{\partial w_{k,j}^o}$$

$$w_{k,j}^o(t + 1) = w_{k,j}^o(t)$$
$$- \eta \, 2(y_i - f_k^o(\text{net}_{i,k}^o)) \left( -\left[ f_k^o(\text{net}_{i,k}^o)(1 - f_k^o(\text{net}_{i,k}^o)) \right] \right) f_j^h(\text{net}_{i,j}^h).$$

Similarly, the Gradient Descent formulated for theta from output neuron $k$ is:

$$\theta_k^o(t+1) = \theta_k^o(t) - \eta \, 2(y_i - f_k^o(\text{net}_{i,k}^o)) \left(-\left[f_k^o(\text{net}_{i,k}^o)(1 - f_k^o(\text{net}_{i,k}^o))\right]\right) 1,$$

which has another term instead of $f_j^h(\text{net}_{i,j}^h)$, once the derivative:

$$\frac{\partial \text{net}_{i,k}^o}{\partial \theta_k^o} = \frac{\partial \left[\sum_j f_j^h(\text{net}_{i,j}^h) w_{k,j}^o + \theta_k^o\right]}{\partial \theta_k^o} = 1,$$

results in the number 1. Finally, we have the update rules for weights and thetas at the output layer.

Next, the corresponding rules for the hidden layer must be found. First we detail the derivative of the squared-error in terms of weights $w_{j,l}^h$:

$$\frac{\partial E_i^2}{\partial w_{j,l}^h} = \frac{\partial \sum_k (y_{i,k} - \hat{y}_{i,k})^2}{\partial w_{j,l}^h}$$

$$= \frac{\sum_k \left(y_{i,k} - f_k^o\left(\sum_j f_j^h(\text{net}_{i,j}^h) w_{k,j}^o + \theta_k^o\right)\right)^2}{\partial w_{j,l}^h}$$

$$= \frac{\sum_k \left(y_{i,k} - f_k^o\left(\sum_j f_j^h\left(\sum_l x_{i,l} w_{j,l}^h + \theta_j^h\right) w_{k,j}^o + \theta_k^o\right)\right)^2}{\partial w_{j,l}^h},$$

given:

$$\text{net}_{i,j}^h = \sum_l x_{i,l} w_{j,l}^h + \theta_j^h. \tag{1.26}$$

Thus, the partial derivative is defined as:

$$\frac{\partial E_i^2}{\partial w_{j,l}^h} = 2\sum_k (y_{i,k} - \hat{y}_{i,k}) \left(\frac{\partial y_{i,k} - \hat{y}_{i,k}}{\partial w_{j,l}^h}\right)$$

$$= 2\sum_k (y_{i,k} - f_k^o(\text{net}_{i,k}^o)) \left(\frac{\partial - f_k^o(\text{net}_{i,k}^o)}{\partial w_{j,l}^h}\right),$$

having:

$$\frac{\partial - f_k^o(\text{net}_{i,k}^o)}{\partial w_{j,l}^h} = -\frac{\partial f_k^o(\text{net}_{i,k}^o)}{\partial \text{net}_{i,k}^o}\frac{\partial \text{net}_{i,k}^o}{\partial w_{j,l}^h} = -f_k^o(\text{net}_{i,k}^o)(1 - f_k^o(\text{net}_{i,k}^o))\frac{\partial \text{net}_{i,k}^o}{\partial w_{j,l}^h},$$

in which:

$$\frac{\partial f_k^o(\text{net}_{i,k}^o)}{\partial \text{net}_{i,k}^o} = f_k^o(\text{net}_{i,k}^o)(1 - f_k^o(\text{net}_{i,k}^o)),$$

for the sigmoid function.

Now we solve:

$$\frac{\partial \text{net}_{i,k}^o}{\partial w_{j,l}^h},$$

which is formulated in terms of the chain rule:

$$\frac{\partial \text{net}_{i,k}^o}{\partial w_{j,l}^h} = \frac{\partial \left[ \sum_j f_j^h (\sum_l x_{i,l} w_{j,l}^h + \theta_j^h) w_{k,j}^o + \theta_k^o \right]}{\partial w_{j,l}^h}$$

$$= \frac{\partial f_j^h(\text{net}_{i,j}^h)}{\partial \text{net}_{i,j}^h} \frac{\partial \text{net}_{i,j}^h}{\partial w_{j,l}^h},$$

having $\text{net}_{i,j}^h$ defined in Eq. (1.26), obtaining:

$$\frac{\partial - f_k^o(\text{net}_{i,k}^o)}{\partial w_{j,l}^h} = -\frac{\partial f_k^o(\text{net}_{i,k}^o)}{\partial \text{net}_{i,k}^o} \frac{\partial \text{net}_{i,k}^o}{\partial w_{j,l}^h}$$

$$= -f_k^o(\text{net}_{i,k}^o)(1 - f_k^o(\text{net}_{i,k}^o))\frac{\partial \text{net}_{i,k}^o}{\partial w_{j,l}^h}$$

$$= -f_k^o(\text{net}_{i,k}^o)(1 - f_k^o(\text{net}_{i,k}^o))\frac{\partial f_j^h(\text{net}_{i,j}^h)}{\partial \text{net}_{i,j}^h} \frac{\partial \text{net}_{i,j}^h}{\partial w_{j,l}^h}$$

$$= -f_k^o(\text{net}_{i,k}^o)(1 - f_k^o(\text{net}_{i,k}^o))\left[ f_j^h(\text{net}_{i,j}^h)(1 - f_j^h(\text{net}_{i,j}^h)) \right]\frac{\partial \text{net}_{i,j}^h}{\partial w_{j,l}^h}$$

$$= -f_k^o(\text{net}_{i,k}^o)(1 - f_k^o(\text{net}_{i,k}^o))\left[ f_j^h(\text{net}_{i,j}^h)(1 - f_j^h(\text{net}_{i,j}^h)) \right] x_{i,l},$$

and, finally:

$$\frac{\partial E_i^2}{\partial w_{j,l}^h} = 2 \sum_k \left( y_{i,k} - f_k^o(\text{net}_{i,k}^o) \right) \left( \frac{\partial - f_k^o(\text{net}_{i,k}^o)}{\partial w_{j,l}^h} \right),$$

$$= 2 \sum_k (y_{i,k} - f_k^o(\text{net}_{i,k}^o))$$

$$\left( -f_k^o(\text{net}_{i,k}^o)(1 - f_k^o(\text{net}_{i,k}^o))\left[ f_j^h(\text{net}_{i,j}^h)(1 - f_j^h(\text{net}_{i,j}^h)) \right] x_{i,l} \right).$$

Connecting all terms, the update rule for weights at the hidden layer is:

$$w_{j,l}^h(t+1) = w_{j,l}^h(t) - \eta \frac{\partial E_i^2}{\partial w_{j,l}^h}$$

$$w_{j,l}^h(t+1) = w_{j,l}^h(t) - \eta \, 2 \sum_k (y_i - \hat{y}_{i,k}) \frac{\partial - \hat{y}_{i,k}}{\partial w_{j,l}^h}$$

$$w_{j,l}^h(t+1) = w_{j,l}^h(t) - \eta \, 2 \sum_k (y_i - f_k^o(\text{net}_{i,k}^o)) \frac{\partial - f_k^o(\text{net}_{i,k}^o)}{\partial w_{j,l}^h}$$

$$w_{j,l}^h(t+1) = w_{j,l}^h(t)$$
$$- \eta \, 2 \sum_k (y_i - f_k^o(\text{net}_{i,k}^o))$$
$$\left( -f_k^o(\text{net}_{i,k}^o)(1 - f_k^o(\text{net}_{i,k}^o)) \left[ f_j^h(\text{net}_{i,j}^h)(1 - f_j^h(\text{net}_{i,j}^h)) \right] x_{i,l} \right),$$

and, because some terms are independent of index $k$, we can simplify the previous formulation as follows:

$$w_{j,l}^h(t+1) = w_{j,l}^h(t)$$
$$- \eta \, 2 \sum_k (y_i - f_k^o(\text{net}_{i,k}^o))$$
$$(-f_k^o(\text{net}_{i,k}^o)(1 - f_k^o(\text{net}_{i,k}^o)) \left[ f_j^h(\text{net}_{i,j}^h)(1 - f_j^h(\text{net}_{i,j}^h)) \right] x_{i,l})$$

$$w_{j,l}^h(t+1) = w_{j,l}^h(t)$$
$$- \eta \, 2 \left[ f_j^h(\text{net}_{i,j}^h)(1 - f_j^h(\text{net}_{i,j}^h)) \right]$$
$$x_{i,l} \sum_k (y_i - f_k^o(\text{net}_{i,k}^o))(-f_k^o(\text{net}_{i,k}^o)(1 - f_k^o(\text{net}_{i,k}^o))).$$

Similarly, the Gradient Descent method formulated for theta of hidden neuron $j$ is:

$$\theta_j^h(t+1) = \theta_j^h(t)$$
$$- \eta \, 2 \left[ f_j^h(\text{net}_{i,j}^h)(1 - f_j^h(\text{net}_{i,j}^h)) \right]$$
$$(1) \sum_k (y_i - f_k^o(\text{net}_{i,k}^o))(-f_k^o(\text{net}_{i,k}^o)(1 - f_k^o(\text{net}_{i,k}^o))),$$

given:

$$\frac{\partial \text{net}_{i,j}^h}{\partial \theta_j^h} = 1.$$

We omit constant 2 in all gradient rules, because the gradient step $\eta$ is already a constant set up by the user. After having all the Gradient Descent rules, the MLP source code, using the R language, is detailed in Listing 1.9.

**Listing 1.9**  The Multilayer Perceptron implementation

```r
# This is the MLP sigmoid activation function
f <- function(net) {
    ret = 1.0 / (1.0 + exp(-net))
    return (ret)
}

# This function is used to build up the MLP architecture, i.
    e.,
# the neurons contained in the hidden and the output layers
# with their respective weights and thetas randomly
    initialized.
mlp.architecture <- function(input.layer.size = 2,
        hidden.layer.size = 2,
        output.layer.size = 1,
        f.net = f) {

    # Here we create a list to contain the layers information
    layers = list()

    # This is the hidden layer in which weights and thetas
    # were initialized in a random manner (using runif) in
    # interval [-1,1]. Term input.layer.size+1 refers to
    # the number of neurons in the input layer (a weight
    # per unit), plus an additional element to define theta
    layers$hidden = matrix(runif(min=-1, max=1,
                    n=hidden.layer.size*(input.layer.size+1)),
                    nrow=hidden.layer.size,
                    ncol=input.layer.size+1)

    # The same as the hidden layer happens here, but for the
    # output layer
    layers$output = matrix(runif(min=-1, max=1,
                    n=output.layer.size*(hidden.layer.size+1)),
                    nrow=output.layer.size,
                    ncol=hidden.layer.size+1)

    # Defining a list to return everything:
    # - the number of units or neurons at the input layer
    # - the number of units at the hidden layer
    # - the number of units at the output layer
    # - layers information (including weights and thetas)
    # - the activation function used is also returned
    ret = list()
    ret$input.layer.size = input.layer.size
    ret$hidden.layer.size = hidden.layer.size
    ret$output.layer.size = output.layer.size
    ret$layers = layers
```

```r
    ret$f.net = f.net

    return (ret)
}

# This function produces the MLP output after providing
    input
# values. Term architecture refers to the model produced by
# function mlp.architecture. Term dataset corresponds to the
# examples used as input to the MLP. Term p is associated to
# the identifier of the current example being forwarded.
forward <- function(architecture, dataset, p) {

    # Organizing dataset as input examples x
    x = matrix(dataset[,1:architecture$input.layer.size],
            ncol=architecture$input.layer.size)
    # Organizing dataset as expected classes y associated to
    # input examples x
    y = matrix(
        dataset[,(architecture$input.layer.size+1):ncol(dataset
            )],
        nrow=nrow(x))

    # Submitting the p-th input example to the hidden layer
    net_h = architecture$layers$hidden %*%
                    c(as.vector(ts(x[p,])), 1)
    f_net_h = architecture$f.net(net_h)

    # Hidden layer outputs as inputs for the output layer
    net_o = architecture$layers$output %*% c(f_net_h, 1)
    f_net_o = architecture$f.net(net_o)

    # Here we have the final results produced by the MLP
    ret = list()
    ret$f_net_h = f_net_h
    ret$f_net_o = f_net_o

    return (ret)
}

# This function is responsible for training, i.e., adapting
# weights and thetas for every neuron (or unit). It
    basically
# applies the Gradient Descent Method.
backpropagation <- function(architecture, dataset,
                        eta=0.1, threshold=1e-3) {

    x = matrix(dataset[,1:architecture$input.layer.size],
            ncol=architecture$input.layer.size)
    y = matrix(
        dataset[,(architecture$input.layer.size+1):ncol(dataset
            )],
        nrow=nrow(x))
```

```r
cat("Input_data...\n")
print(x)

cat("Expected_output...\n")
print(y)

cat("Enter_to_start_running...")
readline()

squared_error = threshold * 2

# This loop will run until the average squared error is
# below some threshold value.
while (squared_error > threshold) {

    # Initializing the squared error to measure the loss
    # for all examples in the training set
    squared_error = 0

    # For every example at index (row) p
    for (p in 1:nrow(x)) {

        # Applying the input example at index p
        f = forward(architecture, dataset, p)

        # Getting results to adapt weights and thetas
        error = (y[p,] - f$f_net_o)

        # Computing term delta for the output layer
        # which simplifies next computations involved
        # in the Gradient Descent method
        delta_o = error * f$f_net_o * (1-f$f_net_o)

        # This is the squared error used as stop criterion.
        # Term sum(error^2) is used because the last layer
        # (i.e., the output layer) may have more than a
        # single neuron. We also use a power of two to
        # ensure negative and positive values do not
        # nullify each other.
        squared_error = squared_error + sum(error^2)

        # Computing term delta for the hidden layer
        w_o = architecture$layers$output[,
                1:architecture$hidden.layer.size]
        delta_h = (f$f_net_h * (1 - f$f_net_h)) *
                sum(as.vector(delta_o) * as.vector(w_o))

        # Adapting weights and thetas at the output layer
        architecture$layers$output =
            architecture$layers$output + eta * delta_o %*%
                c(f$f_net_h, 1)

        # Adapting weights and thetas at the hidden layer
        architecture$layers$hidden =
```

```r
        architecture$layers$hidden + eta * delta_h %*%
            c(x[p,], 1)
    }

    # Dividing the total squared error by nrow to find
    # the average which we decided to use as stop
    #     criterion
    squared_error = squared_error / nrow(x)

    # Printing the average squared error out
    cat("Squared_error_=_", squared_error, "\n")
    }

    # Returning the trained architecture, which can now
    # be used for execution.
    return (architecture)
}

# This function is used to test the MLP
mlp.test <- function(architecture, dataset, debug=T) {

    # Organizing dataset as input examples x
    x = matrix(dataset[,1:architecture$input.layer.size],
                ncol=architecture$input.layer.size)

    # Organizing dataset as expected classes y associated to
    # input examples x
    y = matrix(
        dataset[,(architecture$input.layer.size+1):ncol(dataset
            )],
        nrow=nrow(x))

    cat("Enter_to_start_testing...")
    readline()

    output = NULL

    # For every example at index (row) p
    for (p in 1:nrow(x)) {

        # Applying the input example at index p
        f = forward(architecture, dataset, p)

        # If debug is true, show all information
        # regarding classification
        if (debug) {
            cat("Input_pattern_=_", as.vector(x[p,]),
                "_Expected_=_", as.vector(y[p,]),
                "_Predicted_=_", as.vector(f$f_net_o), "\n")
        }

        # Concatenating all output values as rows in a matrix,
        # so we can check them out later.
        output = rbind(output, as.vector(f$f_net_o))
```

```
    }

    # Returning results
    return (output)
}

# This function is useful to produce a discrete
# (either yes or no) hyperplane to shatter the
# input space of examples
discretize.hiperplane <- function(img, range = c(0.45, 0.55)
    ) {
    ids_negative = which(img < range[1])
    ids_positive = which(img > range[2])
    ids_hiperplane = which(img >= range[1] & img <= range[2])

    img[ids_negative] = 0
    img[ids_positive] = 1
    img[ids_hiperplane] = 0.5

    img
}

# This is a function to train and test the XOR problem
xor.test <- function(eta=0.1, threshold=1e-3) {

    # Loading the dataset "xor.dat"
    dataset = as.matrix(read.table("xor.dat"))

    # Building up the MLP architecture with random weights
    # and thetas. Observe we have two units at the input
    # layer (what is the number of input variables), we
    # have two units at the hidden layer (so we will have
    # two hyperplanes to shatter the space of examples as
    # expected), and we have a single unit at the output
    # layer to provide the answer as 0 or 1 (actually
    # values in range [0,1])
    model = mlp.architecture(input.layer.size = 2,
                             hidden.layer.size = 2,
                             output.layer.size = 1,
                             f.net = f)

    # Now we train the architecture "model" to build up
    # the "trained.model"
    trained.model = backpropagation(model, dataset, eta=eta,
                                    threshold=threshold)

    # Then we test the "trained.model" using the same
    # XOR dataset. For more complex problems, we will use
    # unseen examples.
    mlp.test(trained.model, dataset)

    # Building up hyperplanes to plot
    x = seq(-0.1,1.1,length=100)
    hiperplane_1 = outer(x,x,
```

```
                    function(x,y) {  cbind(x,y,1) %*%
                        trained.model$layers$hidden[1,] } )

    hiperplane_2 = outer(x,x,
                    function(x,y) {  cbind(x,y,1) %*%
                        trained.model$layers$hidden[2,] } )

    cat("Press_enter_to_plot_both_hiperplanes...")
    readline()

    # Plotting the hyperplanes built at the hidden layer
    filled.contour(discretize.hiperplane(hiperplane_1) +
                    discretize.hiperplane(hiperplane_2))
}
```

We approach the problem XOR by calling function xor.test() with the default parameters. However, before invoking it, we need to load its source code using source("mlp.r"); (using the R Statistical Software). After running xor.test(), the reader will observe the input examples and expected output classes, as shown next:

```
> xor.test()
Input data...
      [,1] [,2]
[1,]    0    0
[2,]    0    1
[3,]    1    0
[4,]    1    1
Expected output...
      [,1]
[1,]    0
[2,]    1
[3,]    1
[4,]    0
Enter to start running...
```

then, by typing "enter", messages about the training iterations are shown, until the average squared error converges to some value below the pre-defined threshold:

```
Squared error =   0.00100089
Squared error =   0.001000691
Squared error =   0.001000491
Squared error =   0.001000292
Squared error =   0.001000093
Squared error =   0.0009998944
Enter to start testing...
```

by typing "enter" again, the classification results are provided:

```
Input pattern =  0 0  Expected =  0  Predicted =  0.03463937
Input pattern =  0 1  Expected =  1  Predicted =  0.9699704
Input pattern =  1 0  Expected =  1  Predicted =  0.97008
Input pattern =  1 1  Expected =  0  Predicted =  0.03163266
Press enter to plot both hiperplanes...
```

**Fig. 1.50** The hyperplanes plotted after training the Multilayer Perceptron to solve the problem XOR

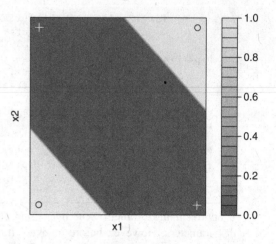

Observe the obtained outputs are not exactly the same as the expected ones, however they close enough. By setting a smaller threshold in Listing 1.9, one can obtain a better approximation to the expected classes. If the user types "enter" again, (s)he will see the two hyperplanes inferred during the training stage, as illustrated in Fig. 1.50.

For illustration purposes, consider the problem of Optical Character Recognition (OCR), in which the characters are represented using binary matrices:

$$
\begin{array}{ccccccc}
0 & 0 & 0 & 0 & 0 & 0 & 0 \\
0 & 0 & 1 & 1 & 1 & 0 & 0 \\
0 & 1 & 1 & 1 & 1 & 1 & 0 \\
0 & 1 & 1 & 0 & 1 & 1 & 0 \\
0 & 1 & 1 & 0 & 1 & 1 & 0 \\
0 & 1 & 1 & 1 & 1 & 1 & 0 \\
0 & 1 & 1 & 1 & 1 & 1 & 0 \\
0 & 1 & 1 & 0 & 1 & 1 & 0 \\
0 & 1 & 1 & 0 & 1 & 1 & 0 \\
0 & 0 & 0 & 0 & 0 & 0 & 0
\end{array}
\quad \text{and} \quad
\begin{array}{ccccccc}
0 & 0 & 0 & 0 & 0 & 0 & 0 \\
0 & 1 & 1 & 1 & 1 & 0 & 0 \\
0 & 1 & 1 & 0 & 1 & 1 & 0 \\
0 & 1 & 1 & 0 & 1 & 1 & 0 \\
0 & 1 & 1 & 1 & 1 & 0 & 0 \\
0 & 1 & 1 & 0 & 1 & 0 & 0 \\
0 & 1 & 1 & 0 & 1 & 1 & 0 \\
0 & 1 & 1 & 0 & 1 & 1 & 0 \\
0 & 1 & 1 & 1 & 1 & 1 & 0 \\
0 & 0 & 0 & 0 & 0 & 0 & 0
\end{array}
,
$$

in which the first represents character "A" and, the second, character "B". So, let MLP be trained to learn those characters. As first step, one needs to set the number of input units (or neurons) for this MLP instance. Observe matrices contain $10 \times 7$ cells, so there will be 70 neurons at the input layer. Since, no one knows[10] how to set the number of hyperplanes, i.e., number of units in the hidden layer, we could guess

---

[10]We can take advantage of the Statistical Learning Theory to set an adequate number of hyperplanes depending on the target problem. This is discussed in Chap. 2.

it by trying some number such as 3 or 5, what is fine for a first attempt. Finally, a single output neuron is used, so MLP could associate "A" to the output value 0 and "B" to 1. That is indeed possible, however what if we decide to "activate" one output neuron for every different example? In such situation, we should define two neurons at the output layer, then when "A" is given as input, MLP would produce the pair $(1, 0)$, and when "B" is received, it would output $(0, 1)$. In this manner, each output neuron represents one input character. By defining the number of output neurons as the number of classes, MLP tends to avoid as much as possible any class overlapping. Such a class mixing may happen due to the output layer combines the hyperplanes built in the hidden layer.

Listing 1.10 includes the previous source code (Listing 1.9) and adds up a new function to train and test the OCR problem. The reader must load this code, run function `ocr.test()` and attempt other parameters instead of the default ones.

**Listing 1.10** Additional function to solve the OCR problem

```
source ("mlp.r")

# Solving the Optical Character Recognition (OCR) problem.
ocr.test <- function(eta =0.1, threshold=1e-3) {

    # Loading the dataset
    dataset = as.matrix(read.table("ocr-asvector.dat"))

    # Loading a test set with unseen examples
    test.dataset = as.matrix(read.table("test-ocr-asvector.
        dat"))

    # Building up the architecture with 70 units at the input
    # layer, 5 units (so 5 hyperplanes) at the hidden layer
    # and 2 at the output layer.
    model = mlp.architecture(input.layer.size = 10*7,
                hidden.layer.size = 5,
                output.layer.size = 2, f.net = f)

    # Training
    trained.model = backpropagation(model,
            dataset,
            eta=eta,
            threshold=threshold)

    # Testing for unseen examples
    mlp.test(trained.model, test.dataset)
}
```

The files used in this problem are `ocr-asvector.dat` and `test-ocr-asvector.dat`. Both transform those previously presented binary matrices into binary vectors, and add the expected output values at the end. After calling function `ocr.test()`, we will see the following output:

```
Input data...
      [,1] [,2] [,3] [,4] [,5] [,6] [,7] [,8] [,9] [,10]
[1,]   0    0    0    0    0    0    0    0    0    1
[2,]   0    0    0    0    0    0    0    0    1    1


      [,11] [,12] [,13] [,14] [,15] [,16] [,17] [,18] [,19]
        [,20]
[1,]     1     1     0     0     0     1     1     1     1
         1
[2,]     1     1     0     0     0     1     1     0     1
         1


      [,21] [,22] [,23] [,24] [,25] [,26] [,27] [,28] [,29]
        [,30]
[1,]     0     0     1     1     0     1     1     0     0
         1
[2,]     0     0     1     1     0     1     1     0     0
         1


      [,31] [,32] [,33] [,34] [,35] [,36] [,37] [,38] [,39]
        [,40]
[1,]     1     0     1     1     0     0     1     1     1
         1
[2,]     1     1     1     0     0     0     1     1     0
         1


      [,41] [,42] [,43] [,44] [,45] [,46] [,47] [,48] [,49]
        [,50]
[1,]     1     0     0     1     1     1     1     1     0
         0
[2,]     0     0     0     1     1     0     1     1     0
         0


      [,51] [,52] [,53] [,54] [,55] [,56] [,57] [,58] [,59]
        [,60]
[1,]     1     1     0     1     1     0     0     1     1
         0
[2,]     1     1     0     1     1     0     0     1     1
         1


      [,61] [,62] [,63] [,64] [,65] [,66] [,67] [,68] [,69]
        [,70]
[1,]     1     1     0     0     0     0     0     0     0
         0
[2,]     1     1     0     0     0     0     0     0     0
         0

Expected output...
      [,1] [,2]
[1,]   1    0
[2,]   0    1
Enter to start running...
```

in which two vectors were provided, the first corresponding to the matrix representing character "A" and the second to "B". Next, the expected values are listed, activating either one of the output neurons according to the input character. After typing "enter", some output information similar to the following is produced:

```
Squared  error  =   0.001000569
Squared  error  =   0.001000473
Squared  error  =   0.001000378
Squared  error  =   0.001000283
Squared  error  =   0.001000188
Squared  error  =   0.001000092
Squared  error  =   0.0009999973
Enter  to  start  testing...
```

which corresponds to the average squared error produced at every training iteration, until the stop criterion is reached, i.e., converging to threshold. Another "enter" will produce a result similar to:

```
Input  pattern  =   0 1 0 0 0 0 1 0 0 1 1 1 0 0 1 1 1 1 1 1 0 0
                    0
                    1 0 1 1 0 0 1 1 0 0 1 0 0 1 1 1 1 1 0 0 1 1 0 1 1 0 0 1
                    1
                    0 1 1 0 0 1 0 0 1 1 0 0 0 1 0 0 0 0
Expected  output  =   1 0
Obtained  output  =   0.9632798  0.03591298

Input  pattern  =   0 0 0 0 0 0 0 0 1 1 1 1 0 0 0 1 1 1 1 1 0 1
                    1
                    0 0 1 1 0 0 1 1 1 1 0 0 0 1 1 1 1 0 0 0 1 1 0 1 1 0 1 1
                    0
                    0 1 1 0 0 1 1 1 1 1 0 0 0 0 0 0 0 0
Expected  output  =   0 1
Obtained  output  =   0.05629617  0.9356224
```

which shows both input examples (in form of vectors), their expected and obtained outputs. The first input example should produce the following output pair $(1, 0)$ as it corresponds to a noisy version of character "A". In fact, it produced $(0.96327978, 0.03591298)$, being very close to the expected values once the first neuron is obviously activated. The second example should produce $(0, 1)$ and, in fact, it generated $(0.05629617, 0.93562236)$, respecting the idea of mostly activating the second neuron.

To improve visualization, those two input vectors are shown next, after being organized as binary matrices:

```
0 1 0 0 0 0 1    0 0 0 0 0 0 0
0 0 1 1 1 0 0    0 1 1 1 1 0 0
1 1 1 1 1 1 0    0 1 1 1 1 1 0
0 0 1 0 1 1 0    1 1 0 0 1 1 0
0 1 1 0 0 1 0    0 1 1 1 1 0 0
0 1 1 1 1 1 0    0 1 1 1 1 0 0
0 1 1 0 1 1 0    0 1 1 0 1 1 0
0 1 1 0 1 1 0    1 1 0 0 1 1 0
0 1 0 0 1 1 0    0 1 1 1 1 1 0
0 0 1 0 0 0 0    0 0 0 0 0 0 0
```

Both test examples contain some noise, however MLP is still capable of learning how to separate them out. We now suggest the reader to extend this problem to cover all English characters and digits, train and test the MLP for unseen examples (with some random noise, for example). We suggest as many neurons at the output layer as the number of possible input characters. About the number of hidden units, the reader may try different values in attempt to produce good results. More information will be provided on that matter throughout this book. Finally, the reader may download the datasets available at the UC Irvine Machine Learning Repository (http://archive.ics.uci.edu/ml), design, train and test different MLP instances on other classification tasks, such as Iris (http://archive.ics.uci.edu/ml/datasets/Iris) and Wine (http://archive.ics.uci.edu/ml/datasets/Wine). The reader may also perform experiments using the MNIST handwritten digit database and compare the MLP results against others reported in literature [7].

## 1.6   Concluding Remarks

This chapter presented a brief review on Machine Learning (ML), mainly focusing on supervised learning algorithms. This type of learning was tackled due to it relies on the theoretical foundation provided by the Statistical Learning Theory (SLT). After introducing the main aspects of the SLT, such as its assumptions, the concept of loss function, the empirical and expected risks, more information was provided about how the Bias-Variance Dilemma is considered in the context of ML and, finally, introduced two well-known supervised learning algorithms (Perceptron and Multilayer Perceptron) for illustration purposes. All those concepts are essential to support the next chapters.

## 1.7   List of Exercises

1. Run the Perceptron on the problem XOR and assess the obtained results. Take your own conclusions why this algorithm cannot learn in such scenario;
2. Randomly select 30% of the examples contained in the Iris dataset to compose your test set. Then use the remaining 70% to form your training set. Start training an Multilayer Perceptron using three hidden neurons and, then, compute the error for the same training examples and, afterwards, the error taking the test examples. Observe how those two errors deviate one another as the number of hidden neurons is increased. A good enough model has to provide close enough errors. When their absolute deviation grows, it means MLP is considering too many (or too few if reducing) hidden units.[11]
3. Using the Wine dataset, train and test the Multilayer Perceptron. Then, assess the obtained results.[12]
4. Based on the previous exercise, notice how the quasi-convexity present in the MLP squared-error function affects training and testing. So, normalize all Wine input attributes in range [0, 1] and observe the new results.
5. Consider the MNIST database to train and test a Multilayer Perceptron algorithm with a varying number of hidden neurons.[13] MNIST is already segmented into training and test sets. Compute errors on the test set and compare to the ones listed in http://yann.lecun.com/exdb/mnist.

## References

1. C.M. Bishop, *Pattern Recognition and Machine Learning*. Information Science and Statistics (Springer, New York, 2006)
2. S. Boyd. L. Vandenberghe, *Convex Optimization* (Cambridge University Press, New York, 2004)
3. G. Carlsson, F. Mémoli, Characterization, stability and convergence of hierarchical clustering methods. J. Mach. Learn. Res. **11**, 1425–1470 (2010)
4. M.E. Celebi, K. Aydin, *Unsupervised Learning Algorithms* (Springer International Publishing, Berlin, 2016)
5. C.M. Grinstead, L.J. Snell, *Grinstead and Snell's Introduction to Probability* (American Mathematical Society, Providence, 2006), Version dated 4 July 2006 edition
6. S. Haykin, *Neural Networks: A Comprehensive Foundation*, 3rd edn. (Prentice-Hall, Upper Saddle River, 2007)
7. Y. LeCun, C. Cortes, MNIST handwritten digit database (2010). http://yann.lecun.com/exdb/mnist/

---

[11]Download the Iris dataset from the UCI Machine Learning Repository available at archive.ics.uci.edu/ml.

[12]Download the Wine dataset from the UCI Machine Learning Repository available at archive.ics.uci.edu/ml.

[13]The MNIST database is available at http://yann.lecun.com/exdb/mnist.

8. T.M. Mitchell, *Machine Learning*, 1st edn. (McGraw-Hill, New York, 1997)
9. H.-L. Nguyen, Y.-K. Woon, W.-K. Ng, A survey on data stream clustering and classification. Knowl. Inf. Syst. **45**(3), 535–569 (2015)
10. R Development Core Team, *R: A Language and Environment for Statistical Computing* (R Foundation for Statistical Computing, Vienna, 2008). ISBN 3-900051-07-0
11. F. Rosenblatt, The perceptron: a perceiving and recognizing automaton, Technical report 85-460-1, Cornell Aeronautical Laboratory, 1957
12. W.C. Schefler, *Statistics: Concepts and Applications* (Benjamin/Cummings Publishing Company, San Francisco, 1988)
13. B. Scholkopf, A.J. Smola, *Learning with Kernels: Support Vector Machines, Regularization, Optimization, and Beyond* (MIT Press, Cambridge, 2001)
14. R.M.M. Vallim, R.F. de Mello, Unsupervised change detection in data streams: an application in music analysis. Prog. Artif. Intell. **4**(1), 1–10 (2015)
15. V.N. Vapnik, *Statistical Learning Theory*. Adaptive and Learning Systems for Signal Processing, Communications, and Control (Wiley, Hoboken, 1998)
16. V. Vapnik, *The Nature of Statistical Learning Theory*. Information Science and Statistics (Springer, New York, 1999)
17. U. von Luxburg, B. Schölkopf, *Statistical Learning Theory: Models, Concepts, and Results*, vol. 10 (Elsevier North Holland, Amsterdam, 2011), pp. 651–706
18. T.J. Ypma, Historical development of the Newton-Raphson method. SIAM Rev. **37**(4), 531–551 (1995)

# Chapter 2
# Statistical Learning Theory

## 2.1 Motivation

This chapter starts by describing the necessary concepts and assumptions to ensure supervised learning. Later on, it details the Empirical Risk Minimization (ERM) principle, which is the key point for the Statistical Learning Theory (SLT). The ERM principle provides upper bounds to make the empirical risk a good estimator for the expected risk, given the bias of some learning algorithm. This bound is the main theoretical tool to provide learning guarantees for classification tasks. Afterwards, other useful tools and concepts are introduced.

As discussed in Chap. 1, there is a great variety of algorithms used to approach classification tasks. In some way, those algorithms divide the input space of examples into different regions, creating a set of decision boundaries according to some supervised learning process. By learning, we mean inferring rules to work on unseen data from examples organized in terms of the pair: input variables (or attributes) and labels (or classes). In order to recall the basic notation, a classifier is a function $f : X \rightarrow Y$, in which[1]:

1. $X = \{x_1, \ldots, x_n\}$ is the input space, in which every $x_i$ is typically in the Euclidean space and may contain multiple dimensions, in form $x_i \in \mathbb{R}^k$;
2. $Y = \{y_1, \ldots, y_n\}$ is the output space, given every $y_i$ is also usually in the Euclidean space and may contain more than a single dimension, in form $y_i \in \mathbb{R}^q$;
3. $f$ is referred to as model, classifier or classification function.

The input space $X$ is composed of variables or attributes associated to the classification task, which must be representative enough to allow the proper learning

---

[1]It is worth to mention that elements in $X$ and $Y$ may be even in another space, such as the Topological, but some mapping is considered to bring them to the Hilbert space in order to respect the definition.

© Springer International Publishing AG, part of Springer Nature 2018
R. Fernandes de Mello, M. Antonelli Ponti, *Machine Learning*,
https://doi.org/10.1007/978-3-319-94989-5_2

of a concept. For instance, in order to classify an object as a domestic cat in contrast, for example, to a horse, the "number of legs" is not a good attribute. Better features would be "has whiskers", "has claws", etc. The output space $Y$ contains all labels, classes or categories to be learned (and later predicted) given examples in $X$. In a binary scenario, it is usual to define $Y \in \{-1, +1\}$, having classes as either positive or negative. Note that $f$ is called a classifier—a model, map or function that is capable of producing a label $y_i$ from some input vector $x_i \in X$—and not a classification algorithm. For illustration purposes, $f$ can be a set of weights and biases after training an instance of the Multilayer Perceptron, or also a set of parameters computed using the Logistic Regression algorithm. Therefore, classifier $f$ is the *result of a supervised learning algorithm after the training stage*.

After this first step, some important questions arise in the context of supervised learning: How to prove that a given algorithm is capable of learning from examples? How good can some classifier $f$ be? Those are the main motivations for the Statistical Learning Theory (SLT) [1, 15, 16], a *theoretical framework designed to understand and assess learning*, under some reasonable assumptions.

**The key notion of learning theory** is to find an algorithm that, provided enough data, outputs a classification hypothesis with a probability close to one, given a small error. In other words, it concerns finding guarantees for classification tasks, and quantifying how much data is required to obtain such learning guarantees. Two main parts are fundamental for the study of this learning theory: (1) the first is the Empirical Risk Minimization (ERM) principle, which approximates the true (and unknown) loss function by taking only the observed examples (from the training set), and uses the hypothesis that minimizes the error inside the training set; (2) the second involves finding a trade-off between the complexity of the hypothesis space, i.e., the bias of the classification algorithm, and the classification error computed on the training data (the empirical risk).

**Assumptions** are necessary to ensure learning in the context of the SLT, as proposed by Vapnik [15, 16]. They are as follows:

1. No assumption is made about the joint probability function $P(X \times Y)$;
2. Examples must be sampled from $P(X \times Y)$ in an independent manner;
3. Labels may assume non-deterministic values due to noise and class overlapping;
4. $P(X \times Y)$ is fixed (static, so it does not change along time);
5. $P(X \times Y)$ is unknown at the training stage.

As a consequence of those assumptions, this chapter introduces more concepts on joint probabilities, loss functions, risk and generalization applied to classifiers.

## 2.2   Basic Concepts

Most of the content of this section is covered in Chap. 1, but here we revisit them under another perspective, making the reader more aware about some specific properties in the light of the learning theory. We begin by defining joint probabilities, data independency, loss functions and risk as well as we raise the question of how to provide learning guarantees through generalization. We then discuss the consistency of classifiers, and look at typical undesired scenarios in which the resulting model fails to learn, i.e., it will overfit or underfit examples.

### 2.2.1 Probability Densities and Joint Probabilities

Probability density functions (PDFs) support us to estimate the likelihood of the occurrence of a given event $X$ assuming values in some interval. This is important because one can determine $P(X)$, for every possible value of $X = x$, only for discrete random variables, while continuous random variables require the definition of intervals. This happens due to the probability of $X$ on any particular value $x$ (there are infinite possible values) tends to 0, and that is why some interval is required.

Given a probability density function (PDF) $f(x)$, the probability of an interval $A$ is given by the area under the function $f(x)$ along $A$, i.e., the integral of $f(x)$ over $A$, written as:

$$P(A) = \int_A p(x)dx.$$

For example, observe the PDF depicted in Fig. 2.1, whose areas under the curve define the probability of randomly selecting a value within intervals $A$ and $B$: the light gray area is $P(0.229 \leq A \leq 0.231) = 0.02$ and the dark gray area is $P(0.249 \leq B \leq 0.251) = 0.05$. Note that, if one chooses a very narrow interval, the area (and thus the probability) will approach zero because there are infinite numbers for $x \in \mathbb{R}$.

Having this notation, the conditions required for a measurable function to be a valid density function are:

**Fig. 2.1** Example of a probability density function over the real line ($x$-axis). Shaded in light gray is the area for the interval $0.229 \leq A \leq 0.231$, which is $P(A) = 0.02$, and, in dark, the area for the interval $0.249 \leq B \leq 0.251$, $P(B) = 0.05$

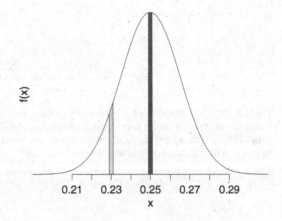

$$f(x) \geq 0 \text{ for all } x$$

$$\int_{\mathbb{R}} f(x)dx = 1,$$

in which the first guarantees that $P(A) \geq 0$ for any set $A$, and, the second, that the probability along the whole real line must sum up to 1 (or 100% as humans usually consider).

To motivate with an example, consider a company sells fresh mushrooms in packs of portions weighing 0.25 kg. One could well figure out that mushroom portions will not weigh exactly 0.25 lb. In fact, even considering a scale of limited precision (say 4 digits), the probability of finding a pack weighing exactly 0.2500 is still low. In practice, to study the probabilities of buying a pack weighing such specific value in kilograms, we estimate a PDF using the following procedure: first, we randomly select 10,000 packs of mushrooms and weigh those, creating a histogram of the resulting weights. A plot of a histogram by grouping the observations into 8 bins would look like Fig. 2.2a. By increasing the number of bins to 30 (and, therefore, decreasing the steps or intervals in the $x$-axis), the histogram would look like Fig. 2.2b. Now let intervals eventually get so small that we would represent the probability distribution of $X$ as a curve (see the curve fitted along the histogram of Fig. 2.2b). We could, by using the data collected weighing 10,000 packs, determine the probability that a randomly selected pack weighs between 0.24 and 0.26 lb, i.e., $P(0.24 < X < 0.26)$, as illustrated through the shaded area shown in Fig. 2.2c.

A classifier is an approximation of a special probability density function referred to as $P(X \times Y)$, also known as the **joint probability density function**, which describes the joint behavior of two variables, having $X$ as the input space of examples (variables or attributes used to compose examples) and $Y$ as the output

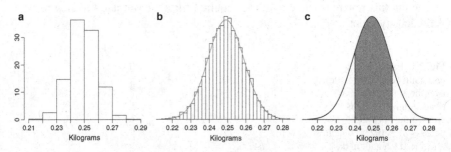

**Fig. 2.2** Practical example of a probability density function. (**a**) By looking at the density frequency of observations, it is possible to picture how the data is distributed; (**b**) with sufficient data and by increasing the resolution of the histogram, we can fit a continuous function as shown in gray; (**c**) PDF to compute probabilities over intervals

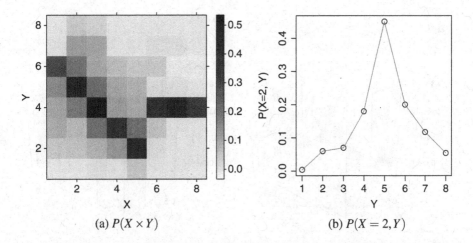

(a) $P(X \times Y)$        (b) $P(X = 2, Y)$

**Fig. 2.3** Example of a joint probability distribution between variables, showing how $X$ and $Y$ depend on each other: (**a**) for each pair $X$, $Y$, the grayscale shows the joint probability value; (**b**) a curve that explains the probabilities for $Y$, when observing a fixed value $X = 2$

space (or classes). In order to be considered a joint probability density function, $P(X \times Y)$ is required to satisfy the following conditions:

$$P(X \times Y) \geq 0 \text{ for all } X, Y$$

$$\int \int P(X \times Y) dX dY = 1.$$

A synthetic example of a joint distribution over two variables $X$ and $Y$ is shown in Fig. 2.3a for 8 possible values of each variable. The grayscale codifies the probability value for each combination $(X, Y)$, which is seen as a probability map. Observe that there is a clear behavior for $X = 1 \cdots 5$, and, then, a different behavior for $X = 6 \cdots 8$. By fixing the event of the variable $X$ and varying $Y$, we have a curve for the joint probability $P(X, Y)$ for a given $X$, for example $P(X = 2, Y)$ as shown in Fig. 2.3b. Fixing $X = 2$, variable $Y$ has its maximum probability at $P(X = 2, Y = 5) = 0.45$.

To improve the comprehension, consider an example of a joint probability density function produced after rolling a die from which two discrete random variables $X$ and $Y$ were obtained, respectively:

1. Let $X = 1$ if an **even** number occurs, i.e., 2, 4 or 6, and $X = 0$ for any **odd** number;
2. Let $Y = 1$ if the number is **prime**, i.e., 2, 3 or 5, and $Y = 0$ otherwise.
3. The joint probabilities for values of $X$ and $Y$ are given by:

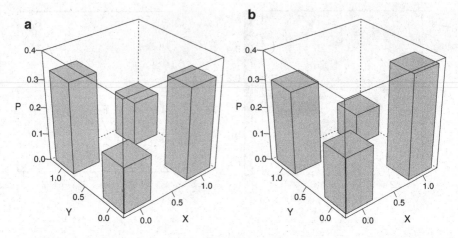

**Fig. 2.4** (**a**) Illustration of the joint probability distribution $P(X \times Y)$ for the problem of rolling a die, and (**b**) a candidate classifier $f_i(X)$ obtained for the same problem

$$P(X = 0, Y = 0) = P\{\{1, 3, 5\} \cap \{1, 4, 6\}\} = \frac{1}{6}$$

$$P(X = 1, Y = 0) = P\{\{2, 4, 6\} \cap \{1, 4, 6\}\} = \frac{2}{6}$$

$$P(X = 0, Y = 1) = P\{\{1, 3, 5\} \cap \{2, 3, 5\}\} = \frac{2}{6}$$

$$P(X = 1, Y = 1) = P\{\{2, 4, 6\} \cap \{2, 3, 5\}\} = \frac{1}{6}.$$

Notice we do not have a continuous problem in this scenario, but a discrete one. If we observe $X = 0$, should we most probably expect to have a prime number $Y = 1$ or a composite (non-prime) number $Y = 0$? Looking at the joint probabilities above, when observing an odd number ($X = 0$), we are most likely to have a prime than a composite one. Figure 2.4a illustrates $P(X \times Y)$ for this problem.

Let us see the same problem in the point of view of some classifier $f$. Note that $f$ is an approximation function for the joint probability density function $P(X \times Y)$. Given we had access to a sample of pairs $\{(x_1, y_1), \ldots, (x_n, y_n)\} \in X \times Y$, we wish to obtain the best approximation that produces:

$$P(X = 0, Y = 0) = \frac{1}{6}$$

$$P(X = 1, Y = 0) = \frac{2}{6}$$

$$P(X = 0, Y = 1) = \frac{2}{6}$$

$$P(X = 1, Y = 1) = \frac{1}{6}.$$

In this case, pairs $(x, y)$ are easy to be listed and obtained, because they are part of a finite set of possibilities, which is not possible when dealing with continuous variables. From this example, when attempting to predict whether the die produced a prime number ($Y = 1$) or not ($Y = 0$) given the number is even ($X = 1$), we have:

$$P(X = 1, Y = 0) = \frac{\frac{2}{6}}{\frac{1}{6} + \frac{2}{6}} \approx 0.66$$

$$P(X = 1, Y = 1) = \frac{\frac{1}{6}}{\frac{1}{6} + \frac{2}{6}} \approx 0.33,$$

and, as consequence, we have most likely $Y = 0$. Similarly, if the die provided an odd number, i.e., $X = 0$, then:

$$P(X = 0, Y = 0) = \frac{\frac{1}{6}}{\frac{1}{6} + \frac{2}{6}} \approx 0.33$$

$$P(X = 0, Y = 1) = \frac{\frac{2}{6}}{\frac{1}{6} + \frac{2}{6}} \approx 0.66,$$

and we should go for $Y = 1$. Although there is a most probable event, even for this simple problem, the random variable $X$ is not enough to provide a single answer (100% sure). This is because the relationship between $X$ and $Y$ may contain uncertainties. In another example (already discussed in Chap. 1), we could have a variable $X$ describing people's heights, while $Y$ is associated to sex. In such a circumstance, just observing a certain height, we would also be unsure whether a person is male or female because $X$ is not sufficient to allow separating sexes.

Looking again at Fig. 2.4a that illustrates the joint probability density function $P(X \times Y)$ for the problem involving a die, now Fig. 2.4b presents a candidate classifier $f_i(x \in X)$, or simply $f(X)$. This classifier $f(X)$ has some divergence or difference when compared to $P(X \times Y)$, which can be measured using an integral as follows:

$$R(f) = \int_{X \times Y} \| P(X \times Y) - (X, f(X)) \| \, dX \times Y,$$

having some pointwise norm of differences between $P(X \times Y)$ and $f(X)$. This divergence defines the expected risk, or simply the risk, of classifier $f(X)$, estimated after a training stage performed on a given sample $\{(x_1, y_1), \ldots, (x_n, y_n)\} \in X \times Y$.

The expected risk of classifier $f(X)$ will be referred throughout this book as $R(f)$. Consequently, the best as possible classifier is the one that minimizes such risk, in form:

$$f_{\text{best}} = \arg\min_{f_i} R(f_i), \quad \forall i. \tag{2.1}$$

If $i$ represents a finite number of possibilities, we can assess all of them to find $f_{\text{best}}$, otherwise we should attempt to obtain an approximation, by iteratively converging to the best classifier.

When one or more random variables, i.e. $X$ and $Y$, are continuous, the risk $R(f)$ cannot be calculated because the fifth assumption states we do not have full access to the joint probability density function $P(X \times Y)$ and, consequently, $f_{\text{best}}$ cannot be found as in Eq. (2.1). Not having full access to $P(X \times Y)$ also implies that this function is not known beforehand, otherwise we could just employ some fitting strategy to find its parameters. For example, if $P(X \times Y)$ is known to be a 2-dimensional Gaussian distribution, the JPD could be estimated via its parameters mean and standard deviation for $X$ and $Y$ in order to find $f_{\text{best}}$.

Instead of assuming knowledge about $P(X \times Y)$, we assume the sampling of pairs $\{(x_1, y_1), \ldots, (x_n, y_n)\} \in X \times Y$ is possible from this joint probability density function, which is then used to estimate candidate classifiers $f_i(X)$ and select the best one. Now the reader may ask how to assess the quality of a given classifier $f$ if $R(f)$ is not computable. In fact, a computable approximation for this risk exists as seen later on this chapter. It is also worth to mention that the discrete set of examples must always be sampled in an independent and identically manner, as described next.

### 2.2.2   Identically and Independently Distributed Data

According to the SLT, we assume examples are sampled in an identically and independently form (or identically and independently distributed—i.i.d.) from the joint probability density function, also referred to as joint probability distribution, $P(X \times Y)$. This basically means that the probability of obtaining a first training example, such as the pair $(x_1, y_1)$, does not affect the probabilities of subsequent drawings, e.g., $(x_2, y_2)$.

Suppose we have an opaque bowl containing four numbered balls: 1, 2, 3 and 4. Let a ball be randomly drawn from the bowl, its number noted, and placed back into the bowl which is then shaked before the next draw. We know the probability of getting a ball with a particular number tag is always $\frac{1}{4}$: if we first draw the ball number 1, there is no change in the probability for future drawings. This is often called drawing *with replacement*. When drawing *without replacement*, we do not put back the ball after it was drawn. Instead, we randomly choose from the remaining balls, changing the probabilities after every draw. Before drawing any ball, the probability of getting ball #1 is $\frac{1}{4}$. Suppose we first draw ball number 1. Then, in the next round, the probability of drawing ball #1 is zero and the remaining others now have a probability of $\frac{1}{3}$. In this situation, we say the events are dependent on each other.

Note that it is rare to be able to perform a census for a given task, that is, to completely collect data from some target population. That is why in practice data analysis is usually carried out using a *sample* that represents a subset of the individuals (commonly referred to as examples or observations), and it is often a small fraction of the population. When i.i.d. sampling is used, each example in the population has the same chance of being observed. If during the data collection, one was permitted to choose the examples to be sampled, it is most likely that such a sample would be skewed towards some subset of the universe of possible observations. Another possible scenario is called convenience sample, in which those easily accessible individuals/examples are more likely to be sampled. In those two cases the i.i.d. assumption is not valid.

**Independent and Dependent Sampling**   In summary, the samples are dependent if the values in one sample affect the values in the other; the samples are independent if the values in one sample reveal no information about samples drawn next.

**Examples of Sampling in Practice**

- *Face detection*:

  - A sample of face images coming from a random group of different ethnicities, sex and age can be considered an *independent* sample in terms of human faces;
  - In the same application, a sample of face images coming from a group of graduate students from a research institute or laboratory is likely to be *dependent*, since they may have biased characteristics.

- *Handwritten recognition*:

  - In character recognition, it is safe to assume that randomly selected characters written by a large number of people are *independent* in terms of the whole population of handwritten characters;
  - In a system that attempts to predict a word in a given language, e.g. English, the sample (sequence of characters) is *dependent* due to the current observation affects the following ones, for example, when a consonant is written, say "z", it reveals information about the next character, for example the probability of observing "j" will be close to zero, while observing vowels will have greater likelihoods.

- *Drug effectiveness to blood pressure reduction*:

  - If we sample the blood pressure of the same group of people before and after they are medicated, there is *dependency* once measurements were taken from the same people: the ones with the highest blood pressure in the first measure will likely have the highest blood pressure in the second one as well;

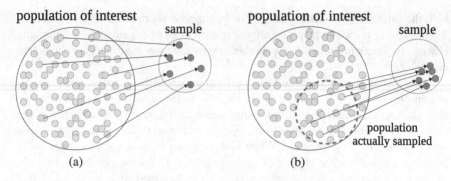

**Fig. 2.5** An example of sampling resulting in (**a**) i.i.d. and (**b**) non-i.i.d. data

- On the other hand, by giving one group of people an active drug and a placebo to a different group, both randomly assigned, the samples can be considered independent because knowing something about the distribution of values when measuring one group does not provide information about the distribution of values in the second one.

In Fig. 2.5, we show an illustration of sampling that will result in both i.i.d. and non-i.i.d. data: using the face detection example, the scenario in Fig. 2.5a happens when we collect data from a large and random group of people from different ethnicities, ages, and sexes; whereas, in (b), we have a sample that is skewed towards some sub-group of the population.

In data stream problems [4, 14], it is common to have one observation affecting the probability of the next ones. In scenarios such as depicted in Fig. 2.6a, by using a set of current observations it is possible to predict with future data which makes the i.i.d. assumption invalid. Note that it could be the case that a data stream is independent, and then it is not possible to predict future observations based on a set of observed ones, see for example Fig. 2.6b. Therefore, while in the second example we consider the data to be i.i.d., in the first one, the i.i.d. assumption is invalid.

In another practical scenario, consider the Sunspot [19] time series illustrated in Fig. 2.7. Observe there is a trend component with sinusoidal characteristics, indicating some dependency among data observations. If we use the same observations from this time space to form the training and test sets, examples will be dependent. The Statistical Learning Theory is not adequate to deal with such type of data.

In the case of Sunspot, it is not difficult to check that if some value in range [150, 160] is observed at the current time instant, there is a high probability that the next value will also lie in the same interval or around it. On the other hand, the probability of drawing a next far value, e.g. one in range [0, 10], is close to zero. Here, the time plays a very important role to define the dependencies among data observations (or data examples). To better understand this scenario, Fig. 2.8a shows a histogram that provides a simplified view for the distribution of the Sunspot time series. Next, we compare how knowledge about the current value $x(t)$ influences

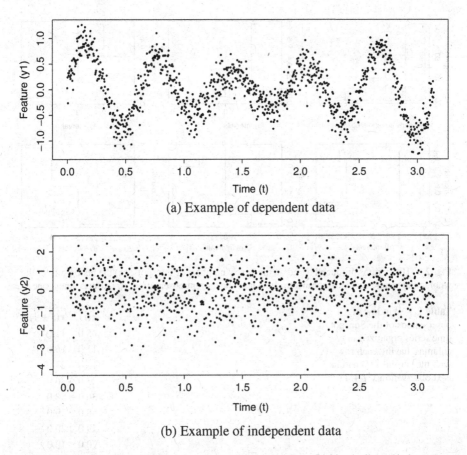

(a) Example of dependent data

(b) Example of independent data

**Fig. 2.6** Examples of data collected over time: (**a**) dependent data; (**b**) identically and independent distributed data

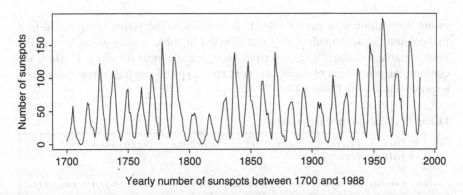

Yearly number of sunspots between 1700 and 1988

**Fig. 2.7** Yearly numbers of sunspots, as from the World Data Center (SIDC). Obtained from the package "datasets" of the R Statistical Software

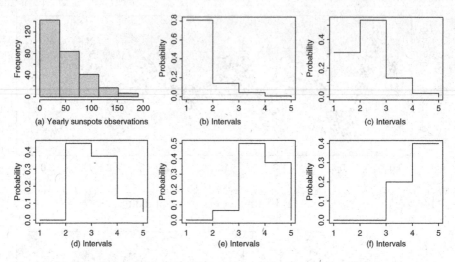

**Fig. 2.8** Histograms produced based on the yearly numbers of Sunspots: (**a**) the histogram for the whole time series; (**b–f**) histograms to show how the current observation $x(t)$ affects a next one

**Table 2.1** Window of observations of the Sunspot time series organized in two columns, having each row with the current $x(t)$ and its next observation $x(t + 1)$

| $x(t)$ | $x(t + 1)$ |
|--------|-----------|
| 5.0 | 11.0 |
| 11.0 | 16.0 |
| 16.0 | 23.0 |
| 23.0 | 36.0 |
| 36.0 | 58.0 |
| 58.0 | 29.0 |
| 29.0 | 20.0 |
| 20.0 | 10.0 |
| 10.0 | 8.0 |
| 8.0 | 3.0 |

in the distribution of a next $x(t + 1)$. The Sunspot time series is organized in a two-column dataset, a subset of those is listed in Table 2.1, in which $x(t)$ is the input variable $X$ and $x(t + 1)$ is seen as the expected output (or class) $Y$. Then, we compute the probability of finding the next $x(t+1)$ providing the current value $x(t)$, as presented in code Listing 2.1.

**Listing 2.1** Read and show regression data

```
# This package is necessary to execute function embedd()
require(tseriesChaos)

# Producing a 2-column dataset with examples (x(t), x(t+1))
X = embedd(ts(sunspot.year), m=2, d=1)
```

```r
# Finding the minimal and maximal values for this time
    series
minValue = min(sunspot.year)
maxValue = max(sunspot.year)

# Splitting the values into 5 intervals so we will use each
    of
# them to compute probabilities
nIntervals = 5

# Compute the probability for every interval
intervals = seq(minValue, maxValue, length=nIntervals+1)
for (i in 1:nIntervals) {
    startInterval = intervals[i]
    endInterval = intervals[i+1]

    cat("Interval [", startInterval, ", ", endInterval,
        "] has probabilities for a next value\n")
    for (j in 1:nIntervals) {
        # Defining the current value for
        # startInterval <= x(t) < endInterval
        ids = which(X[,1] >= startInterval & X[,1] <
            endInterval)

        # Counting occurrences inside every interval for
        x(t+1)
        inside = sum(X[ids,2] >=
                        intervals[j] & X[ids,2] < intervals[j+1])

        # Estimating the probabilities for x(t+1)
        probability = inside / nrow(X[ids,])

        cat("\trange [", intervals[j], ", ",
            intervals[j+1], "] = ", probability, "\n")
    }
}
```

The following output is obtained after running Listing 2.1:

```
Interval [0, 38.04] has probabilities for a next value
    range [0, 38.04] = 0.8098592
    range [38.04, 76.08] = 0.1408451
    range [76.08, 114.12] = 0.04225352
    range [114.12, 152.16] = 0.007042254
    range [152.16, 190.2] = 0
Interval [38.04, 76.08] has probabilities for a next value
    range [0, 38.04] = 0.3095238
    range [38.04, 76.08] = 0.5357143
    range [76.08, 114.12] = 0.1309524
    range [114.12, 152.16] = 0.02380952
    range [152.16, 190.2] = 0
Interval [76.08, 114.12] has probabilities for a next value
    range [0, 38.04] = 0
    range [38.04, 76.08] = 0.45
```

```
        range  [76.08,   114.12]  =  0.375
        range  [114.12,  152.16]  =  0.125
        range  [152.16,  190.2]   =  0.05
Interval  [114.12,  152.16]  has  probabilities  for  a  next  value
        range  [0,  38.04]   =  0
        range  [38.04,  76.08]   =  0.0625
        range  [76.08,  114.12]  =  0.5
        range  [114.12,  152.16]  =  0.375
        range  [152.16,  190.2]   =  0
Interval  [152.16,  190.2]  has  probabilities  for  a  next  value
        range  [0,  38.04]   =  0
        range  [38.04,  76.08]   =  0
        range  [76.08,  114.12]  =  0.2
        range  [114.12,  152.16]  =  0.4
        range  [152.16,  190.2]   =  0.4
```

Notice probabilities change for different intervals of $x(t)$. Those changes occur due to some level of dependency between variables $x(t)$ and $x(t+1)$, matching $X = x(t)$ and $Y = x(t + 1)$. Figure 2.8 shows the data distribution for the entire time series. Figure 2.8b–f illustrate the distributions while taking some current observation $x(t)$ to predict a next $x(t + 1)$. Observe how the current observation, drawn from one of the five intervals, affects the probability of a next.

To complement, let a time series be represented by the random variable $S$ produced using a Normal distribution with mean equals to 0 and variance equals to 1, i.e., $\mathcal{N}(\mu = 0, \sigma^2 = 1)$, as shown in Fig. 2.9. By executing the same code on this new time series, we obtain histograms such as presented in Fig. 2.10a–e. In this situation, observe distributions are very similar, meaning that $s(t)$ has no influence in a next observation, once data examples are independent from each other.

By analyzing the Sunspot and the Normal distributed series through the Auto-Correlation Function [3], one can assess dependencies, as shown in Fig. 2.11. Figure 2.11a confirms dependencies among data observations for Sunspot due to the great correlation values. The second series does not contain relevant dependencies, once correlations are below 10% (dashed line in Fig. 2.11b).

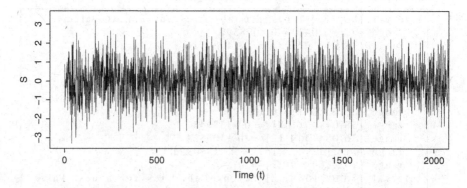

**Fig. 2.9** First 2000 instances of the randomly generated time series

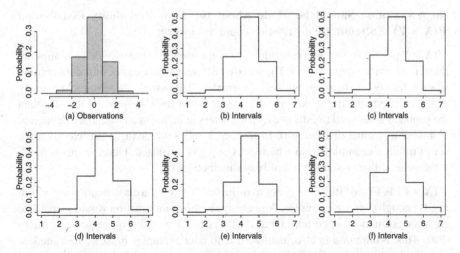

**Fig. 2.10** Histograms based on the random variable $S$ following a Normal distribution: (**a**) the histogram for the entire time series; (**b–f**) histograms to show how the current observation $s(t)$ affects a next one

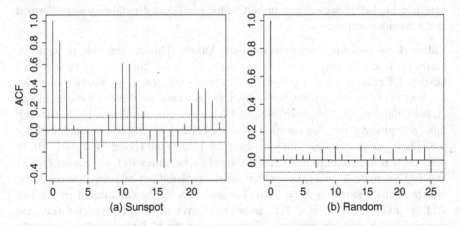

**Fig. 2.11** Autocorrelation function applied on: (**a**) the Sunspot time series; and (**b**) the Normal distributed time series

## 2.2.3 Statistical Learning Theory Assumptions

Provided the foundations, we detail the assumptions taken by the Statistical Learning Theory (SLT) in order to ensure supervised learning:

**Examples Must be Sampled in an Independent Manner** By assuming independency, SLT is sure the data distribution never changes so $P(X \times Y)$ can be estimated using a sample;

**No Assumption Should be Made About the Joint Probability Distribution**
$P(X \times Y)$ Therefore the distribution could be any one;

$P(X \times Y)$ **Is Unknown at Training** As a consequence of the previous assumption, there is no prior about $P(X \times Y)$, what is different than considering data coming from a family of distributions (e.g. Normal and Poisson), as required by many statistical methods. If one has any knowledge about the family of $P(X \times Y)$, then the problem is reduced to estimate its parameters (e.g., mean and standard deviation, if it were a Normal distribution). Instead, SLT relies on having a sufficiently large set of training examples from which $P(X \times Y)$ is estimated. Observe this is much more general than assuming a family of distributions;

$P(X \times Y)$ **Is Fixed/Static** In order to estimate $P(X \times Y)$, a classification algorithm needs enough training examples. Any change in this joint function would jeopardize such an estimation, once additional data examples would require some learning drift. This assumption is also justified due to data examples must be independent from each other, otherwise $P(X \times Y)$ would change along samplings, as illustrated in the previous section. As a consequence, SLT is not suitable for problems involving dependent data, as typical for real-world time series and with concept drift detection [4, 14]. However, it is important to assess data dependency as performed in the previous section;

**Labels Can Assume Non-deterministic Values** This assumption is justified, firstly because data may contain noisy labels, i.e., some class $y_i$ may be incorrect labeled. Of course, a small portion of labels is expected to be wrong, otherwise learning would be impossible. Secondly, it is fair to assume there is some degree of class overlapping, as previously discussed in the problem of rolling a die. Illustrating this concept using the classification of sex according to people's heights, it is not possible to assign a unique label for someone 1.70-m tall (given by $X = 1.70$). In practice, it is important to find the conditional probabilities $P(Y = \text{female}|X = x)$ and $P(Y = \text{male}|X = x)$, i.e., which are the probabilities of such a person to be female or male knowing $X = 1.70$? For instance, let the conditional probability $P(Y = \text{male}|X = 1.7) = 0.6$, an average error of 40% is expected (i.e., the complement of such probability). That concludes the SLT requirements to ensure supervised learning. Next sections formalize this very useful framework.

### 2.2.4   Expected Risk and Generalization

SLT intends to obtain a **measure of performance** for any classifier $f$ built upon some sample $\mathscr{D} = \{(x_1, y_1), (x_2, y_2), \cdots, (x_n, y_n)\}$ from a fixed joint probability density function $P(X \times Y)$. This measure, referred to as **expected risk**, is defined using the expected value of the loss or risk of classifier $f$, when every possible example from $P(X \times Y)$ is evaluated:

$$R(f) := E(\ell(X, Y, f(X)),$$

in which $\ell(.)$ is a loss function. This risk quantifies the integral of divergences between the expected outcomes and those obtained after applying $f(X)$ $\forall X$, given some $P(X \times Y)$.

As discussed in Chap. 1, 0–1 and squared losses are common functions to quantify classification and regression error, as defined in Eqs. (2.2) and (2.3), respectively.

$$\ell_{0-1}(X, Y, f(X)) = \begin{cases} 1, & \text{if } f(X) \neq Y, \\ 0, & \text{otherwise.} \end{cases} \tag{2.2}$$

$$\ell_{\text{squared}}(X, Y, f(X)) = (Y - f(X))^2 \tag{2.3}$$

Given $f$ is linear, the squared-loss function produces a convex optimization problem, making possible the use of the Gradient Descent method to adapt learning parameters, as previously discussed in Chap. 1.

The **expected risk** should not be confused with the **average error** for classifier $f$. In fact, they are related but different: the average error is calculated on a sample, while the expected risk assumes some joint probability density function $P(X \times Y)$, so the integral of divergences between $P(X \times Y)$ and its estimator $f$ can be computed, in form:

$$\int_{X \times Y} \| \ell(P(X \times Y) - (X, f(X))) \| \, dX \times Y,$$

in which $\ell(.)$ is the selected loss function.

The expected risk considers all possible data examples, including the ones **out of the training set** (defined as population in Statistics [11]). This is only available in simplistic scenarios, such as the rolling of a die (Chap. 1), for which $P(X \times Y)$ is known. To contrapose, take the problem of playing soccer according to the assessment of temperature and humidity, as mentioned in Chap. 1. Because those variables are continuous, we cannot know $P(X \times Y)$ for every possible pair. Observe the input variable $X$ is given by a pair of temperature and humidity values, while the output class $Y$ can assume "yes" or "no". This would be the same as having a space with two axes associated to real values and a third to a binary (discrete) variable.

For realistic classification tasks, a finite set of $n$ examples $(x, y)$ sampled from $P(X \times Y)$ is often insufficient to compute the expected risk. On the other hand, those training examples can be used to compute the **empirical risk** of some classifier $f$:

$$R_{\text{emp}}(f) = \frac{1}{n} \sum_{i=1}^{n} \ell(x_i, y_i, f(x_i)).$$

The empirical risk assesses only a finite training set. That might not be enough, since the best classifier estimated using those examples could perform poorly when applied on unseen examples. This context emphasizes the importance of having the empirical risk as a good estimator for the expected risk.

**Generalization** is the concept relating the empirical and the expected risks, as follows:

$$G = |R_{\text{emp}}(f) - R(f)|.$$

This difference allows to understand how some classifier $f$ performs on unseen examples, i.e. how such classifier behaves provided new data. A classifier is said to generalize when the difference between those risks is sufficiently small, meaning it performs similarly over seen and unseen examples. This implies the empirical risk must be a good estimator of the expected risk, so that we use it to select the best among all candidate classifiers.

An important theoretical step may be noticed: if generalization is ensured, one could use empirical risks of classifiers $f_0, \ldots, f_k$ to select the best for new data. The proof of this claim is **the central subject for the Statistical Learning Theory**, detailed later in this chapter. Also observe **any classifier presenting good generalization does not necessarily imply risks are small**. Generalization simply informs us when the empirical risk, i.e., $R_{\text{emp}}(f)$, is a good estimator of $R(f)$, supporting the selection of the best classifier using $R_{\text{emp}}(f)$. The concept of generalization is paramount for the area of Machine Learning, since it ensures learning from some finite sample.

## 2.2.5  Bounds for Generalization: A Practical Example

Let a binary classification task in which two vectors $\mathbf{w}^+$ and $\mathbf{w}^-$ correspond to the average of positive and negative training instances, respectively. The difference vector:

$$\mathbf{w} = \mathbf{w}^+ - \mathbf{w}^-,$$

is normal, i.e., orthogonal, to the hyperplane separating examples from both classes [12].

Listing 2.2 illustrates this problem with 200 random numbers drawn from two 2-dimensional Gaussian distributions, whose mean vectors are $\mu^+ = [0.9, 0.5]$ for the positive class, and $\mu^- = [0.3, 0.3]$ for the negative. This instance used a diagonal covariance matrix with $\sigma = 0.1$ to represent the linear dependencies for such 2-dimensional vectors.

**Listing 2.2** Read and show regression data

```
# Creating the data positive and negative examples
n <- 200
sigma <- 0.1
m_p <- c(0.9, 0.5)
m_n <- c(0.3, 0.3)
posc <- rbind(rnorm(n, m_p[1], sigma), rnorm(n, m_p[2],
    sigma))
negc <- rbind(rnorm(n, m_n[1], sigma), rnorm(n, m_n[2],
    sigma))

# Scatterplot of data
plot(t(posc), pch=1, xlim=c(0,1.2), ylim=c(0,1.0))
par(new = TRUE)
plot(t(negc), pch=5, axes=F, xlab="", ylab="")
```

We can then compute and plot the average vectors for each class, as well as the hyperplane, which is the eigenvector orthogonal to **w**. It crosses **w** at its central point as seen in Fig. 2.12, and exemplified in Listing 2.3.

**Listing 2.3** Read and show regression data

```
# Computing the average for each class
w_p <- c(mean(posc[1,]), mean(posc[2,]))
w_n <- c(mean(negc[1,]), mean(negc[2,]))

# plotting the average vectors
arrows(0,0, w_p[1],w_p[2])
arrows(0,0, w_n[1],w_n[2])

# Computing the vector of the difference 'w'
w <- w_p - w_n

# the vector to be plotted is more interesting if we
# translate its origin to the point where w_n ends
wvec <- c(w_n[1], w_n[2], w[1]+w_n[1], w[2]+w_n[2])
arrows(wvec[1], wvec[2], wvec[3], wvec[4], col=2)

# Computing the central point of 'wvec'
# Obs: wvec[1] = wvec x_1 point; wvec[2] = wvec y_1 point
#      wvec[3] = wvec x_2 point; wvec[4] = wvec y_2 point
wmid <- c( mean(c(wvec[1],wvec[3])), mean(c(wvec[2],wvec[4])
    ))

# The hyperplane is orthogonal to wvec, so we compute two
    points
# by rotating the wvec 90 degrees in both directions and
    adding
# the central point of vector 'w'
hx <- c(-(wvec[4]-wvec[2])+wmid[1], (wvec[4]-wvec[2])+wmid
    [1])
hy <- c( (wvec[3]-wvec[1])+wmid[2], -(wvec[3]-wvec[1])+wmid
    [2])
```

```
# finally , we plot the resulting hyperplane
lines(hx, hy, col=3)
```

In Fig. 2.12, we see the decision boundary, i.e., hyperplane, is not perfect because there is some degree of class overlapping, thus the empirical risk is not zero. In fact, this classifier is very sensitive to outliers: as an exercise, try to include a single outlier example, then compute the hyperplane again to see how it changes.

Despite its limitations, this classifier is yet representative in the point of view of risk, because it always outputs the same hyperplane for a given training set, consequently its empirical risk is deterministic. Considering vectors $\mathbf{w}_D^+$ and $\mathbf{w}_D^-$, computed using a training dataset $\mathbf{x} \in D$, the empirical risk of classifier $f$ is defined as:

$$
\begin{aligned}
E\left[f(D)\right] &= E_D\left[\mathbf{w}_s^+ - \mathbf{w}_s^-\right] \\
&= E_D\left[\mathbf{w}_s^+\right] - E_D\left[\mathbf{w}_s^-\right] \\
&= E_{y=+1}\left[\mathbf{x}\right] - E_{y=-1}\left[\mathbf{x}\right].
\end{aligned}
$$

Let us see, in practice, how the expected value of the classifier changes for samples under different sizes. This allows to analyze how likely a classifier has been misled by the training set. In particular, we are interested in how to bound the probability of some classifier being misled, i.e., what we believe to be the best classifier $f$, serving the lowest empirical risk, may fail for unseen examples, as confirmed by a greater expected risk.

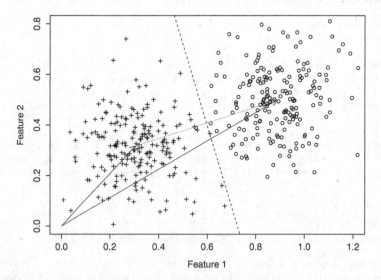

**Fig. 2.12** Example of a two-class dataset: positive and negative. The black arrows represent the average vectors computed for each class, the light gray arrow depicts $\mathbf{w}$, and the dark gray dashed line is the decision boundary, i.e., hyperplane inferred by this classification algorithm

We then wish to bound the probability of error and ask the question: how likely have we been misled by the worst possible function $f$, i.e., the function that provides $R_{emp}(f) \sim 0$ (looks good on the training set), while, in fact, the expected risk is $R(f) > \varepsilon$? Let $n$ be the sample size, this can be written for a single function as $P(|R_{emp}(f) - R(f)| > \epsilon)$, and given the sample was independently and uniformly drawn, then the probability would be $(1 - R(f))$ for the first example.[2] Due to the sample independence, the probabilities are multiplied, yielding:

$$P(|R_{emp}(f) - R(f)| > \epsilon) = (1 - R(f))^n,$$

since we are assuming an error equal or greater than $\epsilon$, it is possible to rewrite as an inequality:

$$P(|R_{emp}(f) - R(f)| > \epsilon) \leq (1 - \epsilon)^n,$$

finally, we approximate it to an exponential using Stirling's formula [5, 20], producing a slightly larger term:

$$P(|R_{emp}(f) - R(f)| > \epsilon) \leq \exp(-\epsilon n).$$

In this case, the probability a classifier $f$ produces an empirical risk equals to zero, while the expected risk is greater than $\epsilon$, becomes exponentially smaller as the sample size $n \to \infty$. To find $\epsilon$ that satisfies some divergence $t$, we set:

$$\exp(-\epsilon n) = t$$

$$-\epsilon n = \ln(t)$$

$$n = \frac{1}{\epsilon} - \ln(t)$$

$$n = \frac{1}{\epsilon} \ln\left(\frac{1}{t}\right)$$

$$\epsilon = \frac{1}{n} \ln\left(\frac{1}{t}\right).$$

Thus, $\epsilon = \frac{1}{n} \ln\left(\frac{1}{t}\right)$ ensures a right-side probability equals to $t$. For example, let us set it as 1%, i.e., $t = 0.01$, what produces:

$$\epsilon = \frac{1}{n} \ln(100),$$

---

[2] Remember that $R(f)$ represents how likely it is for a randomly selected sample to be misclassified by $f$.

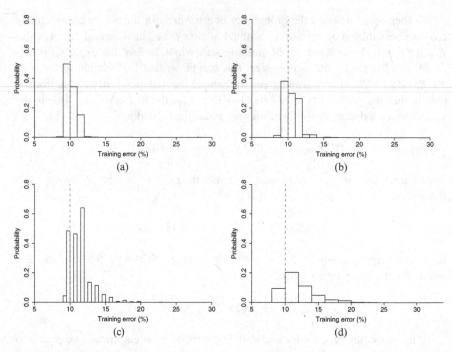

**Fig. 2.13** Density distributions of training errors over 1000 random experiments using different sample sizes $n =$ (**a**) 64, (**b**) 32, (**c**) 16, (**d**) 8. The training error, a.k.a., the empirical risk, computed for the whole dataset with $n = 400$ examples is shown as a dashed vertical line ($\sim$10%)

therefore the upper bound can be computed by the logarithm of 100 divided by the number of examples. By using this formulation, a sample of $n = 450$ would be enough to achieve $\epsilon \approx 0.01$.

To illustrate how the sample size influences the error estimates, we produce a series of simulations with a reduced number of examples $n_r$ used to compute average vectors $\mathbf{w}^+$ and $\mathbf{w}^-$ (randomly choosing $n_r < n$, in which $n$ is the total number of training examples). By measuring the error estimates in the whole dataset, we observed an error of $\sim 5\%$. In order to proceed with the analysis, we draw 1000 training sets $\mathcal{D}_{n_r}$, with sizes $n_r = [64, 32, 16, 8]$, ensuring class balance, and compute a histogram of errors to show how the empirical risk deviates as the training set size is reduced. Those results, shown in Fig. 2.13, empirically confirm that error estimates are jeopardized when relying on small samples.

Unfortunately, such analysis is not as simple in practice. The scenario described is feasible when having a single or a small set of functions, but, depending on the target problem, classification algorithms may produce infinite functions to tackle a task. From that, we define the set or class of functions considered by a classification algorithm as:

$$\mathscr{F} = \{f_1, f_2, \cdots, f_n, \cdots\}.$$

From this definition, our new problem deals with a subspace $\mathcal{F}$ of functions from which we wish the empirical risk to be a good estimator for the expected risk taking any function in $\mathcal{F}$, as the sample size $n$ is increased. By holding this property, we could select the best classifier in $\mathcal{F}$ by only assessing empirical risks to find $R_{\text{emp}}(f_{\text{best}}) \leq R_{\text{emp}}(f_i)$, for all $i = 1, \ldots, n$.

We remember that this set of admissible functions $\mathcal{F}$ represent the bias of some classification algorithm, for example: (1) the Perceptron contains all possible linear functions inside such bias, as discussed in Chap. 1; and (2) the Multilayer Perceptron has even more functions (linear and nonlinear ones), when the number of neurons at the hidden layer is greater than one. Now we discuss about the concepts of Bayes risk and Consistency.

## 2.2.6 Bayes Risk and Universal Consistency

Considering a space of admissible functions $\mathcal{F}$, an algorithm is consistent with respect to such a bias if it converges to the best classifier in $\mathcal{F}$, as the sample size increases. **Consistency** is a property associated to a set of functions (not just one particular), allowing us to study the asymptotic convergence behavior, i.e., as the sample size tends to infinite.

In order to ensure generalization, Vapnik [16] relied on consistency concepts to prove that a classification algorithm starts with some initial function and converges, as the sample size increases, to the best classifier in $\mathcal{F}$. We here recall each classification algorithm may have its own bias $\mathcal{F}$, which may overlap for different algorithms. For instance, the Perceptron has every linear function in $\mathcal{F}_{\text{Perceptron}}$, while the Multilayer Perceptron potentially has more functions in $\mathcal{F}_{\text{MLP}} \supseteq \mathcal{F}_{\text{Perceptron}}$. The only situation in which $\mathcal{F}_{\text{MLP}} = \mathcal{F}_{\text{Perceptron}}$ is when MLP has a single neuron at the hidden layer. MLP forms more complex functions, as more neurons are added into its hidden layer.

There are different types of consistencies defined in the literature. As described in Chap. 1, consider the best possible classifier for a problem, referred to as the **Bayes classifier** $f_{\text{Bayes}}$, which is certainly contained in the space of all functions $\mathcal{F}_{\text{all}}$.[3] The Bayes consistency states that, as the sample size increases, the classification algorithm must approach the best classifier, i.e.:

$$\lim_{n \to \infty} E\left[\ell(X, Y, f(X))\right] = R(f_{\text{Bayes}}),$$

where space $\mathcal{F}_{\text{all}}$ is considered, and $R(f_{\text{Bayes}})$ is the Bayes risk which assumes the lowest possible loss for a given problem.

In practical scenarios, classification algorithms do not span the whole space of functions $\mathcal{F}_{\text{all}}$, but a subspace instead. In this case, we must define the property of

---

[3]This space contains every possible function to tackle any problem.

consistency with respect to a subspace of functions $\mathscr{F} \subset \mathscr{F}_{all}$; thus, a classification algorithm is consistent when it converges to the best classifier in $\mathscr{F}$, i.e.:

$$\lim_{n \to \infty} E\left[\ell(X, Y, f(X))\right] = R(f_{\mathscr{F}}).$$

By using the notions of consistency, the best classifier inside $\mathscr{F}$ is:

$$f_{\mathscr{F}} = \arg\min_{f \in \mathscr{F}} R(f),$$

and the Bayes classifier is:

$$f_{Bayes} = \arg\min_{f \in \mathscr{F}_{all}} R(f).$$

By considering that samples are uniformly drawn from some joint probability density function $P(X \times Y)$, we say a classification algorithm is consistent with respect to Bayes if the expected risk $R(f_i)$ of some classifier $f_i$, inferred by this algorithm, converges to $R(f_{Bayes})$:

$$P(R(f_i) - R(f_{Bayes}) > \epsilon) \to 0 \text{ as } n \to \infty.$$

Likewise, a classification algorithm is said to be consistent with respect to $\mathscr{F}$, if the expected risk $R(f_i)$ of some classifier $f_i$, inferred using this algorithm, converges to $R(f_{\mathscr{F}})$, then:

$$P(R(f_i) - R(f_{\mathscr{F}}) > \epsilon) \to 0 \text{ as } n \to \infty.$$

In summary, those definitions explore classification biases having a fixed but unknown joint density distribution $P(X \times Y)$. There is still another type of consistency called the **universal consistency**. A classification algorithm is said to be universally consistent if it is consistent with respect to subspace $\mathscr{F}$ for *any possible joint probability density function $P(X \times Y)$. This consistency is fundamental to the Statistical Learning Theory*, given $P(X \times Y)$ is assumed to be unknown.

### 2.2.7   Consistency, Overfitting and Underfitting

It is possible to relate consistency and learning errors by looking at how far our solution is from some target classifier. For instance, let $f_i$ be some current solution, $f_{\mathscr{F}}$ the best classifier in $\mathscr{F}$, and $f_{Bayes}$ the best classifier in $\mathscr{F}_{all}$. Two learning errors can be defined:

- **Estimation error**: represents how far our solution $f_i$ is from the best possible classifier $f_{\mathscr{F}} \in \mathscr{F}$. This error is resultant of the uncertainty present in training

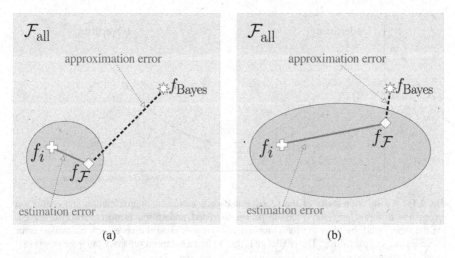

**Fig. 2.14** Illustration of scenarios having the current classifier $f_i$ (cross), the best classifier inside the subspace of functions $f_{\mathscr{F}}$ (diamond), and the best classifier $f_{\text{Bayes}}$ (star) in the space containing all functions $\mathscr{F}_{\text{all}}$: (**a**) small subspace of functions often privileges a small estimation error, while it produces a greater approximation error; and (**b**) large subspace of functions that will likely have a smaller approximation error, but often a greater estimation error

data (e.g., class overlapping or mislabeled examples) and it can be seen as the statistical **variance**;

- **Approximation error**: represents how far $f_{\mathscr{F}}$ is from the best classifier $f_{\text{Bayes}}$ given the whole space of functions $\mathscr{F}_{\text{all}}$. This error is resultant of the bias imposed by the algorithm and might be interpreted as the statistical **bias**.

The total error, i.e., how far our solution $f_i$ is from the best classifier $f_{\text{Bayes}}$, is defined in terms of approximation and estimation errors:

$$R(f_i) - R(f_{\text{Bayes}}) = \underbrace{(R(f_i) - R(f_{\mathscr{F}}))}_{\text{estimation error}} + \underbrace{(R(f_{\mathscr{F}}) - R(f_{\text{Bayes}}))}_{\text{approximation error}}. \tag{2.4}$$

Figure 2.14 illustrates two scenarios according to Eq. (2.4): (1) it is easier to converge to the best classifier inside $\mathscr{F}$ using a restricted subspace, but it may span a space far from $f_{\text{Bayes}}$; on the other hand, (2) a wider subspace is likely to contain a solution closer to $f_{\text{Bayes}}$, but it also faces greater difficulties to convergence to the best solution $f_{\mathscr{F}} \in \mathscr{F}$.

By defining a subspace $\mathscr{F}$, we set a balance between estimation and approximation errors: (1) a **stronger bias** produces lower variances, or small estimation errors, but leads to greater approximation errors; and (2) a **weaker bias** produces greater variances, or greater estimation errors, however leading to smaller approximation errors. From this perspective, we revisit the definitions of underfitting and overfitting in the context of Machine Learning (depicted in Fig. 2.15):

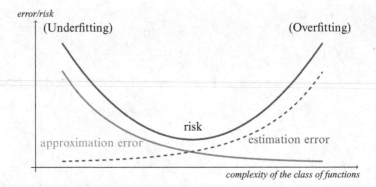

**Fig. 2.15** As the complexity of the function subspace increases, approximation and estimation errors show different behavior. If the subspace is restricted, underfitting is most probably observed. On the other hand, by allowing the complexity to grow beyond the necessary, estimation error increases causing overfitting. The challenge is then to find a balanced subspace (adapted from [18])

- **Underfitting**: for a small $\mathscr{F}$, estimation error is small but approximation error is large;
- **Overfitting**: for a large $\mathscr{F}$, estimation error is large but approximation error is small.

Given those two definitions, let us recall the problem **XOR**, as discussed in Chap. 1. Employing the Perceptron results in **underfitting** since $f_{\mathscr{F}}$ is not even enough to represent the training data. By considering MLP with 2 neurons at the hidden layer, the bias becomes sufficient for the task. It is small enough to prevent a large estimation error, but big enough to ensure convergence to $f_{\mathscr{F}} = f_{\text{Bayes}}$. To contrast, if one guesses a larger number of neurons at the hidden layer, it would result in a much larger subspace than necessary, leading to **overfitting**.[4] In fact, the convergence to $f_{\text{Bayes}}$ becomes harder due to estimation error: the resulting function is likely to be much more complex than the problem needs, producing small empirical risk but great expected risk when unseen examples present any small data variation. This simple example shows an important issue when designing classification algorithms: we should use a subspace which is sufficient and necessary for a given problem, because by using either an insufficient an overcomplex subspace, we may incur in under or overfitting, respectively.

---

[4]In case of **XOR**, the dataset has a finite number of possibilities. Thus, considering all of them were provided, overfitting is not verifiable in practice because memorization is indeed enough when new examples are equal to the ones in the training set.

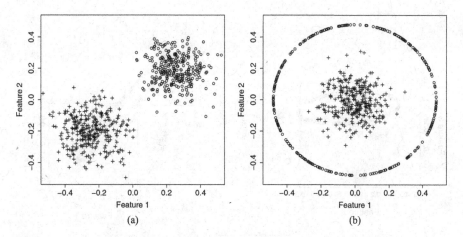

**Fig. 2.16** Two types of binary classification tasks (**a**) linearly separable, and (**b**) nonlinearly separable

### 2.2.8 Bias of Classification Algorithms

The bias of a classification algorithm is defined in terms of a subspace $\mathscr{F} \subseteq \mathscr{F}_{all}$ (most likely $\mathscr{F} \subset \mathscr{F}_{all}$), as discussed before. How is such set of admissible functions influenced? First of all, that depends on the input space. For example, consider a linearly separable problem, as illustrated in Fig. 2.16a. Any classification algorithm considering linear hyperplanes would be enough to provide good results. In a more complex scenario, such as shown in Fig. 2.16b, we need a more complex bias to tackle the problem such as a subspace $\mathscr{F}$ containing Gaussian functions.

Although such Gaussian bias works, one could instead apply the following nonlinear transformation on every input vector $(x_1, x_2)$ from Fig. 2.16b:

$$T\left(\begin{bmatrix} x_1 \\ x_2 \end{bmatrix}\right) = \begin{bmatrix} x_1^2 \\ \sqrt{2}x_1x_2 \\ x_2^2 \end{bmatrix},$$

obtaining a 3-dimensional space, in which examples are linearly separable (see Fig. 2.17). In that sense, instead of increasing the subspace of classification functions, one can transform the input examples in order to simplify the space of functions and, thus, solve such a task.

The subspace of functions, often referred to as the algorithm bias, has also other terms: (1) the **hypothesis bias** is defined by the representation language of hypothesis or models, e.g.: if-then-else rules, decision trees, networks, graphs, topological spaces, etc. For example, MLP uses weights to represent linear hyperplanes shattering the input space, while C4.5 and J48 employ tree nodes to orthogonally shatter every variable in the input space of examples [2]; (2) the **preference bias**

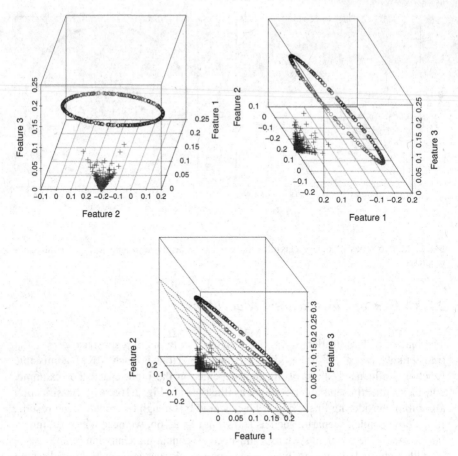

**Fig. 2.17** 3D views of the space obtained after the nonlinear transformation on training examples. The last also depicts a hyperplane allowing linear separation of the classes

defines the conditions under which the algorithm prefers one classifier over another. This is common when, given the current classifier $f_i$, the algorithm needs to opt for a next candidate classifier; and (3) the **search bias** defines some heuristic or search criterion to look for more solutions inside $\mathscr{F}$. In this book, we simply refer to all those biases as the algorithm bias.

## 2.3   Empirical Risk Minimization Principle

The Empirical Risk Minimization (ERM) principle, a central subject to SLT, provides the foundation for ensuring learning generalization. ERM relies on the concept of **universal consistency** and on the **Law of Large Numbers** to provide an upper bound so the training error (empirical risk) is a good estimator for the true error (expected risk).

As previously discussed, all consistency definitions are based on the expected risk of classifiers. However, there is no way to compute it without knowing the joint probability density function $P(X \times Y)$. How could we rely on the empirical risk as an estimator given possible generalization issues related to under and overfitting? This question motivated Vapnik [16] to define relationships and bounds to allow the use of $R_{emp}$ as a sufficient estimator for the expected risk through the concept of universal consistency with respect to a subspace or algorithm bias $\mathscr{F}$.

By assuming a subspace of functions $\mathscr{F}$, a loss function, and the universal consistency, SLT attempts to converge to $f_i \in F$ such that:

$$f_i = \arg \min_{f \in F} R_{emp}(f).$$

According to the Law of Large Numbers, assuming data (examples) are independent and identically distributed (i.i.d.), the average of a sample $\xi_1, \ldots, \xi_n$ converges to the expected value of variable $\xi$ as the sample size increases, i.e.:

$$\frac{1}{n} \sum_{i=1}^{n} \xi_i \rightarrow E[\xi] \text{ with } n \rightarrow \infty,$$

which was considered to assume the empirical risk asymptotically converges to the expected risk:

$$R_{emp}(f) = \frac{1}{n} \sum_{i=1}^{n} \ell(x_i, y_i, f(x_i)) \rightarrow E[\ell(x_i, y_i, f(x_i))] \text{ with } n \rightarrow \infty.$$

However, the Law of Large Numbers could only be used if the joint probability density function $P(X \times Y)$ is kept static and data are i.i.d., that is why both properties are mandatory assumptions.

The Chernoff inequality [6], later extended by Hoeffding, is already an upper bound for the Law of Large Numbers. Let $\xi_i$ be random values in interval $[0, 1]$, the empirical risk approximates the expected risk given a divergence of less than $\epsilon$ with the following probability:

$$P\left( \left| \frac{1}{n} \sum_{i=1}^{n} \xi_i - E[\xi] \right| > \epsilon \right) \leq 2 \exp(-2n\epsilon^2). \tag{2.5}$$

Here the Chernoff inequality provides the right-side term $2 \exp(-2n\epsilon^2)$ to bound the probability that such approximation differs at least by $\epsilon$. Observe this negative exponential function provides smaller values as the sample size $n$ increases. Rewriting it in terms of risks:

$$P(|R_{\text{emp}}(f) - R(f)| > \epsilon) \leq 2\exp(-2n\epsilon^2), \tag{2.6}$$

as a consequence, more training examples facilitate convergence to the best classifier $f \in \mathscr{F}$.

Besides the i.i.d. and the fixed joint probability density function assumptions, there is another limitation to be addressed in this formulation: the Chernoff inequality is valid for the Law of Large Numbers if and only if classifier $f$ is set without any a priori knowledge about data. This is a very tricky issue. In fact Inequality (2.5) holds only when the function being used as estimator is data-independent. However, that does not happen a priori, since the classifier is chosen according to training examples. Therefore, a method had to be designed to surpass such issue.

### 2.3.1   Consistency and the ERM Principle

Let us consider a space of examples versus class labels $(X \times Y)$, which is produced using a deterministic function:

$$Y = \begin{cases} -1 & \text{if } X < 0.5, \\ 1 & \text{if } X \geq 0.5. \end{cases}$$

A classifier $f$ with zero training error, i.e. memorizing $y_i$ for every input sample $x_i$, can be defined as follows:

$$f(x) = \begin{cases} y_i & \text{if } x = x_i \text{ for some } i = 1, \cdots, n, \\ 1 & \text{otherwise.} \end{cases}$$

Although training error is zero, it assigns a fixed label 1 to unseen examples, consequently it works by guessing such label for examples outside the training set. After training, consider new examples with equal class probability, thus $f$ will misclassify 50% of those future instances. Its misclassification is comparable with the flipping of a fair coin.

In this scenario, we have what is called a **memory-based classifier**, which is an extreme situation of **overfitting**. For every new data point $x$, it is optimal while assigning labels to memorized training instances, but behaves randomly for unseen examples. Therefore, $f$ has the poorest as possible generalization, i.e., empirical risk does not approximate the expected risk. As a consequence, ERM Principle is inconsistent.

At this point, we need to evaluate whether the **class of functions** in $\mathscr{F}$ provides universal consistency to the ERM Principle. Thus, if $\mathscr{F}$ contains any memory-based classifier, the ERM principle is **always inconsistent**. On the other hand, if the classification algorithm bias is **restricted** to a particular subspace $\mathscr{F} \subset \mathscr{F}_{\mathrm{all}}$ **not containing any memory-based classifier**, the ERM Principle is consistent and, therefore, learning can be ensured.

## 2.3.2 Restriction of the Space of Admissible Functions

As the consistency guarantee of the ERM Principle can be viewed as restricting the algorithm bias, SLT always evaluates the space of admissible functions $f_i \in \mathscr{F}$ for the worst-case scenario. It considers any classifier $f_i$ can be selected, even including the worst possible $f_w$. So, if the worst classifier has an empirical risk $R_{\mathrm{emp}}(f_w)$ converging to expected risk $R(f_w)$ as the sample size increases, then any other classifier $f_i \in \mathscr{F}$ will also converge. Consequently, this classification algorithm is capable of learning. In that way, we can set the function for the Law of Large Numbers as the worst classifier which does not depend on that training data anymore, but simply on a definition. As consequence, an important issue was solved.

The **uniform convergence** ensures, for every function in $\mathscr{F}$, the divergence $|R_{\mathrm{emp}}(f) - R(f)|$ decreases as $n \to \infty$. As a result, the following inequality is valid for a small epsilon:

$$|R_{\mathrm{emp}}(f) - R(f)| \leq \epsilon \text{ for all } f \in \mathscr{F}, \mathscr{F} \subset \mathscr{F}_{\mathrm{all}}, \text{ as } n \to \infty.$$

Therefore, the divergence between empirical and expected risks will never be greater than a given $\epsilon$, given the whole $\mathscr{F}$. As this is a space of functions, the worst-case scenario is mathematically represented as the supreme of differences [7]:

$$\sup_{f \in \mathscr{F}} |R_{\mathrm{emp}}(f) - R(f)| \leq \epsilon.$$

Figure 2.18 illustrates the risk curves from a hypothetical restricted space of functions $\mathscr{F}$, in which $f_{\mathscr{F}}$ is the best function in $\mathscr{F}$. Note that for any $f \in \mathscr{F}$, there will be a maximum distance $\epsilon$ between the curves of the expected $R(.)$ and the empirical risk $R_{\mathrm{emp}}(.)$.

As the supreme represents an upper bound for differences related to every function $f \in \mathscr{F}$, it is possible to write:

$$|R_{\mathrm{emp}}(f) - R(f)| \leq \sup_{f \in \mathscr{F}} |R_{\mathrm{emp}}(f) - R(f)|.$$

We need to reduce the probability of having classifiers $f_i$ producing errors close to zero for the training set, i.e., $R_{\mathrm{emp}}(f_i) \sim 0$, while the true error is in fact large, say $R(f_i) > R_{\mathrm{emp}}(f_i)$. Thus, we rewrite the previous inequality as follows:

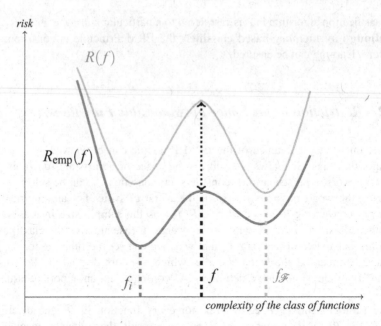

**Fig. 2.18** Expected (light gray) and empirical (dark gray) risk curves traced according to the classifiers contained in a restricted space of functions $\mathscr{F} \subset \mathscr{F}_{\text{all}}$: the arrow represents the maximum distance $\epsilon$ between curves

$$P\left(|R_{\text{emp}}(f_i) - R(f_i)| > \epsilon\right) \le P\left(\sup_{f \in \mathscr{F}} |R_{\text{emp}}(f) - R(f)| > \epsilon\right). \qquad (2.7)$$

Now we have, on the right side of Inequality (2.7), the same scenario considered by the Law of Large Numbers, thus such theoretical foundation can be applied to provide convergence as the sample size increases. This formulation is valid if and only if a fixed set of restricted functions $f \in \mathscr{F}$ is taken into account. Thus, instead of employing the Law of Large Numbers for a given classifier $f$ inferred from training examples, making this law inconsistent, SLT considers all classifiers inside the algorithm bias. Consequently, the Law of Large Numbers is consistent so the uniform convergence of probabilities is ensured as $n \to \infty$.

In summary, the Law of Large Numbers is uniformly valid over a static set of functions for every $\epsilon > 0$ (remember $\epsilon$ represents a distance), as follows:

$$P\left(\sup_{f \in \mathscr{F}} |R_{\text{emp}}(f) - R(f)| > \epsilon\right) \to 0 \text{ as } n \to \infty,$$

in which this probability is an upper limit for the worst possible classifier $f_w \in \mathscr{F}$, and thus for every $f \in \mathscr{F}$.

Considering the distance between the expected risk of any classifier $f_i$ versus the best classifier $f_{\mathscr{F}}$, both in $\mathscr{F}$, it is now clear that:

$$|R(f_i) - R(f_{\mathscr{F}})| > 0,$$

because the expected risk of the best classifier is always smaller or equal to any other. This makes the absolute operator unnecessary:

$$|R(f_i) - R(f_{\mathscr{F}})| = R(f_i) - R(f_{\mathscr{F}}).$$

By adding the empirical risks, we have:

$$
\begin{aligned}
&R(f_i) - R(f_{\mathscr{F}}) \\
&= R(f_i)\left[-R_{\text{emp}}(f_i) + R_{\text{emp}}(f_i) - R_{\text{emp}}(f_{\mathscr{F}}) + R_{\text{emp}}(f_{\mathscr{F}})\right] - R(f_{\mathscr{F}}),
\end{aligned}
\tag{2.8}
$$

in which the term inside square brackets sums up to zero. From this equation, SLT defines upper bounds useful to study the consistency of a classification algorithm regarding its worst-case scenario.

We know $R_{\text{emp}}(f_i) - R_{\text{emp}}(f_{\mathscr{F}}) \leq 0$, given $f_i$ is optimized using the training set, so it will produce a lower empirical risk, even when compared to the best classifier. This confirms that the learning process relies on training set to minimize error. See Section 5.4 of [12] for more details. Then, we write the following upper bound:

$$
\begin{aligned}
&R(f_i) - R_{\text{emp}}(f_i) \underbrace{+ R_{\text{emp}}(f_i) - R_{\text{emp}}(f_{\mathscr{F}})}_{A} + R_{\text{emp}}(f_{\mathscr{F}}) - R(f_{\mathscr{F}}) \\
&\leq R(f_i) - R_{\text{emp}}(f_i) + R_{\text{emp}}(f_{\mathscr{F}}) - R(f_{\mathscr{F}}),
\end{aligned}
\tag{2.9}
$$

in which the right-side term of the inequality was produced by removing term A, given it is always less or equal to zero.

Observe that the right-side term of Inequation (2.9) considers the sum of distances between the expected and the empirical risks for $f_i$ and $f_{\mathscr{F}}$. If we say the worst classifier is the one with the maximum distance, then it is possible to obtain:

$$R(f_i) - R_{\text{emp}}(f_i) + R_{\text{emp}}(f_{\mathscr{F}}) - R(f_{\mathscr{F}}) \leq 2 \sup_{f \in \mathscr{F}} |R_{\text{emp}}(f) - R(f)|, \tag{2.10}$$

which is an upper bound for the worst classifier in $\mathscr{F}$.

Interestingly, all work presented so far was dedicated to provide upper limits. This is the way SLT addresses the problem: by fixing and ensuring consistency for the worst classifier, every other scenario will also be consistent. Then, by using Inequation (2.10), it is possible to write:

$$P\left(|R(f_i) - R(f_{\mathscr{F}})| > \epsilon\right) \leq P\left(\sup_{f \in \mathscr{F}} |R_{\text{emp}}(f) - R(f)| > \epsilon/2\right),$$

which is an enough condition to ensure consistency to the ERM Principle.

Some important remarks follows this result:

1. We need to restrict the space of admissible functions of a classification algorithm in order to make the ERM Principle consistent. The inherent nature of a problem sets up the necessary bias constraints. For example, the Perceptron bias is not enough to learn the problem *XOR* leading to underfitting;
2. The more functions $\mathscr{F}$ contains, the greater is the supreme of the distance: $\sup_{f \in \mathscr{F}} |R_{\text{emp}}(f) - R(f)|$, allowing the algorithm to select more complex functions;
3. If available, those more complex functions are likely to be selected because supervised learning algorithms are driven by the optimization of some loss function over the training data;
4. As more functions are added to $\mathscr{F}$, it is harder to ensure consistency for the ERM Principle and therefore learning;
5. Since the uniform convergence relies on restricting the subspace of functions, **without a sufficiently restricted bias, there is no learning guarantee**;
6. We cannot say there is no learning when the ERM principle is inconsistent, but no learning guarantee is provided according to SLT.

### 2.3.3  Ensuring Uniform Convergence in Practice

We now define the properties $\mathscr{F}$ must hold to ensure uniform convergence. Let the bias of some classification algorithm be defined by a finite subspace $\mathscr{F} = \{f_1 \cdots f_m\}$ of admissible functions. Every individual function in that subspace respects the Law of Large Numbers, having the Chernoff bound as follows:

$$P(|R_{\text{emp}}(f_i) - R(f_i)| > \epsilon) \leq 2\exp(-2n\epsilon^2). \tag{2.11}$$

We need to ensure the same for all functions in $\mathscr{F}$, rewriting Inequation (2.11) to include the probabilities of all individual functions:

$$P\left(\sup_{f \in \mathscr{F}} |R_{\text{emp}}(f) - R(f)| > \epsilon\right)$$
$$= P(|R_{\text{emp}}(f_1) - R(f_1)| > \epsilon \vee |R_{\text{emp}}(f_2) - R(f_2)| > \epsilon \vee \cdots$$
$$\vee |R_{\text{emp}}(f_m) - R(f_m)| > \epsilon), \tag{2.12}$$

in which $\vee$ is a logical OR operator. Let a set of events whose associated probabilities are given by disjunct elements $\{P_1, P_2, \ldots, P_m\}$. As a consequence, the probability of observing any event is:

$$\{P_1, P_2, \ldots, P_m\} = \sum_{i=1}^{m} P_i.$$

Now suppose some events are dependent, i.e. there is some degree of intersection such as in a Venn diagram, therefore the union of all elements represent a lower bound for such sum, in form:

$$\{P_1, P_2, \ldots, P_m\} < \sum_{i=1}^{m} P_i.$$

Consequently, it is possible to found the following upper bound for Inequality (2.12):

$$P(|R_{\text{emp}}(f_1) - R(f_1)| > \epsilon P(|R_{\text{emp}}(f_2) - R(f_2)| > \epsilon \vee \cdots$$
$$\vee |R_{\text{emp}}(f_m) - R(f_m)| > \epsilon)$$
$$\leq \sum_{i=1}^{m} P(|R_{\text{emp}}(f_i) - R(f_i)| > \epsilon),$$

Given:

$$P(|R_{\text{emp}}(f_i) - R(f_i)| > \epsilon) \leq 2 \exp(-2n\epsilon^2),$$

we have:

$$P(|R_{\text{emp}}(f_1) - R(f_1)| > \epsilon) \leq 2 \exp(-2n\epsilon^2)$$

$$P(|R_{\text{emp}}(f_2) - R(f_2)| > \epsilon) \leq 2 \exp(-2n\epsilon^2)$$

$$\vdots$$

$$P(|R_{\text{emp}}(f_m) - R(f_m)| > \epsilon) \leq 2 \exp(-2n\epsilon^2)$$

Therefore, summing all such inequations, we have the uniform convergence considering all functions $f_1 \cdots f_m$ inside the **finite** subspace $\mathscr{F}$ as:

$$P\left(\sup_{f \in \mathscr{F}} |R_{\text{emp}}(f) - R(f)| > \epsilon\right) \leq \sum_{i=1}^{m} P(|R_{\text{emp}}(f_i) - R(f_i)| > \epsilon)$$

$$\leq 2m \exp(-2n\epsilon^2), \quad (2.13)$$

As $\mathscr{F}$ is finite, $m$ is a constant supporting the upper bound at the right-side term, and then the uniform convergence is valid as $n \to \infty$. This solves the consistency for any finite subspace $\mathscr{F}$, but raises the question about how to deal with subspaces containing an infinite number of functions.

## 2.4   Symmetrization Lemma and the Shattering Coefficient

The symmetrization lemma and the shattering coefficient are necessary to ensure the ERM principle in the context of infinite spaces of functions. The **symmetrization lemma** uses a ghost sample, or virtual sample, in order to map infinite spaces into an enumerable space of functions. The **shattering coefficient** is a function relating the sample size with the maximal number of distinct classifications. Therefore, the shattering coefficient is a measurable function of learning capacity, counting the number of different classifiers in the algorithm bias.

A **ghost sample** is an unknown sample with size $n$ drawn from the same joint distribution $P(X \times Y)$ and independently from the training sample. For a single sample:

$$\sup_{f \in \mathcal{F}} |R(f) - R_{\text{emp}}(f)|$$

Adding the second sample with same size $n$:

$$\sup_{f \in \mathcal{F}} |R(f) - R_{\text{emp}}(f)| \leq \sup_{f \in \mathcal{F}} |R(f) - R_{\text{emp}}(f)| + \sup_{f \in \mathcal{F}} |R(f) - R'_{\text{emp}}(f)|.$$

We can simplify this formulation for the worst classifier $f_w \in \mathcal{F}$ as:

$$R_{\text{emp}}(f_w) - R(f_w) \leq R_{\text{emp}}(f_w) - R(f_w) - \left[ R'_{\text{emp}}(f_w) - R(f_w) \right]$$
$$= |R_{\text{emp}}(f_w) - R'_{\text{emp}}(f_w)|,$$

so note the absolute value is needed because we cannot ensure the difference between the empirical risks is positive. In addition, recall $R_{\text{emp}}(f_w) \leq R(f_w)$.

The divergence $|R_{\text{emp}}(f_w) - R'_{\text{emp}}(f_w)|$ can be rewritten in terms of the supreme:

$$|R_{\text{emp}}(f_w) - R'_{\text{emp}}(f_w)| = \sup_{f \in \mathcal{F}} |R_{\text{emp}}(f) - R'_{\text{emp}}(f)|,$$

however this inequality has to be rewritten in terms of $\epsilon$. At first:

$$P\left( \sup_{f \in \mathcal{F}} |R_{\text{emp}}(f) - R(f)| > \epsilon \right) \leq 2m \exp(-2n\epsilon^2).$$

With the addition of the ghost sample, we have $2n$ examples so $2m \exp(-2n\epsilon^2)$ becomes $2m \exp(-2(2n)\epsilon^2)$, simplifying:

$$2m \exp(-2(2n)\epsilon^2) = 2m \exp(-4n\epsilon^2)$$

$$= 2m \exp\left(-n\frac{\epsilon^2}{4}\right)$$

$$= 2m \exp\left(-n\frac{\epsilon^2}{2^2}\right)$$

$$= 2m \exp\left(-n\left(\frac{\epsilon}{2}\right)^2\right)$$

so, finally, we obtain:

$$P(\sup_{f \in \mathcal{F}} |R(f) - R_{\text{emp}}(f)| > \epsilon) \leq 2P(\sup_{f \in \mathcal{F}} |R'_{\text{emp}}(f) - R_{\text{emp}}(f)| > \epsilon/2)$$

$$\leq 2m \exp(-n\epsilon^2).$$

This demonstrates the **symmetrization lemma** [9] which cancels out the expected risk. Note the ghost sample is a conceptual object, and we do not have to draw it, but this step allows us to remove term $R(f)$, which cannot be computed assuming the joint probability distribution is unknown. This will be used later to find bounds for infinite subspaces of functions.

According to the symmetrization lemma, the empirical risks for two different and independent samples from $P(X \times Y)$ approach each other as $n \to \infty$. The symmetrization lemma is necessary to cancel out risk $R(f)$, which is not computable, as well as to prove convergence for the ERM Principle having a subspace $\mathcal{F}$ with infinite functions.

## 2.4.1 Shattering Coefficient as a Capacity Measure

Besides subspace $\mathcal{F}$ contains an infinite number of functions, there is a finite number of classification possibilities for some sample. For instance, in a binary classification problem, i.e., $Y = \{-1, +1\}$, there is an infinite number of available functions to classify data points, however each function $f \in \mathcal{F}$ can only provide at most $2^n$ different classification results over some sample with $n$ examples. Figure 2.19 illustrates how infinite functions may produce the same classification result in terms of the available data.

If two distinct classifiers $f, g \in \mathcal{F}$ produce decision functions with equal classification results, then their empirical risks are the same, i.e., $R_{\text{emp}}(f) = R_{\text{emp}}(g)$. For instance, let $n = 2$, then there are $2^n = 2^2 = 4$ different ways of classifying one sample, and $2^{2n} = 2^4 = 16$ ways of classifying two samples together, each with $n = 2$ examples. Therefore, there are at most $2^{2n}$ different classification results when considering both samples at the same time, as shown in Fig. 2.20.

Let:

1. $Z_{(n)} = \{(x_1, y_1), \cdots, (x_n, y_n),\}$ be a training set with $n$ instances;
2. $|F_{Z_n}|$ be the cardinality of subspace $\mathscr{F}$ for set $Z_n$, which is seen as the number of functions producing different classification results for $Z_n$;

then, the maximum number of distinct functions in $\mathscr{F}$ is:

$$\mathcal{N}(\mathscr{F}, n) = \max\left\{ |F_{Z_n}| \,\middle|\, x_1, \cdots, x_n \in \mathscr{X} \right\}, \tag{2.14}$$

which is a function of $n$ referred to as the **shattering coefficient** for subspace $\mathscr{F}$.

If $\mathcal{N}(\mathscr{F}, n) = 2^n$, then there is at least one sample with $n$ instances that can be classified in all possible ways, having two labels[5] $Y = \{-1, +1\}$. Then, we say that $\mathscr{F}$ is **capable of shattering** at least one sample in all possible ways, as ensured by the maximum operator in Eq. (2.14), provided the data organization in the input space. This does not mean that every sample will be classified (shattered) in all possible ways, but at least one of them.

The shattering coefficient is a capacity measure function for $\mathscr{F}$, as it allows to quantify the number of distinct classification results. Besides being referred to as a

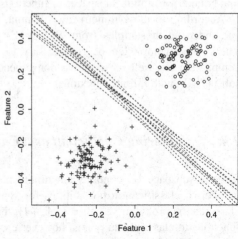

**Fig. 2.19** Given a finite number of input examples, there are infinite functions providing the same classification result

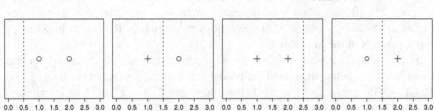

**Fig. 2.20** For $n = 2$, there are 4 different ways of classifying one sample

---

[5]As a binary classifier is considered, this is a power of two.

coefficient, the shattering is a function providing the cardinality for $Z_n$ as $n \to \infty$. Consequently, it outputs the number of admissible functions in $\mathscr{F}$ as $n \to \infty$, which is associated to the **complexity** of such subspace. A larger subspace $\mathscr{F}$ will certainly have a shattering curve growing faster as $n \to \infty$ (see Fig. 2.21).

## 2.4.2 Making the ERM Principle Consistent for Infinite Functions

Now we connect the **symmetrization lemma** to the consistency of the ERM Principle given a finite number of functions. In order to proceed, we should consider $2n$ instances, having half in one sample and the remaining half in a ghost sample. For a problem with $l$ labels, the maximal number of functions producing distinct classification results for both samples combined is given by $l^{2n}$. Assuming a problem with two labels, the shattering coefficient is $\mathscr{N}(\mathscr{F}, 2n) = 2^{2n}$.

From the symmetrization lemma:

$$P(\sup_{f \in \mathscr{F}} |R(f) - R_{\text{emp}}(f)| > \epsilon) \leq 2P(\sup_{f \in \mathscr{F}} |R'_{\text{emp}}(f) - R_{\text{emp}}(f)| > \epsilon/2),$$

Vapnik substituted the supreme over $\mathscr{F}$ for the supreme over $\mathscr{F}_{Z_{2n}}$, in which $Z_{2n}$ represents the union of both samples:

$$2P(\sup_{f \in \mathscr{F}} |R'_{\text{emp}}(f) - R_{\text{emp}}(f)| > \epsilon/2) = 2P(\sup_{f \in \mathscr{F}_{Z_{2n}}} |R'_{\text{emp}}(f) - R_{\text{emp}}(f)| > \epsilon/2),$$

**Fig. 2.21** Illustration of the shattering coefficient for two different subspaces: $\mathscr{F}$ (dashed curve) has a less complex shattering function than $\mathscr{F}'$ (gray curve)

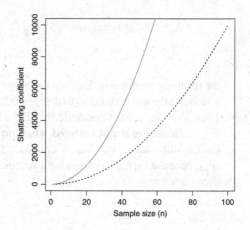

having $\mathscr{F}_{Z_{2n}}$ as the algorithm bias provided $2n$ instances.[6]

As $\mathscr{F}_{Z_{2n}}$ contains at most $\mathscr{N}(\mathscr{F}, 2n)$ distinct classification functions, it is possible to employ the Chernoff bound as follows:

$$2P(\sup_{f \in \mathscr{F}_{Z_{2n}}} |R'_{\text{emp}}(f) - R_{\text{emp}}(f)| > \epsilon/2) \leq 2\mathscr{N}(\mathscr{F}, 2n) \exp(-n\epsilon^2/4),$$

in which term $\mathscr{N}(\mathscr{F}, 2n)$ replaces the number of functions $m$, defined in Eq. (2.13).

There are some important remarks about this result. First, consider the shattering coefficient is significantly smaller than $2^{2n}$, i.e., $\mathscr{N}(\mathscr{F}, 2n) \leq (2n)^k$ given some constant $k$:

- Then the shattering coefficient grows in a polynomial way, and by plugging it into the Chernoff bound:

$$2\mathscr{N}(\mathscr{F}, 2n) \exp(-n\epsilon^2/4) = 2(2n)^k \exp(-n\epsilon^2/4)$$
$$= 2 \exp(k \log(2n) - n\epsilon^2/4),$$

the whole expression converges to zero, as $n \to \infty$;
- Therefore, **the Empirical Risk Minimization Principle is consistent with respect to $\mathscr{F}$ when the shattering coefficient grows polynomially**.

Second, considering the full space of functions $\mathscr{F}_{\text{all}}$, and therefore the less restricted bias as possible:

- It should be clear that by having all functions, a sample could be classified in all possible ways independently of its size, i.e., $\mathscr{N}(\mathscr{F}, 2n) = 2^{2n}$.
- By substituting this term in the Chernoff bound:

$$2\mathscr{N}(\mathscr{F}, 2n) \exp(-n\epsilon^2/4) = 2(2^{2n}) \exp(-n\epsilon^2/4)$$
$$= 2 \exp(n(2 \log(2) - \epsilon^2/4)), \qquad (2.15)$$

the resulting expression does not converge to zero as $n$ increases, once $\epsilon \geq 0$ but a sufficiently small value so that $\epsilon^2/4 < 2 \log(2)$;
- Therefore, we cannot conclude the ERM Principle is consistent with respect to $\mathscr{F}_{\text{all}}$, so **learning is not ensured when no restricted bias is set**;
- On the other hand, we cannot conclude the ERM Principle is inconsistent for $\mathscr{F}_{\text{all}}$, because Eq. (2.15) provides a sufficient condition for consistency, but not a necessary one.

---

[6]Remember the number of instances affect the number of distinct admissible functions.

Finally, Vapnik and Chervonenkis proved the following condition is necessary to ensure the consistency of the ERM Principle:

$$\frac{\log \mathcal{N}(\mathscr{F}, n)}{n} \to 0. \tag{2.16}$$

The reason for this condition will become clearer in Sect. 2.5.

It is possible to notice that if $\mathcal{N}(\mathscr{F}, n)$ is polynomial, then the condition is valid. However, if we have an unrestricted space, i.e., $\mathscr{F} = \mathscr{F}_{\text{all}}$, then $\mathcal{N}(\mathscr{F}, n) = 2^n$ for every value of $n$ and:

$$\frac{\log \mathcal{N}(\mathscr{F}, n)}{n} = \frac{\log(2^n)}{n} = \frac{n}{n} = 1.$$

Thus, the ERM Principle is not consistent given $\mathscr{F}_{\text{all}}$.

## 2.5  Generalization Bounds

The SLT bounds the probability of having "bad" classifiers, i.e., the ones with large divergences from their expected risks. We can interpret this bound like in a statistical test: for instance, by setting a confidence level of 0.01, the chance of observing a divergence above $\epsilon$ in between the empirical and expected risks is less than 1% over random samples. This is also the source of the Valiants's Probably Approximately Correct (PAC) [13] term: the confidence parameter $\delta$ represents how probable an algorithm has been misled by the training set.

In fact, SLT is interested in bounding the probability of having an error around zero for the training set, while the true error is in fact large. From that idea, we take the Chernoff bound:

$$P(\sup_{f \in \mathscr{F}} |R(f) - R_{\text{emp}}(f)| > \epsilon) \le 2\mathcal{N}(\mathscr{F}, 2n) \exp(-n\epsilon^2/4),$$

to be studied in terms of $\delta > 0$, which is an acceptable probability of divergence between risks:

$$P(\sup_{f \in \mathscr{F}} |R(f) - R_{\text{emp}}(f)| > \epsilon) \le \delta,$$

and, then solving it for $\epsilon$:

$$2\mathcal{N}(\mathcal{F}, 2n)\exp(-n\epsilon^2/4) = \delta$$

$$\exp(-n\epsilon^2/4) = \frac{\delta}{2\mathcal{N}(\mathcal{F}, 2n)}$$

$$\log(\exp(-n\epsilon^2/4) = \log(\delta) - \log(2\mathcal{N}(\mathcal{F}, 2n))$$

$$-n\epsilon^2/4 = \log(\delta) - \log(2\mathcal{N}(\mathcal{F}, 2n))$$

$$\epsilon^2 = -\frac{4}{n}(\log(\delta) - \log(2\mathcal{N}(\mathcal{F}, 2n)))$$

$$\epsilon = \sqrt{-\frac{4}{n}(\log(\delta) - \log(2\mathcal{N}(\mathcal{F}, 2n)))}.$$

Since the term is upper bounded by a probability of error $\delta$, it is possible to substitute $\epsilon$:

$$\sup_{f \in \mathcal{F}} |R(f) - R_{\text{emp}}(f)| > \epsilon$$

$$\sup_{f \in \mathcal{F}} |R(f) - R_{\text{emp}}(f)| > \sqrt{-\frac{4}{n}(\log(\delta) - \log(2\mathcal{N}(\mathcal{F}, 2n)))}.$$

If $\delta$ represents the probability that $|R(f) - R_{\text{emp}}(f)|$ is above some $\epsilon$, i.e., the probability of being misled by the training set, then the complement $1 - \delta$ is the hit chance or the probability that such divergence is less than or equal to $\epsilon$:

$$\sup_{f \in \mathcal{F}} |R(f) - R_{\text{emp}}(f)| \leq \epsilon$$

$$\sup_{f \in \mathcal{F}} |R(f) - R_{\text{emp}}(f)| \leq \sqrt{-\frac{4}{n}(\log(\delta) - \log(2\mathcal{N}(\mathcal{F}, 2n)))}.$$

Pay attention to the relational operator, which was inverted due to the focus on the hitting probability $1 - \delta$.

As in Sect. 2.3.2, we assume the empirical risk is a lower bound for the expected risk, i.e., $R_{\text{emp}}(f) \leq R(f)$. Also remember the subspace may contain the memory-based classifier, i.e., the worst function $f_w \in \mathcal{F}$, so:

$$R(f_w) - R_{\text{emp}}(f_w) \leq \sqrt{-\frac{4}{n}(\log(\delta) - \log(2\mathcal{N}(\mathcal{F}, 2n)))}.$$

and, finally, the **generalization bound allowing the selection of the worst classifier** is:

$$R(f_\text{w}) \leq \underbrace{R_\text{emp}(f_\text{w})}_{\text{training error}} + \underbrace{\sqrt{-\frac{4}{n}\left(\log(\delta) - \log(2\mathcal{N}(\mathscr{F}, 2n))\right)}}_{\text{divergence factor } \epsilon}. \tag{2.17}$$

This bound is an important result to ensure generalization in case subspace $\mathscr{F}$ is simple but enough to tackle some classification problem. Enough because such space is required to admit functions (bias) capable of classifying the training sample while, at the same time, it is expected to have a polynomial shattering coefficient.

In Eq. (2.17), terms "training error" $R_\text{emp}(f)$ and "divergence factor" $\epsilon$ control the trade-off between the empirical risk and the generalization. Observe two scenarios: first, when the training error is equal to 0, but the divergence factor is equal to $\infty$, resulting in **overfitting**:

$$R(f) \leq \underbrace{0}_{\text{training error}} + \underbrace{\infty}_{\text{divergence factor } \epsilon} = \infty.$$

Second, when the divergence factor is equal to 0, but the training error is $\infty$, leading to **underfitting**:

$$R(f) \leq \underbrace{\infty}_{\text{training error}} + \underbrace{0}_{\text{divergence factor } \epsilon} = \infty.$$

In this sense, finding a good balance between those terms is the key issue to ensure learning.

By looking at Eq. (2.17), it is now clear why Eq. (2.16) from Sect. 2.4.2 is a condition for consistency:

$$\frac{\log \mathcal{N}(\mathscr{F}, n)}{n} \to 0.$$

The generalization bound also allows the intuitive analysis of further scenarios. When empirical risk and divergence factor are both small, the expected risk is also small with a high probability if:

- the subset of functions $\mathscr{F}$ is restricted, with a finite number of functions producing distinct classification results (there is a restricted bias);
- and, despite restricted, $\mathscr{F}$ is still capable of representing the training set, avoiding underfitting.

In this scenario, there is a high probability of learning a concept from data.

On the other hand, in a more complex classification problem—e.g., a feature space with unclear structure in terms of the class distribution—we need to relax the bias, and therefore consider a greater subspace of admissible functions $\mathscr{F}$, so classifiers will be capable of representing viable solutions. As this subspace tends to the full space $\mathscr{F}_\text{all}$, we have $\mathcal{N}(\mathscr{F}, n) \to 2^n$ for all $n$, reducing the probability of learning a concept eventually to zero. In such scenario, it is not possible to define an upper bound, and therefore we cannot ensure the empirical risk is a good estimator for the expected risk, which is the main point of the ERM Principle.

Fortunately, even for complex problems, it is possible to restrict $\mathcal{F}$, and make the ERM Principle consistent. That is why it is important to assess different techniques under different biases to address the same problem, justifying ensembles on complex scenarios. By defining a set of classification algorithms with restricted and diverse subspaces, it is still possible to ensure learning. Ensembles are out of the scope of this book, but we suggest the reader to refer to [8, 10] for further details.

## 2.6 The Vapnik-Chervonenkis Dimension

The **shattering coefficient** can be estimated for some space **F** in order to study its influences on the Chernoff bound. However, in order to simplify learning guarantees, Vapnik and Chervonenkis [15, 16] proposed a capacity measure to characterize the exponential growth of the shattering coefficient. This measure is known as the Vapnik-Chervonenkis (VC) dimension.

They assumed that a sample $Z_n$ containing $n$ examples is shattered by a class of functions $\mathcal{F}$ if it can produce all possible binary classifications for $Z_n$, i.e., the cardinality is $|\mathcal{F}_{Z_n}| = 2^n$. VC dimension of $\mathcal{F}$ is defined as the largest integer $n$ such that $Z_n$ is shattered in all possible ways. Mathematically:

$$VC(\mathcal{F}) = \max\left[n \in \mathbb{Z}^+ \,\middle|\, |\mathcal{F}_{Z_n}| = 2^n \text{ for some } Z_n\right].$$

If the VC dimension is finite for the class of functions in $\mathcal{F}$, then the shattering coefficient grows polynomially as the sample size goes to infinity. This implies in consistency for the ERM Principle, and therefore ensures learning. Otherwise, if there is no such maximum, the VC dimension is infinite, meaning there is at least one sample that can be shattered in all possible ways, producing $2^n$ different binary classifications. As previously seen, this is the case in which consistency is not ensured for the ERM Principle, consequently there is no learning guarantee according to such principle.

For example, let a sample having 3 examples in $\mathbb{R}^2$ and $\mathcal{F}$ containing only linear functions. If this sample is distributed along a line, as in Fig. 2.22a, we could shatter this sample in only 6 different ways, which is less than $2^n = 2^3 = 8$. But using a different setting as in Fig. 2.22b, $\mathcal{F}$ is capable of shattering the sample in all $2^n$ possible ways, and therefore $\mathcal{F}$ has a VC dimension at least equal to 3—note that the value of 3 refers to the sample size $n$. We say "at least", because we should still evaluate it for the next sample size. Observe it is not necessary that all 3-point samples to be classified in all possible ways, but having **just one sample** is already enough. Observe that depends on the organization of points in the input space.

Following the same example, for a classification problem in which $Z_n \in \mathbb{R}^2$ with $\mathcal{F}$ containing only linear functions, $\mathcal{F}$ is capable of shattering samples in at least

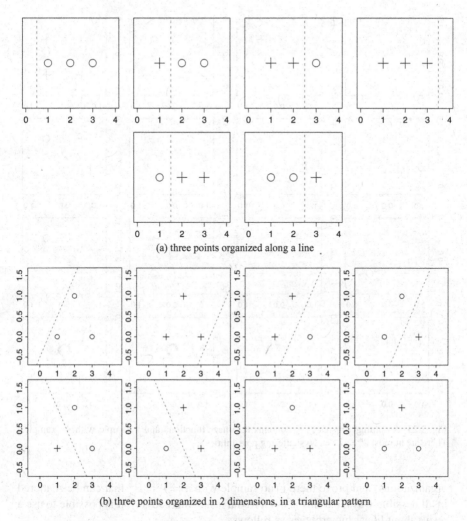

(a) three points organized along a line

(b) three points organized in 2 dimensions, in a triangular pattern

**Fig. 2.22** Assuming subspace $\mathscr{F}$ contains all linear functions, and a sample of 3 examples is organized along a line (**a**). Then, there are 6 different ways to shatter those points. But if those 3 points are organized in a different way (**b**), then there are $2^n = 2^3 = 8$ shattering possibilities

$2^3 = 8$ ways, and we know that $VC(\mathscr{F}) \geq 3$. But so far we only know that its minimal value is equal to 3, and not the actual VC dimension. This is because VC is a function of $n$ that provides a maximum value for which the cardinality $|\mathscr{F}_{Z_n}| = 2^n$ still holds. We already tried $n = 3$, but we still do not know the behavior when increasing $n$: Will the VC dimension grow for the same subspace of functions $\mathscr{F}$?

By considering $n = 4$ and the same subspace of linear functions (see Fig. 2.23), one might conclude it is not possible to shatter those points in all $2^4 = 16$ ways. In order to be sure, we have to verify this for all possible sample settings (input space

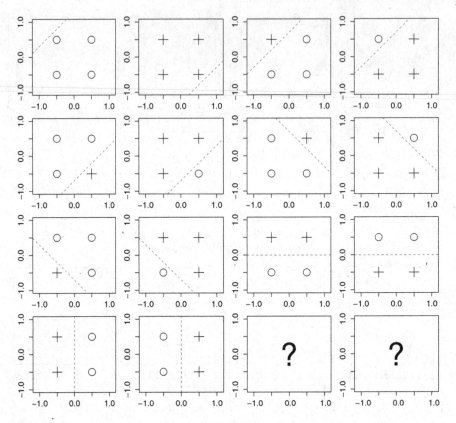

**Fig. 2.23** Assuming subspace $\mathscr{F}$ contains all linear functions and a sample with 4 examples. There are at most $2^n = 2^4 = 16$ shattering possibilities

organizations) and prove there is no sample with $n = 4$ in $\mathbb{R}^2$ that can be shattered in all possible ways, but this is not a trivial task. Alternatively, it is possible to use a result about linear hyperplanes as follows.

**Linear Hyperplanes and VC Dimension** If subspace $\mathscr{F}$ contains all possible $(n - 1)$-linear hyperplanes to classify points in $\mathbb{R}^n$, its VC dimension is $n + 1$. For example, considering an input space $\mathbb{R}^2$ and having 1-dimensional hyperplanes (lines), the VC dimension is equal to 3.

One way to interpret the VC dimension is via the behavior of $\mathscr{N}(\mathscr{F}, n)$ as a function of $n$. Considering the input examples are in $\mathbb{R}^2$ (see Figs. 2.22 and 2.23), note the shattering coefficient has an exponential behavior up to the VC dimension, i.e., for $n = 1, 2, 3$, then it becomes polynomial for $n \geq 4$. Therefore, $2^n = 2^3 = 8$ produces the maximum size $n$ for which the sample can be shattered in all possible ways. Also observe that, from the algorithm bias, one can find the VC dimension, and, thus, analyze if the ERM Principle is consistent. Finally, it allows to conclude whether learning is guaranteed or if nothing can be said.

**Fig. 2.24** Interpreting the margin bound as main motivation to the Support Vector Machines

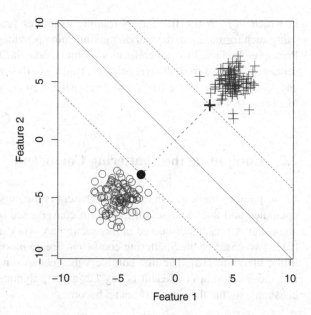

### 2.6.1 Margin Bounds

Consider linear functions are available to shatter a linearly separable sample in input space $\mathbb{R}^2$. The **margin of a classifier** $f$ is the shortest distance from any example to hyperplane, as depicted in Fig. 2.24.

This concept is relevant once there is a proof about the VC dimension for a class of linear functions $\mathscr{F}_p$, confirming margin $\rho$ is bounded by the ratio of the radius $R$ of the **smallest hypersphere enclosing all data points** (see Fig. 2.24):

$$VC(\mathscr{F}_p) \leq \min\left(d, \frac{4R^2}{\rho^2}\right) + 1,$$

in which $d$ is the dimension of the input space, i.e., $\mathbb{R}^d$.

Consequently, by maximizing margin $\rho$, we minimize the VC dimension. In this sense, the margin can be used as a capacity measure to a class of functions, which is the main motivation for the Support Vector Machines (SVM). Thus, the correspondent shattering coefficient grows slower, so the term associated with the **divergence factor** in Inequality (2.17) becomes smaller, supporting the faster convergence of the empirical risk to the expected one.

Vapnik also connected the margin bound to the ERM Principle by finding:

$$R(f) \leq v(f) + \sqrt{\frac{c}{n}\left(\frac{R}{\rho^2}\log(n^2) + \log(1/\delta)\right)},$$

in which $v(f)$ is the fraction of training examples presenting margin $\leq \rho$. By using such formulation, the margin maximization provides consistency for the ERM Principle. In fact, this generalization bound finds the best decision function to separate examples from different classes, leading to the lowest shattering coefficient and, consequently, to the best possible classification algorithm: the Support Vector Machines (SVMs).

## 2.7 Computing the Shattering Coefficient

In our point of view, the Shattering coefficient is the most important function to be computed and used to assess the uniform convergence of any supervised learning algorithm All results discussed in this section are work in progress, last updates in [21]. Two cases of the Shattering coefficient are commonly studied and illustrated in the literature [18]. The first considers the scenario in which $\mathcal{N}(\mathcal{F}, 2n) = n^k$, i.e., the Shattering coefficient is any $k$th-order polynomial function (given $k$ is a constant), so that the Chernoff bound becomes:

$$2 \exp \left(\log \mathcal{N}(\mathcal{F}, 2n) - n\epsilon^2/4\right)$$

$$2 \exp \left(\log n^k - n\epsilon^2/4\right)$$

$$2 \exp \left(k \log n - n\epsilon^2/4\right),$$

allowing us to conclude that the linear term $-n\epsilon^2/4$ will asymptotically dominate $k \log n$ as $n \to \infty$. As consequence, we have a negative exponential function which certainly converges to zero. Such convergence ensures the probability term on the right-side of the Empirical Risk Minimization principle also converges to zero (see Sect. 2.3 for more details), so that the empirical risk $R_{\text{emp}}(f)$ is a good estimator for the expected risk $R(f)$, as we desire.

The second scenario considers the exponential function $\mathcal{N}(\mathcal{F}, 2n) = 2^n$:

$$2 \exp \left(\log \mathcal{N}(\mathcal{F}, 2n) - n\epsilon^2/4\right)$$

$$2 \exp \left(\log 2^n - n\epsilon^2/4\right)$$

$$2 \exp \left(n \log 2 - n\epsilon^2/4\right),$$

from which learning is not guaranteed, given $\log 2 > \epsilon^2/4$ once we always set $\epsilon < 1$ to measure the divergence between the empirical and the expected risks.

From this perspective, we conclude the Shattering coefficient $\mathcal{N}(\mathcal{F}, 2n)$ is essential to prove learning guarantees to supervised machine algorithms. In addition, by having such growth function, we can also find out the minimal sample size to ensure $\epsilon$ as divergence factor. In that sense, the Shattering coefficient $\mathcal{N}(\mathcal{F}, 2n)$ of any $d$-dimensional Hilbert space $\mathcal{H}$ being classified with a single $(d-1)$-

dimensional hyperplane is [17]:

$$\mathcal{N}(\mathscr{F}, 2n) = 2 \sum_{i=0}^{d} \binom{n-1}{i},\tag{2.18}$$

for a generalized data organization with sample size equals $n$.

*Proof* Let a sample with $2^d$ instances in general organization in a $d$-dimensional Hilbert space $\mathscr{H}$ which must be classified using a single $(d-1)$-dimensional hyperplane. At first, consider $d=2$ then $2^d=4$ instances projected into a 0-dimensional Hilbert space, forming a single point (see Fig. 2.25). In that scenario, either the point could be classified as laying on one side of the hyperplane or on the other size, composing a total of 2 possibilities (either positive or negative, for example), thus:

$$2\binom{4-1}{0} = 2.$$

Then, consider the same $2^d=4$ instances are now projected into an 1-dimensional Hilbert space. In such data organization, we have six possible classifications in addition to the previous space dimension (see Fig. 2.25). Therefore, until now, we have 8 possible classifications when combining both spaces, in form:

$$\bullet \quad 2\left(\sum_{i=0}^{1}\binom{4-1}{i}\right) = 8.$$

Observe that every time we project the points to a greater dimension, we reorganize them into a generalized form, to next analyze all different classifications obtained when compared to previous spaces. Next, we project points into a 2-dimensional Hilbert space which can be classified into six other forms that remain different from the previous projections (see Fig. 2.25), so that:

$$2\left(\sum_{i=0}^{2}\binom{4-1}{i}\right) = 14.$$

This remains valid for any space dimensionality. In case of any other sample with more elements, this is, if we add a single instance into this even sample, it is obvious that the number of possible classifications can only be equal or greater than for this current sample, therefore the Shattering coefficient is a monotonically increasing function. From this, we conclude the proof for any $2^d \times 2^\beta$ sample size for $\beta \in \mathbb{Z}_+$, what is enough to study the Shattering coefficient of hyperplane-based supervised learning algorithms.

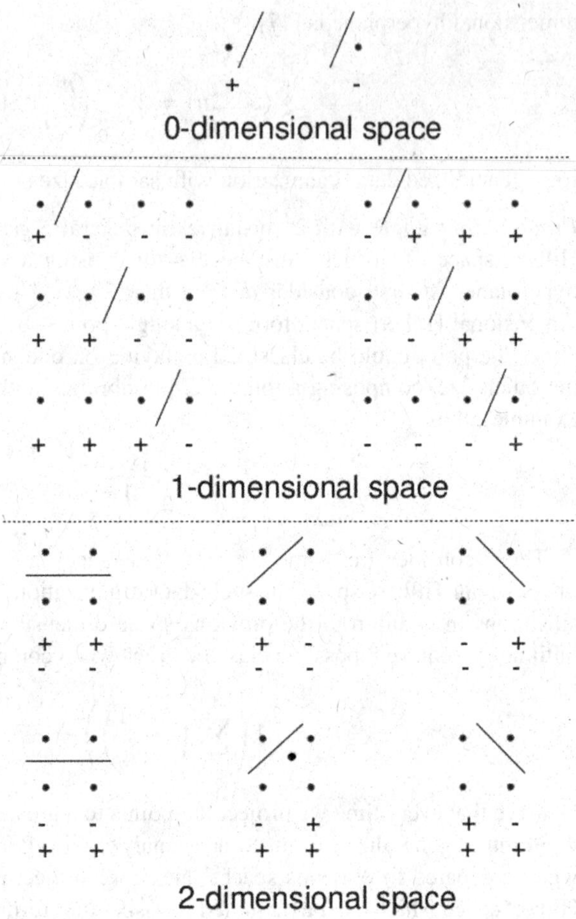

**Fig. 2.25** Illustrating the
proof on Shattering
coefficient

0-dimensional space

1-dimensional space

2-dimensional space

From this, we can conclude that:

$$\mathcal{N}(\mathscr{F}, 2n) = 2 \sum_{i=0}^{d} \binom{n-1}{i} = 2^n - 2 \sum_{i=d+1}^{n} \binom{n-1}{i},$$

therefore, the definition of a $d$-dimensional Hilbert space implies in a reduction of the exponential space of admissible functions $2^n$ of $2 \sum_{i=d+1}^{n} \binom{n-1}{i}$. In such a manner, in addition to characterize the cardinality of the algorithm bias, we can also understand its complement to the space containing all possible classifiers for a sample size with $n$ examples.

From this conclusion, we also notice that whenever $h < n$, there is a reduction in the space containing all admissible functions $2^n$, so that such reduction allows to obtain:

$$2 \exp \left( \log \mathcal{N}(F, 2n) - n\epsilon^2/4 \right)$$

$$= 2 \exp \left( \log \left( 2^n - 2 \sum_{i=d+1}^{n} \binom{n-1}{i} \right) - n\epsilon^2/4 \right),$$

as consequence, learning is ensured if and only if (both terms are always positive):

$$\log \left( 2^n - 2 \sum_{i=d+1}^{n} \binom{n-1}{i} \right) < n\epsilon^2/4.$$

Therefore, when $d \geq n$, we cannot define a more restrictive bias, as consequence the Shattering coefficient is:

$$\mathcal{N}(\mathscr{F}, 2n) = 2 \sum_{i=0}^{d} \binom{n-1}{i} = 2^n$$

and learning cannot be ensured according the ERM principle [16].

All the previous conclusions were drawn for a single hyperplane shattering some input space. If we use multiple indexed hyperplanes to classify a given generalized input space, then Shattering coefficient is:

$$\mathcal{N}(\mathscr{F}, 2n) = 2 \sum_{i=0}^{d} \binom{n-1}{i}^p,$$

due to the direct combination of $p$ $(d-1)$-dimensional hyperplanes used to classify the $d$-dimensional Hilbert space $\mathscr{H}$, and considering each hyperplane is different from any other so that it can be combined to provide the output space $Y$.

As last consequence, we suggest the computation of the Shattering coefficient for the supervised learning algorithms in order to prove their uniform convergences and their minimal training set sizes. For instance, suppose the training of an artificial neural network with $p$ neurons, which must produce a classifier $f$ whose empirical risk $R_{\text{emp}}(f)$ (this may be seen as the risk computed on a test sample) diverges from the expected risk $R(f)$, seen as the risk for unseen data, at most by 5%, so that $\epsilon = 0.05$ and:

$$P(\sup_{f \in F} |R_{\text{emp}}(f) - R(f)| > \epsilon) \leq 2\mathcal{N}(\mathscr{F}, 2n) \exp(-n\epsilon^2/4)$$

$$P(\sup_{f \in F} |R_{\text{emp}}(f) - R(f)| > 0.05) \leq 2 \left( 2 \sum_{i=0}^{d} \binom{n-1}{i}^p \right) \exp(-n \, 0.05^2/4),$$

then we may compute the probability $\delta$ in form:

$$\delta = 2 \left( 2 \sum_{i=0}^{d} \binom{n-1}{i}^{p} \right) \exp\left(-n\, 0.05^2/4\right),$$

thus, by defining some $\delta$ one wishes to ensure, for example, $\delta = 0.01$, one will have a probability of divergence between the empirical and expected risks less than or equal to 0.01 from which the minimal training set size $n$ can be found:

$$0.01 = 2 \left( 2 \sum_{i=0}^{d} \binom{n-1}{i}^{p} \right) \exp\left(-n\, 0.05^2/4\right).$$

In such scenario, in 99% of cases, the empirical risk $R_{emp}(f)$ will be a good estimator for the actual risk $R(f)$, therefore providing strong learning bounds to researchers and machine learning users. For instance, if we consider an input space in $\mathbb{R}^2$, the following bound is found:

$$0.01 = 4 \exp\left(-0.000625n\right)(2^{-p}(n^2 - 3n + 2)^p + (n-1)^p + 1)$$

In order to carry on with this instance, let $p = 5$, so we have:

$$0.01 = 4 \exp\left(-0.000625n\right)(2^{-5}(n^2 - 3n + 2)^5 + (n-1)^5 + 1)n \quad \approx 199{,}281,$$

the number of training examples necessary to ensure such predefined learning guarantee. For the sake of comparison, if we set $\epsilon = 0.1$ and solve for the sample size we obtain $n \approx 43{,}755$, significantly reducing the training set size required, however a greater divergence between risks is acceptable. This assessment of the Shattering coefficient is especially necessary to take conclusions on the current Deep Learning approaches that have been empirically proposed in the literature.

## 2.8  Concluding Remarks

This chapter introduced the main concepts of the Statistical Learning Theory, including the empirical risk, the expected risk, the Empirical Risk Minimization Principle, the Symmetrization lemma, the Shattering coefficient, the Generalization Bound, and the VC dimension. Then, the maximal margin bound was introduced to justify why Support Vector Machines are taken as the most effective classification algorithm from literature. At last, we discuss and formulate the Shattering coefficient for general data organizations in any $d$-dimensional Hilbert space. Relationships among such concepts were discussed in order to provide guarantees for the supervised machine learning scenario. Next chapter employs all the same concepts as tools to assess learning algorithms.

## 2.9 List of Exercises

After reading the paper "Statistical Learning Theory: Models, Concepts, and Results" by von Luxburg, U. and Schölkopf, B. and complement all concepts discussed throughout this chapter, address the following tasks:

1. What is the relation between the Statistical Learning Theory and the Principle of Minimum Description Length?
2. What is the association between the restriction of the space of admissible functions (Sect. 2.3.2) and the No Free Lunch Theorem?
3. Is the Generalization Bound a regularization? How do you compare it with the Tikonov Regularization?
4. How can you compare the Probably Approximately Correct (PAC) framework with the Statistical Learning Theory?
5. What is the Rademacher complexity and its relation with supervised learning?

## References

1. M. Anthony, N. Biggs, *Computational Learning Theory*. Cambridge Tracts in Theoretical Computer Science (Cambridge University Press, Cambridge, 1992)
2. C.M. Bishop, *Pattern Recognition and Machine Learning*. Information Science and Statistics (Springer-Verlag New York, Secaucus, 2006)
3. G.E.P. Box, G.M. Jenkins, *Time Series Analysis: Forecasting and Control*, 3rd edn. (Prentice Hall PTR, Upper Saddle River, 1994)
4. F.G. da Costa, R.A. Rios, R.F. de Mello, Using dynamical systems tools to detect concept drift in data streams. Expert Syst. Appl. **60**(C), 39–50 (2016)
5. J.B. Dence, Reply to a letter by Weissman on Stirling's approximation. Am. J. Phys. **51**(9), 776–778 (1983)
6. L. Devroye, L. Györfi, G. Lugosi, *A Probabilistic Theory of Pattern Recognition*. Applications of Mathematics, vol. 31, corrected 2nd edn. (Springer, Berlin, 1997), missing
7. E.L. Lima, *Análise real*. Number v. 1 in Análise real. IMPA (1989)
8. M. Ponti Jr., Combining classifiers: from the creation of ensembles to the decision fusion, in *Graphics, Patterns and Images Tutorials (SIBGRAPI-T), 2011 24th SIBGRAPI Conference on* (IEEE, Piscataway, 2011), pp. 1–10
9. C. Sammut, G.I. Webb, *Encyclopedia of Machine Learning*, 1st edn. (Springer Publishing Company, New York, 2011)
10. R.E. Schapire, Y. Freund, *Boosting: Foundations and algorithms* (MIT press, Cambridge, 2012)
11. W.C. Schefler, *Statistics: Concepts and Applications* (Benjamin/Cummings Publishing Company, San Francisco, 1988)
12. B. Scholkopf, A.J. Smola, *Learning with Kernels: Support Vector Machines, Regularization, Optimization, and Beyond* (MIT Press, Cambridge, 2001)
13. L.G. Valiant, A theory of the learnable. Commun. ACM **27**(11), 1134–1142 (1984)
14. R.M.M. Vallim, R.F. de Mello, Unsupervised change detection in data streams: an application in music analysis. Prog. Artif. Intell. **4**(1), 1–10 (2015)
15. V.N. Vapnik, *Statistical Learning Theory*. Adaptive and Learning Systems for Signal Processing, Communications, and Control (Wiley, Hoboken, 1998)

16. V. Vapnik, *The Nature of Statistical Learning Theory*. Information Science and Statistics (Springer, New York, 1999)
17. V.N. Vapnik, A.Ya. Chervonenkis, On the uniform convergence of relative frequencies of events to their probabilities, in *Measures of Complexity: Festschrift for Alexey Chervonenkis*, ed. by V. Vovk, H. Papadopoulos, A. Gammerman (Springer International Publishing, Cham, 2015), pp. 11–30
18. U. von Luxburg, B. Schölkopf, *Statistical Learning Theory: Models, Concepts, and Results*, vol. 10 (Elsevier North Holland, Amsterdam, 2011), pp. 651–706
19. WDC-SILSO, Solar Influences Data Analysis Center (SIDC), Royal Observatory of Belgium, Brussels (2017). http://www.sidc.be/silso/datafiles
20. Y. Weissman, An improved analytical approximation to n! Am. J. Phys. **51**(1), 9–9 (1983)
21. R.F. de Mello, M.A. Ponti, C.H.G. Ferreira, Computing the shattering coefficient of supervised learning algorithms (2018). http://arxiv.org/abs/1805.02627

# Chapter 3
# Assessing Supervised Learning Algorithms

## 3.1 Practical Aspects of the Statistical Learning Theory

Chapter 2 introduced the concepts and formulation developed in the context of the Statistical Learning Theory. In this chapter, those concepts are illustrated using the following algorithms: Distance-Weighted Nearest Neighbors, Perceptron, Multilayer Perceptron, and Support Vector Machines.

## 3.2 Distance-Weighted Nearest Neighbors

The Distance-Weighted Nearest Neighbors (DWNN) algorithm [1, 2], based on the K-Nearest Neighbors (KNN), defines the number of $k$ closest neighbors considering radial basis functions weighing the influence of training examples. To define DWNN, first consider a training set (a.k.a. knowledge base in this circumstance) composed of $n$ pairs $(x_1, y_1), \ldots, (x_n, y_n) \in X \times Y$, having $X$ as the input space of examples and $Y$ as their class labels. In this scenario, DWNN receives a query point (or unseen example) $x_q$ to compute the classification output as follows:

$$f(x_q) = \frac{\sum_i^n y_i w_i(x_q)}{\sum_i^n w_i(x_q)},$$

in which the weighing function is:

$$w_i(x_q) = \exp\left(\frac{-\left\|x_i - x_q\right\|_2^2}{2\sigma^2}\right),$$

© Springer International Publishing AG, part of Springer Nature 2018
R. Fernandes de Mello, M. Antonelli Ponti, *Machine Learning*,
https://doi.org/10.1007/978-3-319-94989-5_3

**Fig. 3.1** Example of radial
functions assuming different
values for the spreading
parameter $\sigma = \{3, 2, 1, 0.5\}$;
a small $\sigma$ leads to relevant
weights only for nearest
points, while a large $\sigma$
provides relevant weights also
for distant objects

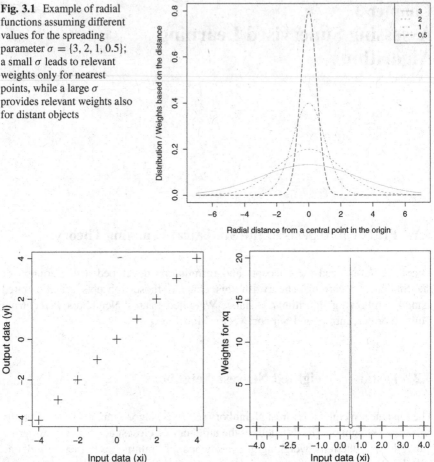

**Fig. 3.2** Examples drawn from an identity function (**a**) input space with 9 points; (**b**) the spread
defined by $\sigma = 0.02$ does not allow any neighbor for $x_q = 0.5$, so we cannot predict its output

having $\left\| x_i - x_q \right\|_2^2$ as the $L_2$-norm (Euclidean norm) between vectors $x_i$ and $x_q$, and
$\sigma$ as the spread of the radial function. Thus, the greater $\sigma$ is, the more open is the
radial function as illustrated in Fig. 3.1.

Figure 3.2a illustrates a training set in $\mathbb{R}^2$, in which there is a linear association
between every input variable $x_i$ and output class $y_i$. For instance, the example or
point $x_i = 4$ is associated to the output class $y_i = 4$, and so on. Notice, we have
more than two classes in this scenario, in fact this is a regression problem in which
there is a linear association between $x_i$ and $y_i$.

Consider data were produced using an identity function, and let $\sigma \geq 0$ (given
$\sigma \in \mathbb{R}_+$). Now let the smallest possible value for $\sigma$, which makes it greater than
zero but small enough to avoid nearest neighbors $x_i$, as illustrated in Fig. 3.2b.

Still in this context, let $\sigma = 0.01$ and the query point $x_q = 0.5$, resulting in no closest neighbor for all training examples, as computed next:

$$w_{x_i=0}(x_q) = \exp\left(\frac{-\|x_i - x_q\|_2^2}{2\sigma^2}\right) = \exp\left(\frac{-\|(0) - x_q\|_2^2}{2\sigma^2}\right)$$

$$= \exp\left(\frac{-\|(0) - (0.5)\|_2^2}{2\sigma^2}\right) = \exp\left(\frac{0.25}{2 \times 0.01^2}\right) \approx 0$$

$$w_{x_i=1}(x_q) = \exp\left(\frac{-\|x_i - x_q\|_2^2}{2\sigma^2}\right) = \exp\left(\frac{-\|(1) - x_q\|_2^2}{2\sigma^2}\right)$$

$$= \exp\left(\frac{-\|(1) - (0.5)\|_2^2}{2\sigma^2}\right) = \exp\left(\frac{0.25}{2 \times 0.01^2}\right) \approx 0.$$

In this manner, the output class for $x_q = 0.5$ is:

$$f(x_q = 0.5) = \frac{\sum_i^n y_i 0}{\sum_i^n 0} = \frac{0}{0}, \tag{3.1}$$

being undefined. This happens because there is no nearby point in the training set. So, having $\sigma = 0.01$, DWNN would classify query points only if they were very close to the training examples. Now consider we make $\sigma$ so small that it will tend to zero by the positive side, i.e., $\sigma \to 0^+$. In such circumstance, DWNN would only produce outputs for query points that coincide with that exact training example. This is the perfect instance to represent the **memory-based classifier** (see Chaps. 1 and 2 for a detailed discussion). Observe this classifier only memorizes training examples, consequently it is incapable of generalizing learning. This is the most representative situation of **overfitting**, i.e., the classifier only memorizes (it does not learn) the training set.

Going to the other extreme in which $\sigma \to +\infty$, the weighing function would produce:

$$w_{x_i=-5}(x_q) = \exp\left(\frac{-\|x_i - x_q\|_2^2}{2\sigma^2}\right) = \exp\left(\frac{-\|(-5) - x_q\|_2^2}{2\sigma^2}\right)$$

$$= \exp\left(\frac{-\|(-5) - (0.5)\|_2^2}{2\sigma^2}\right) = \exp\left(\frac{30.25}{2 \times \infty^2}\right) \approx 1$$

$$w_{x_i=5}(x_q) = \exp\left(\frac{-\|x_i - x_q\|_2^2}{2\sigma^2}\right) = \exp\left(\frac{-\|(5) - x_q\|_2^2}{2\sigma^2}\right)$$

$$= \exp \left( \frac{-\,\|(5) - (0.5)\|_2^2}{2\sigma^2} \right) = \exp \left( \frac{20.25}{2 \times \infty^2} \right) \approx 1.$$

So, weights are equal to 1 for any value of $x_i$, even those far from $x_q = 0.5$. Then, the output class will be:

$$f(x_q = 0.5) = \frac{\sum_i^n y_i 1}{\sum_i^n 1} = \frac{\sum_i^n y_i}{n} = \frac{1}{n} \sum_i^n y_i,$$

i.e., the average value for all $y_i$ taking the training set. In fact, we now tend to the average value given all output classes in the training set. This is the most representative situation of **underfitting**, meaning this classifier is not even capable of modeling the training set.

From those two extreme scenarios, we observe $\sigma$ is the parameter defining the learning bias for DWNN. When $\sigma \to 0^+$, DWNN produces the memory-based classifier (overfitting), and if $\sigma \to +\infty$, it builds up an average-based classifier (underfitting). Notice $\sigma$ must be set so that it provides a representative enough model for this identity function. But what would be the most adequate value for it?

To answer that question, we should investigate the association between the input space and the output classes. After plotting this problem, one could simply decide to set $\sigma$, so every query point $x_q$ would have at least two nearest neighbors. By proceeding with that approach, we would obtain the output results illustrated in Fig. 3.3, in which an affine relationship is defined for the two closest points. As a drawback, we may notice our training set is not enough to characterize this linear association when the query point is significantly smaller than $-4$ or greater than 4, i.e., when $x_q < -4$ or $x_q > 4$.

What does it happen if $\sigma$ is small enough so there is only a single nearest neighbor? In that situation, DWNN would produce outputs as shown in Fig. 3.4, in which query points lead to a discontinuous step function.

Notice the influence $\sigma$ has on a more complex regression between the input and output spaces, such as in Figs. 3.5 and 3.6. If $\sigma$ is enough to have a single neighbor, we would have a function composed of steps; while for $\sigma$ defining two nearest points, we would have a linear approximation; and if $\sigma$ sets between 2 and 4 nearest neighbors, DWNN is capable of outputting something similar to such sinusoidal function. The question that stays is: Is this approximately sinusoidal output the best? That should be evaluated over unseen examples in order to take a final conclusion.

Let us analyze the influence of distinct radial basis functions in the bias of such algorithm, given $\sigma \to 0^+$ and $\sigma \to +\infty$. Consider an infinite number of training points on $x_i$ producing the linear identity function illustrated in Fig. 3.7, showing what happens for a small $\sigma$. Observe an asymptotic infinite number of radial basis functions along $x_i$, every one centered at each one of the query points.

By setting $\sigma \to +\infty$, a single and unique radial basis function is defined along $x_i$ as shown in Fig. 3.7b. Observe it would be impossible to plot such radial function.

Thus, when $\sigma \to 0^+$, DWNN considers an infinite set of all possible radial basis functions, while for $\sigma \to +\infty$, its bias contains a single radial. Both situations

**Fig. 3.3** Output results provided by DWNN given $\sigma = 0.35$, which is enough to set two nearest neighbors for each input example. We highlight a test example at $x_i = 0.5$. Note that, given the limited training set, we can only predict from $-4$ to $4$, also, we just plot a single radial function for clarity

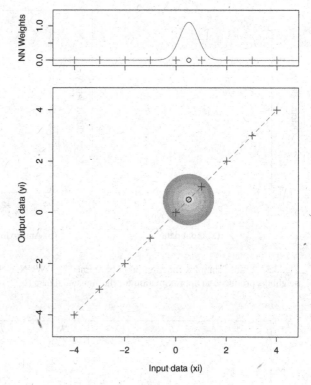

**Fig. 3.4** A discontinuous step function when $\sigma$ provides a single nearest neighbor

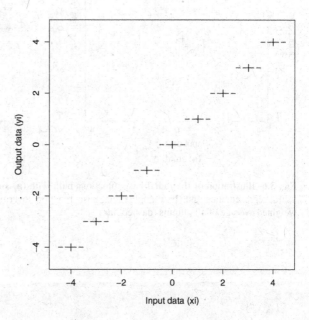

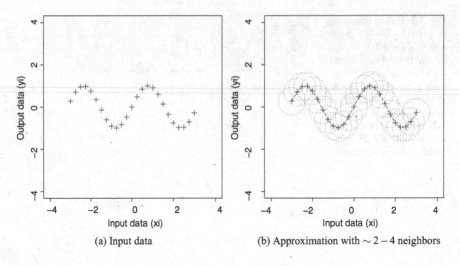

(a) Input data                                (b) Approximation with ∼ 2 − 4 neighbors

**Fig. 3.5** A more complex function (**a**) to be learned by DWNN: a choice of $\sigma$ that includes ∼2–4 neighbors produces an approximation for the sinusoidal data (**b**)

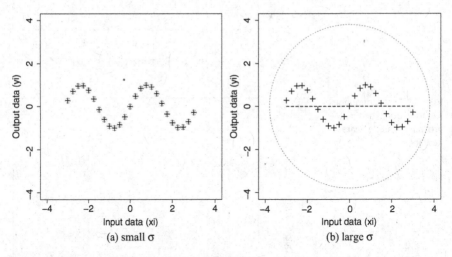

(a) small σ                                        (b) large σ

**Fig. 3.6** Illustration of the radial basis functions built with (**a**) small $\sigma$, insufficient to obtain a useful representation and (**b**) a single radial basis function covering all data points, yielding to a weighted average of all outputs (dashed line)

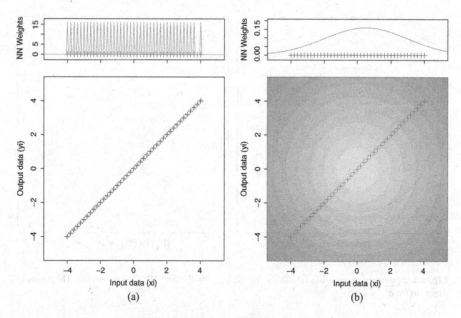

**Fig. 3.7** Illustration of radial basis functions built when (**a**) $\sigma \to 0^+$ and (**b**) $\sigma \to +\infty$

are illustrated side by side in Fig. 3.8, having $\mathscr{F}_{\text{all}}$ as the space containing every possible function, $\mathscr{F}_{\sigma \to 0^+}$ as an illustration for the space containing every possible radial basis function provided by DWNN when $\sigma \to 0^+$, and finally $\mathscr{F}_{\sigma \to +\infty}$ corresponds to a space containing only a single radial basis function, i.e. the one providing the average along every possible query point.

Note the memory-based classifier is most likely to be selected when the space of functions, i.e., the algorithm bias, is less restricted. We may also refer to it as a weaker bias. At the same time, it is possible to understand why a very small space of functions imply underfitting. In this case, we can say it has a strong bias.

Listing 3.1 presents the implementation of the DWNN algorithm using the R Statistical Software, whose main function is dwnn(). The reader is suggested to execute testIdentity() to assess the effect of $\sigma = \{0.01, 0.1, 100\}$. For $\sigma = 0.01$, all DWNN output classes will be NaN, i.e., not a number, due to the division by zero from Eq. (3.1). For $\sigma = 0.1$, the number of neighbors is adequate for this problem. For $\sigma = 100$, DWNN will consider all points as neighbors, tending to the average.

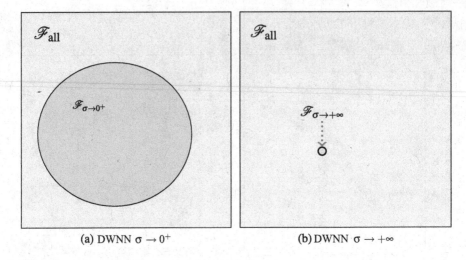

(a) DWNN $\sigma \to 0^+$        (b) DWNN $\sigma \to +\infty$

**Fig. 3.8** Spaces of admissible functions (or biases) for the DWNN: (**a**) $\sigma \to 0^+$; (**b**) a single function for $\sigma \to +\infty$

**Listing 3.1** Distance-Weighted Nearest Neighbor (DWNN) algorithm

```
# Computing the L2-norm between vector x_q and x_i
euclidean <- function(x_i, x_q) {
    sqrt(sum((x_i-x_q)^2))
}

# This is the weighing function
w_i <- function(dist, sigma) {
    exp(-dist^2/(2*sigma^2))
}

# This is the DWNN algorithm. It receives the training set,
# the test set and then sigma.
dwnn <- function(training.set, test.set, sigma = 1) {

    # Number of input attributes (we consider only the
    # last one as the output class)
    nAttrs = ncol(training.set)-1
    class = ncol(training.set)

    obtained = rep(0, nrow(test.set))

    # For every example in the test set
    for (q in 1:nrow(test.set)) {
        x_q = as.vector(test.set[q,1:nAttrs])
        num = 0
        den = 0

        # Computing the output class based on every
        # example i in the training set
```

```
        for (i in 1:nrow(training.set)) {
            # Computing the L2-norm
            dist = euclidean(training.set[i,1:nAttrs], x_q)

            # Computing the weight
            weight = w_i(dist, sigma)
            num = num + weight * training.set[i, class]
            den = den + weight
        }

        # The output class according to DWNN
        produced_output = num / den
        obtained[q] = produced_output
    }

    # List of DWNN results
    ret = list()

    # The obtained class after executing DWNN
    ret$obtained = obtained

    # The absolute error in terms of the expected class
    # versus the obtained one
    ret$absError = abs(test.set[,class] - obtained)

    # Here we save the expected class for later use
    # (if necessary)
    ret$expected = test.set[,class]

    return (ret)
}

# Test the identity function
testIdentity <- function(sigma=0.01) {

    # Defining the training set
    training.set = cbind(seq(-5,5,by=1), seq(-5,5,by=1))

    # Defining the test set
    test.set = cbind(seq(-5.5,5.5,by=1), seq(-5.5,5.5,by=1))

    results = dwnn(training.set, test.set, sigma)

    # Plotting the training set
    plot(training.set, xlab="x_i_(input_value)",
                       ylab="y_i_(expected_class)")
    obtained.result = cbind(test.set[,1], results$obtained)
    # Plotting the DWNN results for the unseen example (in
        red)
    points(obtained.result, col=2)

    return (results)
}
```

Similarly, Listing 3.2 presents another problem involving a sinusoidal function with some noise added. We also invite the reader to execute function `testSin()` for different values of $\sigma$. Of course, if $\sigma$ is too small, NaNs will be produced. If $\sigma$ is too large, results will tend to the average of $y_i$. A fair value for $\sigma$ is in range $[2, 5]$ as the reader may conclude.

**Listing 3.2** Using DWNN on the input examples produced using a sinusoidal function with added noise

```r
source("dwnn.r")

# Test the sinusoidal function
testSin <- function(sigma=0.01) {

    # Producing data
    data = sin(2*pi*seq(0,2, length=100)) +
              rnorm(mean=0, sd=0.1, n=100)
    training.ids = sample(1:length(data), size=50)
    test.ids = setdiff(1:length(data), training.ids)

    # Defining the training set
    training.set = cbind(training.ids, data[training.ids])

    # Defining the test set
    test.set = cbind(test.ids, data[test.ids])

    # Running DWNN
    results = dwnn(training.set, test.set, sigma)

    # Plotting the training set
    plot(training.set, xlab="x_i (input value)",
                       ylab="y_i (expected class)")
    obtained.result = cbind(test.set[,1], results$obtained)

    # Plotting the DWNN results for unseen examples (in red)
    points(obtained.result, col=2)

    return (results)
}
```

## 3.3  Using the Chernoff Bound

As approached in Sect. 2.4.2, the Statistical Learning Theory allows us to prove:

$$P\left(\sup_{f \in \mathscr{F}} |R(f) - R_{\text{emp}}(f)| > \epsilon\right) \leq 2P\left(\sup_{f \in \mathscr{F}} |R'_{\text{emp}}(f) - R_{\text{emp}}(f)| > \epsilon/2\right)$$

$$= 2P\left(\sup_{f \in \mathscr{F}_{z_{2n}}} |R'_{\text{emp}}(f) - R_{\text{emp}}(f)| > \epsilon/2\right) \leq 2\mathscr{N}(\mathscr{F}, 2n)\exp\left(-n\epsilon^2/4\right),$$

having the right-side term provided by the Chernoff bound and, in particular, term $\mathcal{N}(\mathcal{F}, 2n)$ is the Shattering coefficient for two samples with $n$ examples each. We here remind the reader that this coefficient is indeed a function of $n$.

For instance, consider we have three arbitrary classification algorithms $A_1$, $A_2$ and $A_3$, each one with the following Shattering coefficients as $n$ increases:

$$\mathcal{N}(\mathcal{F}, 2n)_{A_1} = n^2$$

$$\mathcal{N}(\mathcal{F}, 2n)_{A_2} = n^4$$

$$\mathcal{N}(\mathcal{F}, 2n)_{A_3} = 2^n.$$

Plugging those in $2\mathcal{N}(\mathcal{F}, 2n)\exp(-n\epsilon^2/4)$, for $\epsilon = 0.1$, and plotting it for $n$ from 1 to 20,000 we can analyze the algorithms convergence as shown in Fig. 3.9a, b. This analysis is produced using Listing 3.3.

**Listing 3.3** Assessing three Shattering coefficients according to the number of training examples $n$

```
epsilon = 0.1
n = 1:20000
N_F_2n_A1 = n^2
N_F_2n_A2 = n^4
N_F_2n_A3 = 2^n

Upper_bound_A1 = 2*N_F_2n_A1*exp(-n*epsilon^2/4)
Upper_bound_A2 = 2*N_F_2n_A2*exp(-n*epsilon^2/4)
Upper_bound_A3 = 2*N_F_2n_A3*exp(-n*epsilon^2/4)

par(mfrow=c(1,2))
plot(Upper_bound_A1, col=1, t="1",
             xlab="2 Samples with n examples each",
             ylab="Probability bound")
lines(Upper_bound_A2, col=2, )
lines(Upper_bound_A3, col=3)

plot(log(Upper_bound_A1), col=1, t="1",
             ylim=log(range(Upper_bound_A2)),
             xlab="2 Samples with n examples each",
             ylab="Natural Logarithm of the Probability bound"
             )
lines(log(Upper_bound_A2), col=2)
lines(log(Upper_bound_A3), col=3)
```

We observe algorithm $A_1$ converges faster to zero, with the following upper bound for 10,000 training examples:

$$P(\sup_{f \in \mathcal{F}} |R(f) - R_{\text{emp}}(f)| > \epsilon) \leq 0.00278,$$

meaning the empirical risk is a good estimator for the expected risk, given an acceptable divergence of $\epsilon = 0.1$ with the probability less than or equal to 0.00278. Observe the need for defining a given acceptable divergence $\epsilon$ between the estimator and the expected value, so we can obtain the upper-bound limit for such probability.

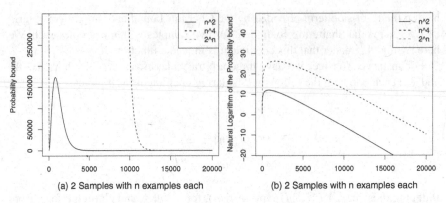

(a) 2 Samples with n examples each          (b) 2 Samples with n examples each

**Fig. 3.9** Assessing three different Shattering coefficients according to the number of training examples $n$

In the same scenario, we may notice $A_2$ requires much more training examples to provide a similar upper-bound, because it has a more complex space of admissible functions, therefore its Shattering coefficient is greater. Finally, $A_3$ cannot be even seen in the figure, because it is exponential and produces points off the chart.

By having this piece of information, one can better choose a classification algorithm. Certainly not $A_3$, because it would never converge to an acceptable upper-bound probability. We still have $A_1$ and $A_2$ to choose from: so which is the best? For now we only know the empirical risk is a good estimator for both $A_1$ and $A_2$. So, let us investigate those in more detail.

Let the empirical risks of $A_1$ and $A_2$ be equal, $\sup_{f \in \mathscr{F}_{A_1}} R_{emp}(f) = \sup_{f \in \mathscr{F}_{A_2}} R_{emp}(f) = 0.05$. Now we are sure about selecting $A_1$ instead of $A_2$, as it provides the same empirical risk and it converges faster according to the Chernoff bound, requiring less training examples to learn. In this scenario, we ensure:

$$P(\sup_{f \in \mathscr{F}_{A_1}, \mathscr{F}_{A_2}} |R(f) - R_{emp}(f)| > \epsilon) \leq 0.00278,$$

so that $A_1$ requires 10,000, while $A_2$ needs 18,339 examples to provide the same bound.

This comparison relied on the Shattering coefficients which must be computer in some manner, as discussed in Sect. 2.7. Besides such theoretical approach, Listing 3.4 introduces an empirical manner to estimate the Shattering coefficient as $n$ increases. It is useful to users understand what happens in terms of counting the number of different functions admitted by some classification algorithm. It considers $p$ hyperplanes classifying some $R$-dimensional input space for $R \geq 2$. In our code, `shattering.coefficient.estimator()` produces two columns the first with the sample size $n$, and the second with the estimated number of distinct classifications for each particular sample size. Observe that the number of iterations `iter` should become excessively large in order to obtain a good approximation for the theoretical Shattering coefficient (see Sect. 2.7).

**Listing 3.4** Shattering estimation using random input examples

```
# Estimating the Shattering coefficient (or function) for an
    input space given
# by the cartesian product of R real lines. For instance,
    for R^2 you use R=2.
# Parameter iter means the number of iterations used to
    assess every linear
# hyperplane in order to check out how many different
    classifications are found.
# Parameters n.start and n.end set the sample size n for the
    estimation.
# Term p sets the number of hyperplanes. Observe the range
    for vector w, term b
# and to generate the data sample (matrix sample) is fixed,
    but the user is suggested
# to adapt it to analyze a broader space.
shattering.coefficient.estimator <- function(iter=1000, n.
    start=1, n.end=100, p=1, R=2) {

        shatter = NULL
        cat("#Sample_size\tNumber_of_different_
            classifications_found...\n")

        # For every sample size
        for (i in n.start:n.end) {
            sample = NULL

                # Produce some random data in the input
                    space
                for (j in 1:R) {
                    sample = cbind(sample, rnorm(mean=0,
                        sd=1, n=i))
                }

            shatter.ways = list()

                # Attempt to find different classifications
                # provided by a single linear hyperplane
                for (j in 1:(i^2*iter)) {

                        combined.labels = rep(0, nrow(sample
                            ))
                        for (k in 1:p) {
                                # Randomly sets vector w
                                    which is normal to the
                                    hyperplane
                                w = runif(min=-10, max=10, n
                                    =R)
```

```
                    # Randomly sets term b to
                        define the intersection
                        of the
                    # hyperplane with the input
                        variables
                    b = runif(min=-5, max=5, n
                        =1)

                    # Providing the outcomes
                        giving this random
                        hyperplane
                    labels = sample %*% w + b

                    # If the outcome is equal to
                        zero or greater
                    # we will assume the
                        positive class (or label
                        )
                    id = which(labels >= 0)

                    # Otherwise the negative
                        class
                    nid = which(labels < 0)

                    # Setting the positive and
                        negative outcomes
                    labels[id] = 2^k-2
                    labels[nid] = 2^k-1

                    # Combining hyperplanes
                    combined.labels = combined.
                        labels + labels
                }

                # Defining a key such as in a
                    hashtable so we
                # can inform that this particular
                    classification happened
                key = paste(combined.labels, sep="#"
                    , collapse="")
                shatter.ways[[key]] = 1
            }

            # Printing results out
            cat(i, " ", length(shatter.ways), "\n")
            shatter = rbind(shatter, c(i, length(shatter
                .ways)))
        }

    return (shatter)
}
```

Results in Listing 3.5 shows outputs for the following setting *shattering.coefficient.estimator(iter=1000, n.start=1, n.end=100, p=1, R=2)* (input space is $\mathbb{R}^2$). To estimate the Shattering coefficient, we take this output and produce some regression to best fit it.

**Listing 3.5** Output provided by the estimation of the Shattering coefficient using Listing 3.4

```
#Sample size    Number of different classifications
1    2
2    4
3    8
4    14
5    22
6    32
7    42
8    56
9    74
10   92
11   109
12   132
13   156
14   180
15   206
16   237
17   266
18   298
19   337
20   373
21   414
22   459
23   490
24   535
25   583
26   635
27   682
28   738
29   787
30   840
```

The results of Listing 3.6 are saved in a file r2.dat, and then loaded in Gnuplot, a command line graphing tool for Linux, to estimate the shattering coefficient via regression. As output, we obtained something similar to Fig. 3.10, in which points correspond to the observed data and the curve is the regression. This empirical estimation approach is useful so the reader can picture we must assess all possible but different classification results produced given some input space and a single hyperplane.

**Listing 3.6** Gnuplot script to estimate and plot the Shattering coefficient

```
f(x)=a*x**2+b*x+c
fit f(x) "r2.dat" via a,b,c
plot "r2.dat" with points, f(x)
```

**Fig. 3.10** The estimated
Shattering coefficient and its
regression function obtained
from Gnuplot

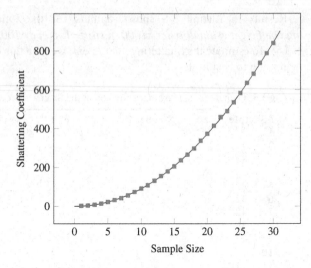

**Table 3.1** The regression
coefficients and respective
errors as estimated by
Gnuplot

| Coefficient | Value | Error | Error (%) |
|---|---|---|---|
| a | 0.952914 | $\pm\,0.006659$ | (0.6988%) |
| b | −0.475372 | $\pm\,0.2128$ | (44.75%) |
| c | 0.474878 | $\pm\,1.431$ | (301.3%) |

Regression results may change slightly because Gnuplot randomly selects the
seed for solving this fitting problem. The polynomial coefficients $a$, $b$ and $c$ and their
respective errors are listed in Table 3.1. For any classification algorithm building up
a single hyperplane in $\mathbb{R}^2$ to separate two classes ($\{-1, +1\}$), note the Shattering
coefficient function can be approximated by $f(n) = 0.952914n^2 - 0.475372n +
0.474878$. It is relevant to compare this result against the theoretical:

$$\mathcal{N}(\mathscr{F}, 2n) = \sum_{i=0}^{2} \binom{n-1}{i}^{1} = n^2 - n + 2,$$

from which we certainly notice our estimator is a infimum function. In any case,
this would be a fair estimation/approximation for the Shattering coefficient of
the Perceptron or the single-hyperplane Multilayer Perceptron working on a 2-
dimensional input space.

What does it happen when more linear hyperplanes are included in an algorithm
bias? As consequence, its Shattering coefficient will combine, or multiply, functions. We here invite the reader to use our estimator and compare its results with the
theoretical Shattering coefficient (see Sect. 2.7).

From now on, consider the theoretical coefficient to study the training set
illustrated in Fig. 3.11 and let us assess the following classification algorithms for
this task: the Perceptron, an MLP with 5 hyperplanes and another MLP with 10
hyperplanes.

**Fig. 3.11** Input space requiring more hyperplanes to proceed with the classification

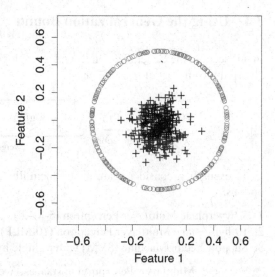

From Sect. 2.7, the Shattering coefficient of the Perceptron is $f(n) = n^2 - n + 2$, while for the 5-hyperplane MLP in a 2-dimensional input space is:

$$g(n) = \sum_{i=0}^{2} \binom{n-1}{i}^5 = 2\left(\frac{1}{32}(n-2)^5(n-1)^5 + (n-1)^5 + 1\right), \qquad (3.2)$$

and for the 10-hyperplane MLP:

$$h(n) = \sum_{i=0}^{2} \binom{n-1}{i}^{10} = 2\left(\frac{(n-2)^{10}(n-1)^{10}}{1024} + (n-1)^{10} + 1\right). \qquad (3.3)$$

Let the empirical risks for the algorithms be:

$$\sup_{f \in \mathscr{F}_{\text{Perc}}} R_{\text{emp}}(f) = 0.75$$

$$\sup_{f \in \mathscr{F}_{\text{5-MLP}}} R_{\text{emp}}(f) = 0.05$$

$$\sup_{f \in \mathscr{F}_{\text{10-MLP}}} R_{\text{emp}}(f) = 0.05.$$

When assessing those classification algorithms, the 5-hyperplane MLP is easily selected as the most adequate, because it converges faster to the upper-bound probability while also having a good enough empirical risk. In order to check how many training examples are required, we suggest the reader to follow the same steps provided in Sect. 2.7.

## 3.4   Using the Generalization Bound

In this section, the Generalization Bound is employed to select the best classification algorithm from a set of possible ones:

$$R(f) \leq \underbrace{R_{\text{emp}}(f)}_{\text{training error}} + \underbrace{\sqrt{-\frac{4}{n}\left(\log(\delta) - \log(2\mathcal{N}(\mathscr{F}, 2n))\right)}}_{\text{divergence factor}}$$

To exemplify, consider again the problem illustrated in Fig. 3.11. Assume three options:

1. 5-hyperplane Multilayer Perceptron (5-MLP);
2. 10-hyperplane Multilayer Perceptron (10-MLP);
3. Support Vector Machine (SVM) with a single hyperplane.

For both Multilayer Perceptron instances, we use the Shattering coefficient previously estimated. The Shattering coefficient for the 5-MLP and the 10-MLP are $g(n)$ and $h(n)$ as defined in Eqs. (3.2) and (3.3), respectively.

Then, let the SVM be an algorithm producing a single linear hyperplane. Could it divide such an input space? Obviously not. As matter of fact, some embedding (mapping) to another space would be necessary so that SVM can shatter examples. In this scenario, the following nonlinear kernel function is used:

$$k\left(\begin{bmatrix} x_{i,1} \\ x_{i,2} \end{bmatrix}\right) = \begin{bmatrix} x_{i,1}^2 \\ \sqrt{2}x_{i,1}^2 x_{i,2}^2 \\ x_{i,2}^2 \end{bmatrix},$$

to embed every example $x_i$ from $\mathbb{R}^2$ into $\mathbb{R}^3$, in which indices $i, 1$ and $i, 2$ refer to both dimensions for every example $i$, so they will be reorganized as discussed later in this chapter (see Fig. 3.19).

After applying such a kernel, the space $\mathbb{R}^3$ becomes linearly separable so that SVM can be used. In such situation, the Shattering coefficient will be:

$$\mathcal{N}(\mathscr{F}, 2n) = 2\sum_{i=0}^{3}\binom{n-1}{i}^1 = \frac{1}{3}n(n^2 - 3n + 8)$$

as discussed in Sect. 2.7.

According to the Chernoff bound, SVM converges much faster so that less examples are needed. In addition, supposing the empirical risks are:

$$\sup_{f \in \mathscr{F}_{5\text{-MLP}}} R_{\text{emp}}(f) = 0.05$$

$$\sup_{f \in \mathcal{F}_{10\text{-MLP}}} R_{\text{emp}}(f) = 0.04$$

$$\sup_{f \in \mathcal{F}_{\text{SVM}}} R_{\text{emp}}(f) = 0.1,$$

which is the best algorithm for this problem? This is analytically answered using the Generalization Bound:

1. Given the 5-hyperplane Multilayer Perceptron:

$$R(f) \leq \underbrace{R_{\text{emp}}(f)}_{\text{training error}} + \underbrace{\sqrt{-\frac{4}{n}\left(\log(\delta) - \log(2\mathcal{N}(\mathcal{F}, 2n))\right)}}_{\text{divergence factor}}$$

$$R(f) \leq 0.05 + \sqrt{-\frac{4}{n}\left(\log(\delta) - \log(5 - \text{MLP}(n))\right)},$$

2. Given the 10-hyperplane Multilayer Perceptron:

$$R(f) \leq \underbrace{R_{\text{emp}}(f)}_{\text{training error}} + \underbrace{\sqrt{-\frac{4}{n}\left(\log(\delta) - \log(2\mathcal{N}(\mathcal{F}, 2n))\right)}}_{\text{divergence factor}}$$

$$R(f) \leq 0.04 + \sqrt{-\frac{4}{n}\left(\log(\delta) - \log(10 - \text{MLP}(n))\right)},$$

3. Given the Support Vector Machine:

$$R(f) \leq \underbrace{R_{\text{emp}}(f)}_{\text{training error}} + \underbrace{\sqrt{-\frac{4}{n}\left(\log(\delta) - \log(2\mathcal{N}(\mathcal{F}, 2n))\right)}}_{\text{divergence factor}}$$

$$R(f) \leq 0.1 + \sqrt{-\frac{4}{n}\left(\log(\delta) - \log(\text{SVM}(n))\right)}.$$

Figure 3.12 illustrates the right-side term of the Generalization Bound for all three classification algorithms, providing an upper-limit for $R(f)$. We conclude SVM converges faster to zero than the other options, so it defines a tighter (more precise) upper-bound for the expected risk, making it more robust to classify unseen examples. Figure 3.12 was produced, having $\delta = 0.01$ in:

$$P(\sup_{f \in \mathcal{F}} |R(f) - R_{\text{emp}}(f)| > \epsilon) \leq \delta.$$

**Fig. 3.12** Generalization
Bounds for all three
classification algorithms
under analysis

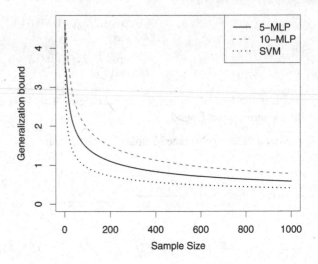

**Listing 3.7** Computing the Generalization Bound for all three classification algorithms under analysis

```
# Sample size variation
n = 1:1000
delta =0.01

# Empirical risks
R_emp_f_5_MLP = 0.05
R_emp_f_10_MLP = 0.04
R_emp_f_SVM = 0.1

# Shattering coefficients
Shattering_5_MLP = 2 *(1/32 *(n − 2)^5 * (n − 1)^5 + (n − 1)
    ^5 + 1)
Shattering_10_MLP = 2 *(((n − 2)^10 * (n − 1)^10)/1024 + (n
    − 1)^10 + 1)
Shattering_SVM = 1/3 * n * (n^2 − 3*n + 8)

# Computing the Generalization Bounds
R_f_5_MLP = R_emp_f_5_MLP + sqrt(−4/n * (log(delta) −
    log(Shattering_5_MLP)))
R_f_10_MLP = R_emp_f_10_MLP + sqrt(−4/n * (log(delta) −
    log(Shattering_10_MLP)))
R_f_SVM = R_emp_f_SVM + sqrt(−4/n * (log(delta) −
    log(Shattering_SVM)))

plot(R_f_5_MLP, t="l", col=1, ylim=c(0, max(c(R_f_5_MLP, R_f
    _10_MLP)))))
lines(R_f_10_MLP, col=2)
lines(R_f_SVM, col=3)
```

**Fig. 3.13** Assessing the
Generalization Bound using
greater sample sizes

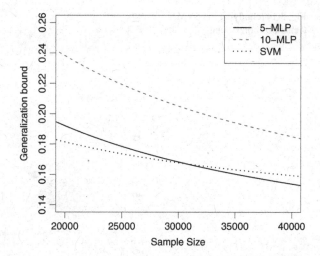

By analyzing a greater sample (e.g., $n > 30,000$), we observe the 5-hyperplane
MLP crosses the convergence of SVM when $n$ is around 31,000 (Fig. 3.13). Thus, if
one has that many examples available, the 5-hyperplane MLP would perform better.
Otherwise, SVM is the best choice.

According to this classification task, we conclude the empirical risk is not the
only factor to rely on when taking decisions. We should consider the Generalization
Bound as well. Despite this fact, many studies do not report the Generalization
Bounds to ensure learning. Some of them neither present results using the $k$-fold
cross validation strategy [1], which is a way to approximate such a bound. We
encourage the reader to analyze the Shattering coefficient for any input space (s)he
is working on.

## 3.5 Using the SVM Generalization Bound

The previous sections considered the most common approach to compute the
Generalization Bound, which is defined as follows:

$$R(f) \leq R_{\mathrm{emp}}(f) + \sqrt{-\frac{4}{n}\left(\log(\delta) - \log(2\mathcal{N}(\mathscr{F}, 2n))\right)},$$

in which $R(f)$ is the expected risk, $R_{\mathrm{emp}}(f)$ is the empirical risk, $n$ is the training
set size (or sample size), and $\mathcal{N}(\mathscr{F}, 2n)$ is the Shattering coefficient. However,
there is a tighter bound for Support Vector Machines:

**Fig. 3.14** Studying the SVM
Generalization Bound

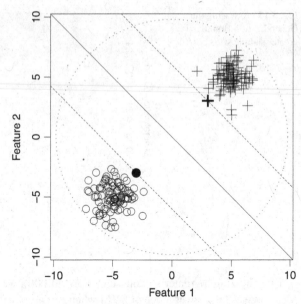

$$R(f) \leq \nu(f) + \sqrt{\frac{c}{n}\left(\frac{R}{\rho^2}\log(n^2) + \log(1/\delta)\right)}, \qquad (3.4)$$

in which $\rho$ is the maximal margin, $R$ as the radius of the smallest open ball capable of containing all training examples in the input space, $c$ is a constant depending on the target scenario, and, finally, $\nu(f)$ is the fraction of the training samples lying on the margin limits.

This bound is exemplified through the training set depicted in Fig. 3.14, having the smallest open ball circumscribing all training examples with radius $R = 7.386189$, the maximal margin as $\rho = 4.242641$, and $\nu(f) = 0$ due to no example is located within the support hyperplanes of each class.

From such information, we compute the SVM Generalization Bound as follows (considering $c = 4$ and $\delta = 0.01$ for convenience):

$$R(f) \leq \nu(f) + \sqrt{\frac{c}{n}\left(\frac{R}{\rho^2}\log(n^2) + \log(1/\delta)\right)}$$

$$R(f) \leq 0 + \sqrt{\frac{4}{n}\left(\frac{7.386189}{4.242641^2}\log(n^2) + \log(1/0.01)\right)}$$

$$R(f) \leq \sqrt{\frac{4}{n}\left(0.4103438\log(n^2) + 4.60517\right)}$$

**Fig. 3.15** The common versus the more precise SVM Generalization Bound

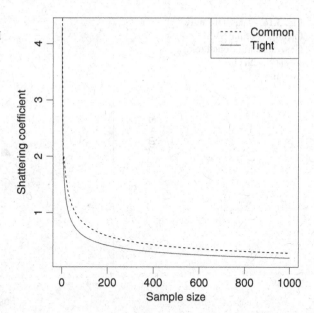

$$R(f) \leq \sqrt{\frac{1.641375}{n} \log(n^2) + \frac{18.42068}{n}}$$

$$R(f) \leq \sqrt{\frac{3.28275}{n} \log(n) + \frac{18.42068}{n}}.$$

This is a tighter (more precise) bound when compared to the common Generalization Bound used in previous section, as illustrated in Fig. 3.15 (based on Listing 3.8). However, this bound requires more knowledge about the input data organization. Thus, given there are many situations parameters $R$ and $\rho$ are unknown, we suggest the use of the common bound instead.

**Listing 3.8**  Analyzing the common versus the tighter (more precise) SVM Generalization Bound

```
n = 1:1000
delta = 0.01

# Using the SVM Generalization Bound
Radius = 7.386189
rho = 4.242641
R_f_SVM_Generalization_Bound = sqrt(3.28275/n * log(n) +
    18.42068/n)

# Approximating the Shattering coefficient using the common
    Generalization Bound
Shattering_SVM = 1/3 * n * (n^2 - 3*n + 8)
R_f_common_Generalization_Bound = sqrt(-4/n * (log(delta) -
    log(Shattering_SVM)))
```

**Fig. 3.16** Dataset to study the impact of the SVM Generalization Bound for a scenario with class overlapping

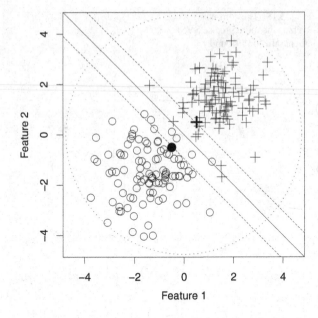

```
plot(R_f_SVM_Generalization_Bound, t="1")
lines(R_f_common_Generalization_Bound, col=2)
```

To better understand the SVM Generalization Bound, Fig. 3.16 shows a problem instance with $n = 200$ examples, having $R = 3.886189$ and $\rho = \frac{\sqrt{2}}{2} = 0.7071068$. Parameters $c = 4$ and $\delta = 0.01$ are again assumed. In this situation, 15 training examples lie within the margin so that $\nu(f) = \frac{15}{200} = 0.075$.

Thus, we have the following SVM Generalization Bound:

$$R(f) \leq \nu(f) + \sqrt{\frac{c}{n}\left(\frac{R}{\rho^2}\log(n^2) + \log(1/\delta)\right)}$$

$$R(f) \leq 0.075 + \sqrt{\frac{4}{n}\left(\frac{3.886189}{0.7071068^2}\log(n^2) + \log(1/0.01)\right)}$$

$$R(f) \leq 0.075 + \sqrt{\frac{4}{n}\left(7.772378\log(n^2) + 4.60517\right)}$$

$$R(f) \leq 0.075 + \sqrt{\frac{31.08951}{n}\log(n^2) + \frac{18.42068}{n}}$$

$$R(f) \leq 0.075 + \sqrt{\frac{62.17902}{n}\log(n) + \frac{18.42068}{n}},$$

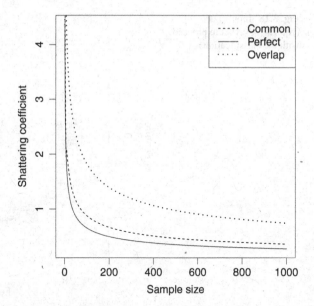

**Fig. 3.17** Analyzing the impact of the SVM Generalization Bound for a scenario with class overlapping

which is illustrated and compared against the perfect linearly separable case (Fig. 3.14) as well as against the common Generalization Bound for a single linear hyperplane, as seen in Fig. 3.17. As we may notice, the convergence is slower when some class overlapping is present, as expected. Listing 3.9 shows the script to plot such a figure.

**Listing 3.9** Assessing the SVM Generalization Bound for perfectly separable versus class overlapping sets

```
n = 1:1000
delta = 0.01
nu_f = 0.075

# Using the SVM Generalization Bound for the perfectly
    separable training set
Radius = 7.386189
rho = 4.242641
R_f_SVM_Generalization_Bound_Perfect = nu_f + sqrt(3.28275/n
    * log(n) + 18.42068/n)

# Using the SVM Generalization Bound for the training set
    with some class overlapping
Radius = 3.886189
rho = 0.7071068
R_f_SVM_Generalization_Bound_Mix = nu_f + sqrt(62.17902/n *
    log(n) + 18.42068/n)

# Approximating the Shattering coefficient using the common
    Generalization Bound
```

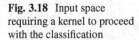

**Fig. 3.18** Input space
requiring a kernel to proceed
with the classification

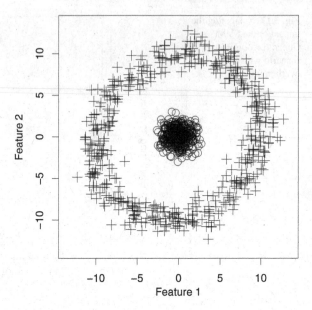

```
Shattering_SVM = 1/3 * n * (n^2 - 3*n + 8)
R_f_common_Generalization_Bound = nu_f + sqrt(-4/n * (log(
    delta) - log(Shattering_SVM)))

plot(R_f_SVM_Generalization_Bound_Perfect, t="1")
lines(R_f_SVM_Generalization_Bound_Mix, col=2)
lines(R_f_common_Generalization_Bound, col=3)
```

We then proceed to the last problem instance to be approached in this section,
which deals with the training set shown in Fig. 3.18.

To solve this problem, we must apply the following nonlinear kernel function:

$$k\left(\begin{bmatrix} x_{i,1} \\ x_{i,2} \end{bmatrix}\right) = \begin{bmatrix} x_{i,1}^2 \\ \sqrt{2}x_{i,1}^2 x_{i,2}^2 \\ x_{i,2}^2 \end{bmatrix},$$

to obtain a third-dimensional feature space, as depicted in Fig. 3.19. This new space
allows the perfect linear separation between classes, having $v(f) = 0$.

In this circumstance, $R = 104.4398$ and $\rho = 41.08596$. Assuming $c = 4$ and
$\delta = 0.01$:

$$R(f) \leq v(f) + \sqrt{\frac{c}{n}\left(\frac{R}{\rho^2}\log(n^2) + \log(1/\delta)\right)}$$

**Fig. 3.19** Feature space after applying the nonlinear kernel function. We show a hyperplane separating the classes

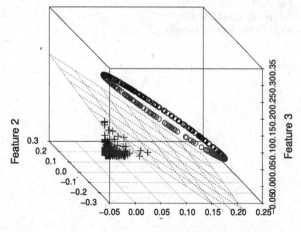

$$R(f) \leq 0 + \sqrt{\frac{4}{n}\left(\frac{104.4398}{41.08596^2}\log(n^2) + \log(1/0.01)\right)}$$

$$R(f) \leq \sqrt{\frac{4}{n}\left(0.06186986\log(n^2) + 4.60517\right)},$$

$$R(f) \leq \sqrt{\frac{0.2474794}{n}\log(n^2) + \frac{18.42068}{n}}$$

$$R(f) \leq \sqrt{\frac{0.4949589}{n}\log(n) + \frac{18.42068}{n}},$$

which is compared to the 5-hyperplane MLP (capable of classifying the original input space $\mathbb{R}^2$), as seen in Fig. 3.20 (based on Listing 3.10). After applying the nonlinear kernel function, the feature space is linearly separable and, in addition, SVM converges much faster than MLP. However, if we attempt to apply SVM directly on the original 2-dimensional input space, classification results are poor.

**Listing 3.10** Comparing the 5-hyperplane MLP versus the SVM Generalization Bound

```
n = 1:1000
delta = 0.01
nu_f = 0

# Using the SVM Generalization Bound for the perfectly
    separable training set
Radius = 104.4398
rho = 41.08596
R_f_SVM_Generalization_Bound_Perfect = nu_f + sqrt(0.4949589
    /n * log(n) + 18.42068/n)
```

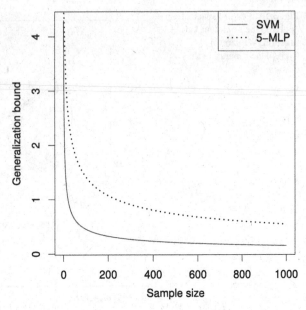

```
# Using the common Generalization Bound for the 5−hyperplane
   MLP on the original 2−dimensional input space
Shattering_5_MLP = 2 *(1/32 *(n − 2)^5 * (n − 1)^5 + (n − 1)
   ^5 + 1)
R_f_5_MLP_common_Generalization_Bound = nu_f + sqrt(−4/n * (
   log(delta) − log(Shattering_5_MLP)))

plot(R_f_SVM_Generalization_Bound_Perfect, t="l")
lines(R_f_5_MLP_common_Generalization_Bound, col=2)
```

SVM may be naively compared against other classification algorithms based
solely on empirical risks. It will provide bad results when the input space is not
adequate to shatter examples, requiring some space transformation. If such ideal
transformation is found, no other algorithm can outperform SVM. This is the main
reason to discuss about kernel functions in a following chapter. One should study
the input space and possible kernel-based transformations rather than investing time
in designing new classification algorithms that do not have the same aforementioned
tight learning guarantees.

## 3.6 Empirical Study of the Biases of Classification Algorithms

We can use the Shattering coefficient and the empirical risk to illustrate the biases of classification algorithms. In summary, both pieces of information are part of the Generalization Bound:

$$R(f) \leq \underbrace{R_{\mathrm{emp}}(f)}_{\text{training error}} + \underbrace{\sqrt{-\frac{4}{n}\left(\log(\delta) - \log(2\mathcal{N}(\mathscr{F}, 2n))\right)}}_{\text{divergence factor}}.$$

In this section, we analyze three binary classification problems, having a 2-dimensional input space, in which:

1. classes are linearly separable, such as in Fig. 3.14;
2. there is a low degree of class overlapping, as seen Fig. 3.16;
3. examples under a given class are surrounded by another, as in Fig. 3.18.

In the first situation, given classes are linearly separable, the Perceptron could be used instead of the Multilayer Perceptron, once a single hyperplane is enough. Thus, if we consider the Perceptron, the Shattering coefficient will be:

$$f(n) = n^2 - n + 2.$$

If we take a $k$-hyperplane MLP (for $k > 1$), the coefficient is unnecessarily complex. From this information, the biases for the Perceptron and for the $k$-hyperplane MLP are illustrated in Fig. 3.21.

**Fig. 3.21** Illustrating the biases for the Perceptron, 1-MLP and $k$-MLP for $k > 1$

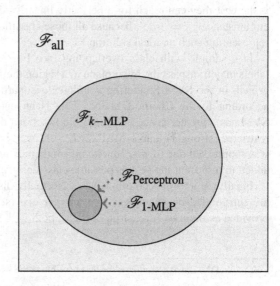

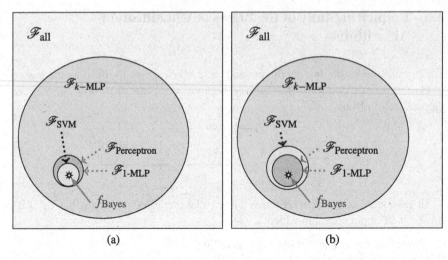

(a)                                    (b)

**Fig. 3.22** Illustrating the SVM bias in comparison with Perceptron and MLP, according to possible sample organizations in terms of $R$ and $\rho$. (**a**) SVM with a stronger bias, (**b**) SVM with a weaker bias

Bias $\mathscr{F}_{\text{Perceptron}} = \mathscr{F}_{1-\text{MLP}}$, otherwise MLP contains more admissible functions as represented by a greater space. The best classifier for Perceptron and $k$-MLP is within the same region as seen in Fig. 3.21. Approximately, we may say the SVM bias is similar to the Perceptron, but not exactly. As seen in the previous section, the SVM generalization bound may change depending on terms $R$ and $\rho$ (Eq. (3.4)), thus its bias may be more (stronger) or less (weaker) restricted depending on the sample under analysis (see Fig. 3.22).

Observe that SVM, Perceptron and $k$-MLP share a common classifier $f$, which is the best they can reach for a perfectly linearly separable task (Fig. 3.14). In this circumstance, $f = f_{\text{Bayes}}$ because all those classification algorithms are capable of representing such an ideal solution.[1]

In a sample with class overlapping (see Fig. 3.16), examples under different labels may transpass the hyperplane to a region they do not belong to. This causes a growth in the SVM Shattering coefficient, jeopardizing the learning convergence according to the Chernoff bound. This is an enough evidence to confirm that SVM has a greater space of admissible functions than Perceptron, but being more restricted (stronger) than $k$-MLP for $k > 1$, as in Fig. 3.23. Notice classifier $f_{\text{Bayes}}$ was suppressed due to we cannot confirm there is no other best solution, but $f$ was added to represent the best as possible classifier given such biases.

Finally, in a third scenario, a set of Normally distributed examples from a class are surrounded by examples from another one (see Fig. 3.18). If such dataset is provided as input to Perceptron or SVM, both with insufficient biases to model this

---

[1] We remind the Bayes classifier is the best possible in the whole space of functions $\mathscr{F}_{\text{all}}$.

**Fig. 3.23** Analyzing the
SVM bias in a
class-overlapping scenario: $f$
corresponds to the best as
possible classifier

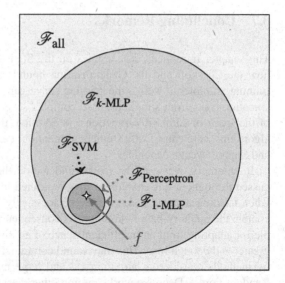

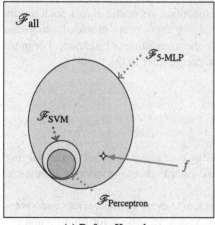

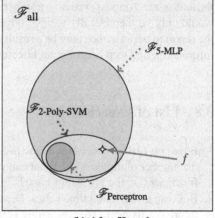

(a) Before Kernel                    (b) After Kernel

**Fig. 3.24** Analyzing algorithm biases given a dataset in which Normally distributed examples
are surrounded by examples from another class. In (**a**) the original 2-dimensional input space is
provided to all algorithms, while in (**b**) a 2-order polynomial kernel is only employed to support
SVM classification (see Fig. 3.19)

task, no feasible solution would be ever found. In comparison, the 5-MLP would
provide a more than enough solution. The best classifier $f$ for this task is only
within the 5-MLP bias, as depicted in Fig. 3.24a.

By applying the nonlinear kernel function discussed in the previous section, SVM
is then sufficient to model a feasible decision function. As consequence, 5-MLP and
SVM now share the best classifier $f$, thus SVM bias is stronger, ensuring faster
learning convergence (Fig. 3.24b). From this illustrative point of view, we conclude
the Shattering coefficient is directly associated with the number of admissible
functions.

## 3.7   Concluding Remarks

This chapter introduced case studies on the Statistical Learning Theory, showing how the Chernoff and the Generalization bounds impact the minimal number of training examples as well as the learning convergence. We also discussed about classification tasks from simple to more complex scenarios, motivating the adaptation of the space of admissible functions. In addition, the biases of some classification algorithms were empirically studied, to mention: Perceptron, Multilayer Perceptron and Support Vector Machines.

In summary, we wish the reader understood the most important subjects discussed throughout this chapter: at first, restricted biases may represent a drawback when tackling nontrial tasks, however excessively weak biases are more likely to contain the memory-based classifier. As consequence, the data scientist is responsible for adapting from an insufficient space of admissible functions to a sufficiently greater and enough bias. This may sound contradictory to the good results reported by methods considering an empirical and large number of hyperplanes, such as Random Forest, Deep Networks, or any other classification algorithm that shatters the data space using an excessive number of functions. As matter of fact, such results would only be theoretically valid if huge training datasets are provided, otherwise the reported performance may be a result of either overfitting or by chance. For more information on Deep Learning architectures, please refer to [3].

## 3.8   List of Exercises

1. Based on Listing 3.4, estimate the Shattering coefficient for input spaces varying the number of dimensions. What can you conclude about the space of admissible functions (i.e. the algorithm bias)?
2. Build up a sinusoidal time series. Next, organize every current series observation to predict its next. Then, separate part of your own dataset for training and the remaining for testing. Notice such data separation is performed along the time axis. Use the Distance-Weighted Nearest Neighbors to address this regression task. Start with a great value for $\sigma$ and reduce it. Analyze error results as such parameter is adapted.
3. Using the same setting of the previous exercise, now attempt to modify DWNN to recurrently predict $k$ further observations. This means you should take the current observation to predict a next, and this next will be used to predict the succeeding, and so on. Analyze how the trajectory of recurrent predictions diverge from expected values.

4. Using the Chernoff Bound, set $\varepsilon = 0.01$, i.e., the maximum acceptable divergence, and analyze the sample size necessary to ensure $P(\sup_{f \in \mathscr{F}} |R(f) - R_{\mathrm{emp}}(f)| \geq \epsilon) \leq 0.1$. Given the last classification task you approached, use this concept to estimate the number of training examples to guarantee learning. As suggestion, try any classification dataset available at the UCI Machine Learning Repository—archive.ics.uci.edu/ml.
5. Estimate the Shattering coefficient for a 3-hyperplane Multilayer Perceptron given the Iris dataset. Is the number of available examples sufficient to draw theoretical conclusions about the classification results obtained?

# References

1. C.M. Bishop, *Pattern Recognition and Machine Learning*. Information Science and Statistics (Springer-Verlag New York, Secaucus, 2006)
2. L. de Carvalho Pagliosa, R.F. de Mello, Applying a kernel function on time-dependent data to provide supervised-learning guarantees. Expert Syst. Appl. **71**, 216–229 (2017)
3. R.F. de Mello, M.D. Ferreira, M.A. Ponti, Providing theoretical learning guarantees to deep learning networks, CoRR, abs/1711.10292 (2017)

# Chapter 4
# Introduction to Support Vector Machines

## 4.1 About this Chapter

This chapter starts by reviewing the basic concepts on Linear Algebra, then we design a simple hyperplane-based classification algorithm. Next, it provides an intuitive and an algebraic formulation to obtain the optimization problem of the Support Vector Machines. At last, hard-margin and soft-margin SVMs are detailed, including the necessary mathematical tools to tackle them both.

## 4.2 Linear Algebra

Some relevant concepts on Linear Algebra are briefly introduced in the next sections: basis, linear transformations and their inverses, dot products, change of basis, orthonormal basis, and finally eigenvalues and eigenvectors.

### 4.2.1 Basis

When we numerically describe some vector, this description depends on a choice of *basis vectors*. In a 2-dimensional space, it is common to use the unit vectors $\mathbf{i} = (1, 0)$ and $\mathbf{j} = (0, 1)$, which form a basis because all other vectors in such space can be represented by scalar multiplications and vector additions between $\mathbf{i}$ and $\mathbf{j}$. This means vectors are produced by linear combinations in form: $a\mathbf{i} + b\mathbf{j}$. The pair $\mathbf{i}$ and $\mathbf{j}$ is called the "canonical" (or standard) basis because they are orthonormal; in fact, the 2-dimensional cartesian plane that we commonly use to draw graphs considers such canonical basis.

© Springer International Publishing AG, part of Springer Nature 2018
R. Fernandes de Mello, M. Antonelli Ponti, *Machine Learning*,
https://doi.org/10.1007/978-3-319-94989-5_4

However, there are other possible basis vectors. By using the formulation: $\mathbf{u} = a\mathbf{x} + b\mathbf{y}$, and varying $a$ and $b$ over all real numbers, we get the set of all possible combinations of *linearly independent* vectors $\mathbf{x}$ and $\mathbf{y}$ in a space $\mathbb{R}^n$. This set is called the *span* of $\mathbf{x}$ and $\mathbf{y}$.

If we take a pair of vectors that are colinear, i.e., they line up, then it means that one of them is redundant: it does not add any information so that we could remove it without reducing the span. We say those vectors are *linearly dependent* and, for this reason the resulting vectors $\mathbf{u}$ will lie on the same space (a.k.a. eigenspace). In addition, if any of those is the zero vector, then they do not form a basis.

Consider two vectors forming a basis $B = \{\mathbf{v}_1, \mathbf{v}_2\}$, as follows:

$$\mathbf{v}_1 = \begin{bmatrix} 2 \\ 1 \end{bmatrix} \quad \mathbf{v}_2 = \begin{bmatrix} 1 \\ 2 \end{bmatrix},$$

and a vector obtained using such basis:

$$\mathbf{u} = 3\mathbf{v}_1 + 2\mathbf{v}_2.$$

If we plot basis $B$ and vector $\mathbf{u}$ in a cartesian plane, we confirm $\mathbf{u}$ is a linear combination of the basis vectors:

$$[\mathbf{u}]_B = \begin{bmatrix} 3 \\ 2 \end{bmatrix}.$$

As illustrated in Fig. 4.1, this is nothing but scaling $\mathbf{v}_1$ and $\mathbf{v}_2$, and summing the resulting scaled vectors. Given such vectors are not colinear, they form a basis, and, thus, it is possible to span the entire 2-dimensional space by using $B$.

We are used to visualize the 2-dimensional plane formed by the canonical basis. However, basis $B$ is not orthogonal and, therefore, we need to draw a grid using parallel and equally spaced lines using $B$ as reference, as illustrated in Fig. 4.2. Then, it is possible to see that the resulting space is bent, but it still spans the entire 2-dimensional plan.

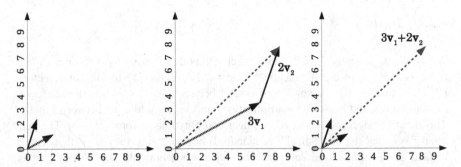

**Fig. 4.1** Representation of a vector using the basis $B$. First, we show the two basis vectors (left), then how we obtain a new vector by using the linear combination of the basis (centre) and its position when overlayed with the plane formed by the canonical vectors $\mathbf{i}, \mathbf{j}$

**Fig. 4.2** Depicting the basis change and how it modifies the shape of the space: the grid lines, however, maintain equally spaced and parallel

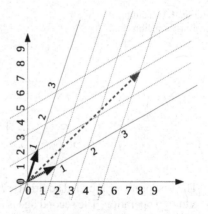

One could unbend the space in order to represent the same vector $[\mathbf{u}]_B$ using the canonical basis instead. By applying the basis vectors and performing the sum, we have:

$$\mathbf{u} = 3 \begin{bmatrix} 2 \\ 1 \end{bmatrix} + 2 \begin{bmatrix} 1 \\ 2 \end{bmatrix} = \begin{bmatrix} 8 \\ 7 \end{bmatrix}.$$

This means the basis can be changed, which is a very useful tool for data analysis. But before talking about the change of basis, we must introduce the concept of linear transformation, once it is fundamental to understand how to transform the entire span of vectors from one basis to another.

### 4.2.2  Linear Transformation

A linear transformation is the result of a matrix multiplied by some vector, mapping an input space into some output space. Let us consider the 2-dimensional space: if we visualize a grid of horizontal and vertical lines defining the orientation of vectors lying in such space, then a linear transformation will always keep grid lines parallel and evenly spaced, as well as a fixed origin point (see Fig. 4.3). This visualization makes easier to understand a linear transformation as a function that takes all possible input vectors (or points in the space represented by position vectors) to produce output vectors while respecting the mentioned constraints.

More formally, let $T : \mathbb{R}^n \to \mathbb{R}^n$ be a transformation, which is linear if and only if the following properties are held:

$$T(\mathbf{a} + \mathbf{b}) = T(\mathbf{a}) + T(\mathbf{b})$$
$$T(c\mathbf{a}) = cT(\mathbf{a}).$$

**Fig. 4.3** Example of a linear
transformation

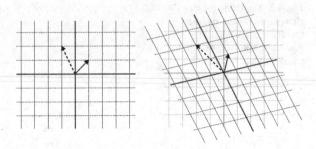

The first property is called additivity, which ensures the transformation preserves the
addition operation. The second one is the homogeneity, which implies scale variance
when a vector is multiplied by a real scalar. Those conditions provide necessary
foundation, relating linear transformations to linear combination of vectors.

Considering the following transformation:

$$T(x_1, x_2) = (x_1 + x_2, 3x_1) \quad \text{for any vector } \mathbf{x} = [x_1, x_2]^T \in \mathbb{R}^2.$$

In order to verify if $T$ is linear, we must first check the additivity property:

$$T(\mathbf{a} + \mathbf{b}) = T(\mathbf{a}) + T(\mathbf{b})$$

$$T\left(\begin{bmatrix} a_1 + b_1 \\ a_2 + b_2 \end{bmatrix}\right) = T\left(\begin{bmatrix} a_1 \\ a_2 \end{bmatrix}\right) + T\left(\begin{bmatrix} b_1 \\ b_2 \end{bmatrix}\right)$$

$$\begin{bmatrix} a_1 + b_1 + a_2 + b_2 \\ 3 \cdot (a_1 + b_1) \end{bmatrix} = \begin{bmatrix} a_1 + a_2 \\ 3a_1 \end{bmatrix} + \begin{bmatrix} b_1 + b_2 \\ 3b_1 \end{bmatrix}$$

$$\begin{bmatrix} a_1 + a_2 + b_1 + b_2 \\ 3a_1 + 3b_1 \end{bmatrix} = \begin{bmatrix} a_1 + a_2 + b_1 + b_2 \\ 3a_1 + 3b_1 \end{bmatrix},$$

which is true, so let us check the second property:

$$T(c\mathbf{a}) = cT(\mathbf{a})$$

$$T\left(c\begin{bmatrix} a_1 \\ a_2 \end{bmatrix}\right) = c\begin{bmatrix} a_1 + a_2 \\ 3a_1 \end{bmatrix}$$

$$\begin{bmatrix} ca_1 + ca_2 \\ 3ca_1 \end{bmatrix} = c\begin{bmatrix} a_1 + a_2 \\ 3a_1 \end{bmatrix},$$

which shows the scale variance is also held, therefore this transformation is linear.

**Fig. 4.4** Example of a linear transformation using basis vectors

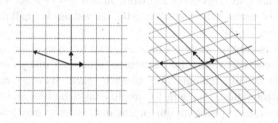

Note that multiplications between vectors, squares and other more complex functions are not linear. For example, by inspecting the transformation below:

$$T(x_1, x_2) = (x_1^2 + \sqrt{2}x_1x_2, x_2^2) \quad \text{for any vector } \mathbf{x} \in \mathbb{R}^2,$$

the additivity and homogeneity constraints are not held.

**Matrix Representation of Linear Transformations** one of the most useful ways to represent a linear transformation is by using a basis matrix, composed of a set of linearly independent vectors that span the whole space.

Considering the 2-dimensional space basis $\mathbf{i} = (1, 0)$ and $\mathbf{j} = (0, 1)$. As depicted in Fig. 4.4, it is possible to write any other vector in this space by linearly combining $\mathbf{i}$ and $\mathbf{j}$. For example, vector $\mathbf{x} = (-3, 1)$ can be written as: $-3\mathbf{i} + 1\mathbf{j}$. If a linear transformation is applied on this space, then it is possible to know $T(\mathbf{x})$ by assessing the transformed versions of the basis vectors. Let the transformed basis vectors be $\hat{\mathbf{i}} = (5/6, 1/3)$ and $\hat{\mathbf{j}} = (-1, 1)$, then:

$$\mathbf{x} = x_1 \cdot \mathbf{i} + x_2 \cdot \mathbf{j}$$

$$T(\mathbf{x}) = x_1 \cdot T(\mathbf{i}) + x_2 \cdot T(\mathbf{j})$$

$$T(\mathbf{x}) = x_1 \begin{bmatrix} 5/6 \\ 1/3 \end{bmatrix} + x_2 \begin{bmatrix} -1 \\ 1 \end{bmatrix}$$

$$T(\mathbf{x}) = \begin{bmatrix} 5/6 \cdot x_1 + (-1) \cdot x_2 \\ 1/3 \cdot x_1 + 1 \cdot x_2 \end{bmatrix}$$

The matrix formed by the transformed basis vectors describes the linear transformation:

$$A = \begin{bmatrix} 5/6 & -1 \\ 1/3 & 1 \end{bmatrix},$$

so transforming $\mathbf{x} = (-3, 1)$ (see Fig. 4.4) results in:

$$T(\mathbf{x}) = A\mathbf{x} = \begin{bmatrix} 5/6 \cdot -3 + (-1) \cdot 1 \\ 1/3 \cdot -3 + 1 \cdot 1 \end{bmatrix} = \begin{bmatrix} -2.5 - 1 \\ -1 + 1 \end{bmatrix} = \begin{bmatrix} -3.5 \\ 0 \end{bmatrix},$$

matching the visualization, in which we overlay the transformed space with the original grid. This linear transformation $T : \mathbb{R}^n \to \mathbb{R}^m$ maps the elements (vectors, in this case), of the first set into the second set. We call the first set *domain*, and the second *co-domain*.

It is easy to see that Linear Algebra has important relationships with the study of kernels, in particular because it formalizes many concepts related to mapping some set of elements into another space. By designing a linear or nonlinear transformation,[1] we aim to simplify tasks by reorganizing data. In the case of a classification problem, this means data is reorganized so that a single hyperplane is sufficient to separate classes. Linear transformations are also widely used in Computer Graphics, as well as a framework for other applications such as to solve differential equations, image restoration, and compute Markov chains.

### 4.2.3  Inverses of Linear Transformations

In many scenarios, it is useful to map the vectors of some transformed space back into the original space. This is often the case when one needs an alternative representation, i.e. the transformed space, to facilitate some operation. Afterwards, in order to bring the resultant vectors back, an inverse transformation is needed. Mathematically, having a linear transformation described by a matrix $A$ so that $T(\mathbf{x}) = \mathbf{y} = A\mathbf{x}$, then, in order to obtain $\mathbf{x}$, we need the inverse matrix $A^{-1}$ in order to compute $\mathbf{x} = A^{-1}\mathbf{y}$.

Let $T : \mathbb{R}^n \to \mathbb{R}^m$ be a linear transformation. The conditions to ensure the inverse of the transformation are:

1. the transformation $T(.)$ must be bijective, meaning it is at the same time injective (maps distinct elements of the domain to also distinct elements in the co-domain), and surjective (every element $\mathbf{y}$ in the co-domain has a corresponding element $\mathbf{x}$ in the domain, such that $T(\mathbf{x}) = \mathbf{y}$);
2. $m \times n$ matrix $A$ has to be square, i.e., $m = n$, otherwise for $m > n$ the number of elements in the co-domain is greater than the domain. In such cases, there are alternative techniques such as pseudo-inverses to approximate results, but for exact inverses both the domain and the co-domain must have the same cardinality;
3. every column vector in the matrix must be linearly independent between each other. In this case, the reduced row echelon form of $A$ is the identity matrix, forming a basis for $\mathbb{R}^n$.

Another way to confirm that a square matrix $A$ is invertible is by computing its determinant: if $\det(A) \neq 0$, then $A$ is invertible. One way to interpret the value of the determinant is to describe how areas in the original space are increased or

---

[1]Nonlinear transformations are typical when designing kernels.

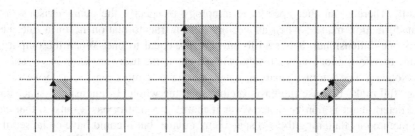

**Fig. 4.5** Area of the rectangle formed by the basis vectors in the original space is 1 (left), after scaling is 8 (center), and after shearing is 1 (right)

decreased. For example, considering again the 2-dimensional space basis vectors $\mathbf{i} = (1, 0)$ and $\mathbf{j} = (0, 1)$ and the following transformation matrices:

$$A = \begin{bmatrix} 2 & 0 \\ 0 & 4 \end{bmatrix} \quad B = \begin{bmatrix} 1 & 1 \\ 1 & 0 \end{bmatrix},$$

then matrix $A$ will scale $\mathbf{i}$ by a factor of 2 and $\mathbf{j}$ by a factor of 4, while matrix $B$ (that produces a shear transformation) will keep $\mathbf{i}$ unaltered, while moving $\mathbf{j}$ to the position $(1, 1)$. If we pay attention on the rectangle formed by the vectors in the original space, and the transformed spaces (see Fig. 4.5), $A$ scales the area of the original rectangle by a factor of 8. The matrix $B$ turns the rectangle into a parallelogram, but it keeps the area unchanged. By computing the determinants, we can see that $\det(A) = 8$ and $\det(B) = 1$.

Notice that in the case of linear transformations, by looking at how the unit rectangle area changes, we can understand the modifications spanned throughout the space. If a determinant of a matrix $A$ is greater than zero, but less than 1, i.e., $0 < det(A) < 1$, then the transformation decreases areas. Negative determinants are possible, indicating the space is flipped over by the transformation, but its absolute value, $|det(A)|$ is still an area scaling indicator.

However, when the determinant is zero, the transformation is mapping the current space into a subspace with lower dimensionality. For example, if the domain is in $\mathbb{R}^2$, a transformation with zero determinant might be mapping the space either into a line or a single point, making impossible to compute an inverse, since distinct vectors in the original space are mapped into the same vector in the target space, i.e., the transformation function is surjective, but not injective. If the original space is $\mathbb{R}^3$, a zero determinant indicates that the transformation is mapping the 3-dimensional space into a plane, a line, or a single point. Once the whole 3-dimensional space is collapsed, for example, into a plane, it would be impossible to unfold it into the whole 3-dimensional space again. In such scenarios, the column vectors of the matrix are linearly dependent. Consequently $det(A) = 0$ and matrix $A$ is referred to as singular or degenerate.

**Rank** There is another specific terminology to specify the characteristics of a transformation matrix: when all vectors after a transformation lie on a line, i.e., it is one-dimensional, it is said to have a *rank* equal to one. When the output of some transformation maps all vectors into a plane, the transformation has a rank of 2. Notice a $3 \times 3$ transformation matrix can have at most rank 3, if so we say it has "full rank" because the basis vectors span the whole 3-dimensional space and the determinant is non-zero. However if that $3 \times 3$ matrix has a rank 2, that can be pictured as flattening the 3-d space into a plane, but it could have collapsed the space even more if it had a rank 1. The rank means the dimensionality of the output of a transformation, more precisely the number of dimensions in the *column space* of the transformation matrix. The set of all outputs of $A\mathbf{x}$ is called the column space of such matrix, since the columns define where the basis vectors will lie after the transformation.

### 4.2.4    Dot Products

Dot products are useful to understand projections and to compare the directions of two vectors. Taking the dot product between $\mathbf{x}$ and $\mathbf{y}$ of the same dimensionality, which is denoted by $\mathbf{x} \cdot \mathbf{y} =< \mathbf{x}, \mathbf{y} >$, is to pair each of their coordinates, multiply the pairs, and add those products. Considering two vectors in $\mathbb{R}^2$, this is defined by: $< \mathbf{x}, \mathbf{y} >= (x_1 \cdot y_1) + (x_2 \cdot y_2)$. Notice the dot product outputs a scalar, a real number, regardless the dimensionality of vectors. Also, the order of the dot product is irrelevant: $< \mathbf{x}, \mathbf{y} >=< \mathbf{y}, \mathbf{x} >$.

A **vector projection** of $\mathbf{y}$ onto $\mathbf{x}$ is computed as:

$$\mathrm{Proj}_{\mathbf{x}}(\mathbf{y}) = \frac{\langle \mathbf{y}, \mathbf{x} \rangle}{||\mathbf{x}||},$$

which orthogonally maps vector $\mathbf{y}$ into the eigenspace of $\mathbf{x}$.

The geometrical interpretation of this operation in $\mathbb{R}^2$ (depicted in Fig. 4.6) can be seen as first projecting vector $\mathbf{y}$ onto the line that passes through the origin and the tip of vector $\mathbf{x}$, and then multiplying the length of this projection, $\mathrm{proj}_{\mathbf{x}} \mathbf{y}$, by the length of $\mathbf{x}$, i.e., $|| \mathrm{proj}_{\mathbf{x}} \mathbf{y}|| \cdot ||\mathbf{x}||$. This leads us to three scenarios: when vectors are pointing to the same direction, their dot product is positive; when vectors are orthogonal, their dot product is zero (the projected vector has length zero); and when vectors are pointing to different directions, their dot product is negative.

As mentioned before, the dot product produces a single value from a pair of vectors. This can be seen as a linear transformation that maps a given multidimensional space into the real line, or a function that takes as input a pair of vectors and outputs an one-dimensional value, i.e. a scalar number. If we have a line of evenly spaced dots in a plane, and apply a linear transformation $T : \mathbb{R}^2 \rightarrow \mathbb{R}^1$, then it remains evenly spaced after transformation. This gives us an interesting interpretation of the dot product: by looking where the unit basis vectors $\mathbf{i}$ and $\mathbf{j}$ will lie on the space $\mathbb{R}^1$, we can compose a transformation matrix.

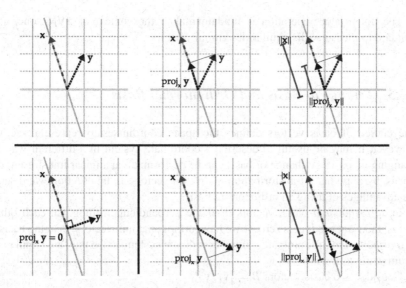

**Fig. 4.6** Geometrical interpretation of the dot product: projection of **y** onto the line passing through the origin and the tip of vector **x**, and the resulting lengths of the projection and vector **x**, which when multiplied provide the dot product between vectors

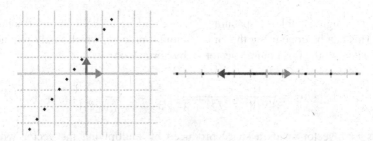

**Fig. 4.7** A linear transformation from a two to one-dimensional space will keep a set of evenly-spaced vectors on a line, also evenly spaced in the output space

In Fig. 4.7, we depict an example in which **i** lies on 1.5, while **j** lies on $-3$, leading to a $1 \times 2$ transformation matrix given by $A = [1.5 \ -3]$. Let a vector $\mathbf{x} = [3 \ 2]^T$, then its transformation can be computed by:

$$[1.5 \ -3] \begin{bmatrix} 3 \\ 2 \end{bmatrix} = (1.5 \cdot 3) + (-3 \cdot 2) = 4.5 - 6 = -1.5,$$

which is computationally similar to perform a dot product (multiply and add). The interesting relationship with the dot product is that such a transformation outputs an one-dimensional space associated with the input vector, so that the transformation matrix corresponds to vector $[1.5 \ -3]$.

The dot product operation is fundamental in the context of SVM, once it a standard to measure (dis)similarities among vectors.

### 4.2.5   Change of Basis and Orthonormal Basis

The choice of basis vectors defines the space coordinates to represent vectors. However, it may be useful to describe a coordinate system in a different way by changing its basis. A change of basis can be performed via a linear transformation, that is, by applying a transformation matrix on vectors. In this section, we discuss the advantages of using an orthonormal basis.

In an *orthogonal* basis, vectors in $B$ are perpendicular/orthogonal each other, that is, their dot product is zero. If we have orthogonal unit vectors, than the basis is also *orthonormal*, forming an optimal coordinate system, since they are simple and intuitive to work with.

Consider two basis vectors $B = \{\mathbf{v}_1, \mathbf{v}_2\}$:

$$\mathbf{v}_1 = \begin{bmatrix} 2 \\ 3 \end{bmatrix} \quad \mathbf{v}_2 = \begin{bmatrix} 7 \\ 4 \end{bmatrix},$$

which, although capable of spanning $\mathbb{R}^2$, it could be adapted to an orthonormal basis. This can be done using the Gram-Schmidt process, which firstly produces a unity vector for the first column vector $\mathbf{v}_1$ by normalizing it:

$$\mathbf{u}_1 = \frac{\mathbf{v}_1}{||\mathbf{v}_1||} = \frac{1}{\sqrt{2^2 + 3^2}} \begin{bmatrix} 2 \\ 3 \end{bmatrix} = \frac{1}{\sqrt{13}} \begin{bmatrix} 2 \\ 3 \end{bmatrix}.$$

This first vector spans a space produced by multiplying the vector with all possible scalars so that $S_1 = \text{span}(\mathbf{u}_1)$, forming an infinite line that goes along the vector (a.k.a. the linear eigenspace). Now we need to transform the second vector so it becomes orthonormal with respect to $\mathbf{u}_1$. In order to do so, we project vector $\mathbf{v}_2$ onto the subspace $S_1$ and take the vector corresponding to the subtraction between $\mathbf{v}_1$ and its projection onto $S_1$, referred to as $\text{Proj}_{S_1}(\mathbf{v}_1)$:

$$\mathbf{y}_2 = \mathbf{v}_2 - \text{Proj}_{S_1}(\mathbf{v}_2).$$

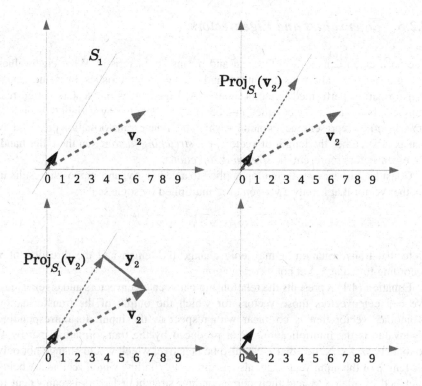

**Fig. 4.8** Finding an orthonormal basis by using the Gram-Schmidt process: first, we have the subspace $S_1$, onto $v_2$ is projected; then by using the vector subtraction, we produce $y_2$ which is orthogonal to $u_1$, and, by normalizing it, we obtain $u_2$ to compose the orthonormal basis

In our example:

$$y_2 = \begin{bmatrix} 7 \\ 4 \end{bmatrix} - \begin{bmatrix} 4 \\ 6 \end{bmatrix} = \begin{bmatrix} 3 \\ -2 \end{bmatrix}.$$

Thus, as $y_2$ is already orthogonal to $u_1$, one just needs to normalize it:

$$u_2 = \frac{y_2}{\|y_2\|} = \frac{1}{\sqrt{13}} \begin{bmatrix} 3 \\ -2 \end{bmatrix}.$$

Figure 4.8 depicts this process. We can now say that $\mathrm{span}(v_1, v_2) = \mathrm{span}(u_1, u_2)$, but the new basis allows the typical representation of vectors, as well as simpler vectorial operations.

## 4.2.6  *Eigenvalues and Eigenvectors*

The term eigen comes from German and means characteristic. What eigenvalues and eigenvectors allow us to understand is the characteristics of some linear transformation performed by some matrix $A$. Eigenvalues are scalars (often real numbers, but can also be complex numbers), often denoted by variables $\lambda$, and, in general terms, represent the amount by which the linear transformation provided by matrix $A$ stretches the length of vectors, i.e. *stretching factors*. On the other hand, the eigenvectors represent the *stretching directions*.

Consider a linear transformation applied to a vector $A\mathbf{v}$, and suppose it results in another vector that is only a version of $\gtrsim$ multiplied by some scalar:

$$A\mathbf{v} = \lambda\mathbf{v}. \tag{4.1}$$

Note that transformation $A$ may only change the length and the direction of $\mathbf{v}$ according to scalar $\lambda$, but not its orientation.

Equation (4.1) represents the relationship between eigenvectors and eigenvalues. We call eigenvectors those vectors for which the output of the transformation is another vector that is co-linear with respect to the input. Its corresponding eigenvalue is the multiplication factor produced by the transformation matrix $A$, so that $A\mathbf{v} = \lambda\mathbf{v}$. Notice that $\lambda$ can also be negative, which would change not only the length of the input vector, but also its direction. Finding which vectors are being stretched by matrix $A$ and their corresponding stretching factors is equivalent to solve the eigenvectors and eigenvalue equation.

Let us approach these concepts by using an example. Consider the following transformation matrix:

$$A = \begin{bmatrix} 1 & 2 \\ 4 & 3 \end{bmatrix}.$$

In Fig. 4.9, we have examples of the following vectors transformed by matrix $A$:

$$\mathbf{v}_1 = \begin{bmatrix} -1 \\ 0 \end{bmatrix}; \mathbf{v}_2 = \begin{bmatrix} 0 \\ 1 \end{bmatrix}; \mathbf{v}_3 = \begin{bmatrix} 1/2 \\ 1 \end{bmatrix}; \mathbf{v}_4 = \begin{bmatrix} -1 \\ 1 \end{bmatrix}.$$

Note that in the case of vectors $\mathbf{v}_1$ and $\mathbf{v}_2$, the transformation changes both scale and orientation, but in the case of $\mathbf{v}_3$ and $\mathbf{v}_4$, the transformed vectors are represented by $T(\mathbf{v}_3) = \lambda\mathbf{v}_3$ and $T(\mathbf{v}_4) = \lambda\mathbf{v}_4$, respectively. Those two last vectors are important because they define *reference axes* for the linear transformation. Remembering the concepts of basis and linear transformations (see Figs. 4.2, 4.3, and 4.4), the eigenvectors would define the axes in which the transformed space grid lines are drawn.

Observe each of those vectors span infinite lines forming axes. The linear transformation $A$ produces changes in vectors using as reference axes $\lambda\mathbf{v}_3$ and

**Fig. 4.9** Vectors produced
after the linear transformation

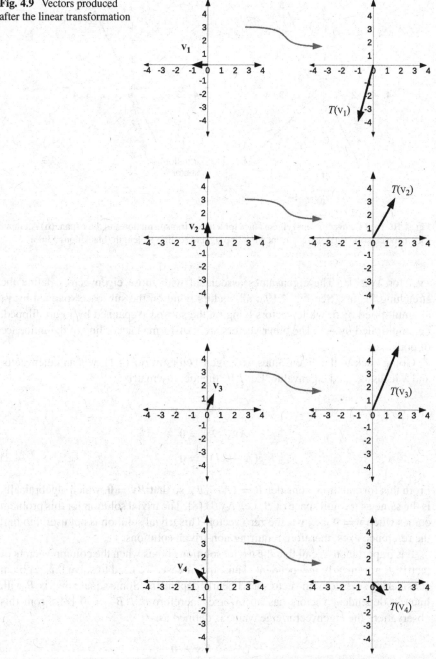

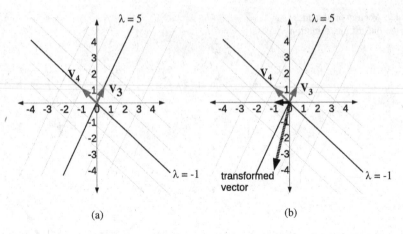

**Fig. 4.10** (**a**) Eigenvectors and eigenvalues for a linear transformation and their span; (**b**) An input vector (black solid line) is transformed by flipping and scaling its length (black dotted line)

$\lambda \mathbf{v}_4$, for $\lambda \in \mathbb{R}$. The eigenvalues associated with those eigenvectors define the stretching factors. See Fig. 4.10a: all vectors lying on the subspace spanned by $\mathbf{v}_3$ are multiplied by 5, while vectors lying on the subspace spanned by $\mathbf{v}_4$ are flipped, i.e., multiplied by $-1$. The other vectors are transformed according to the influence of both axes.

Going back to the eigenvalues and eigenvectors in Eq. (4.1), $\mathbf{v}$ is an eigenvector and $\lambda$ its associated eigenvalue. Let $I$ be the identity matrix[2]:

$$A\mathbf{v} = \lambda\mathbf{v}$$

$$A\mathbf{v} - \lambda\mathbf{v} = \mathbf{0}$$

$$(A - \lambda I)\mathbf{v} = \mathbf{0} \tag{4.2}$$

From this formulation, consider $B = (A - \lambda I)$, so that $B\mathbf{v} = \mathbf{0}$, which algebraically is the same as the null space of $B$, i.e., $N(B)$ [8]. The trivial solution for this problem comes when $\mathbf{v} = \mathbf{0}$, i.e., $\mathbf{v}$ is the zero vector. This trivial solution is not useful to find the reference axes, therefore requiring non-trivial solutions.

It is important to recall that the trivial solution comes when the column vectors of matrix $B$ are linearly independent. Thus, in our case, we need to solve this problem considering such column vectors are linearly dependent. Some square matrix $B$ with linearly dependent vectors has no inverse, therefore $det(B) = 0$ [8]. From this observation, the eigenvector/eigenvalue is defined as:

$$(A - \lambda I)\mathbf{v} = \mathbf{0}$$

---

[2]Remember that $\lambda\mathbf{v} = \lambda I\mathbf{v}$, allowing us the last step.

$$Bv = 0$$

$$det(B) = 0$$

$$det(A - \lambda I) = 0. \tag{4.3}$$

So consider a problem in which matrix $A$ is:

$$A = \begin{bmatrix} 1 & 2 \\ 4 & 3 \end{bmatrix},$$

so that:

$$det\left(\begin{bmatrix} 1 & 2 \\ 4 & 3 \end{bmatrix} - \lambda \begin{bmatrix} 1 & 0 \\ 0 & 1 \end{bmatrix}\right) = 0$$

$$det\left(\begin{bmatrix} 1-\lambda & 2 \\ 4 & 3-\lambda \end{bmatrix}\right) = 0$$

$$(1-\lambda) \times (3-\lambda) - (4 \times 2) = 0$$

$$\lambda^2 - 4\lambda - 5 = 0, \tag{4.4}$$

thus there are two values $\lambda = -1$ and $\lambda = 5$, which are the eigenvalues for matrix $B$. Going back to Eq. (4.2), the eigenvectors are found by solving, first for $\lambda = -1$:

$$\left(\begin{bmatrix} 1 & 2 \\ 4 & 3 \end{bmatrix} - (-1) \begin{bmatrix} 1 & 0 \\ 0 & 1 \end{bmatrix}\right) v = 0$$

$$\left(\begin{bmatrix} 2 & 2 \\ 4 & 4 \end{bmatrix}\right) v = 0.$$

After applying the row-reduced echelon form on such matrix:

From: $\begin{bmatrix} 2 & 2 \\ 4 & 4 \end{bmatrix}$, we divide the first row by 2 $\rightarrow \begin{bmatrix} 1 & 1 \\ 4 & 4 \end{bmatrix}$,

and then the second row receives $-4$ the first row plus the second $\rightarrow \begin{bmatrix} 1 & 1 \\ 0 & 0 \end{bmatrix}$.

From this last result:

$$\begin{bmatrix} 1 & 1 \\ 0 & 0 \end{bmatrix} v = 0$$

$$\begin{bmatrix} 1 & 1 \\ 0 & 0 \end{bmatrix} \begin{bmatrix} v_1 \\ v_2 \end{bmatrix} = \begin{bmatrix} 0 \\ 0 \end{bmatrix}$$

$$v_1 + v_2 = 0$$

$$v_1 = -v_2$$

Therefore, assuming $v_2 = t, t \in \mathbb{R}$:

$$E_{\lambda=-1} = N(A - \lambda I) = \left\{ \begin{bmatrix} v_1 \\ v_2 \end{bmatrix} = t \begin{bmatrix} -1 \\ 1 \end{bmatrix}, t \in \mathbb{R} \right\},$$

in which $E_{\lambda=-1}$ is the eigenspace for the first eigenvector found:

$$\begin{bmatrix} -1 \\ 1 \end{bmatrix}.$$

We still need to find the second eigenvector, by using $\lambda = 5$:

$$\left( \begin{bmatrix} 1 & 2 \\ 4 & 3 \end{bmatrix} - (5) \begin{bmatrix} 1 & 0 \\ 0 & 1 \end{bmatrix} \right) \mathbf{v} = 0$$

$$\left( \begin{bmatrix} -4 & 2 \\ 4 & -2 \end{bmatrix} \right) \mathbf{v} = 0.$$

Applying the row-reduced echelon form on such matrix:

From: $\begin{bmatrix} -4 & 2 \\ 4 & -2 \end{bmatrix}$, second row receives itself plus the first row $\rightarrow \begin{bmatrix} -4 & 2 \\ 0 & 0 \end{bmatrix}$,

and then the first row divided by $-4 \rightarrow \begin{bmatrix} 1 & -\frac{1}{2} \\ 0 & 0 \end{bmatrix}$.

From this last result:

$$\begin{bmatrix} 1 & -\frac{1}{2} \\ 0 & 0 \end{bmatrix} \mathbf{v} = \mathbf{0}$$

$$\begin{bmatrix} 1 & -\frac{1}{2} \\ 0 & 0 \end{bmatrix} \begin{bmatrix} v_1 \\ v_2 \end{bmatrix} = \begin{bmatrix} 0 \\ 0 \end{bmatrix}$$

$$v_1 - \frac{1}{2} v_2 = 0$$

$$v_1 = \frac{1}{2} v_2$$

Therefore, assuming $v_2 = t, t \in \mathbb{R}$:

$$E_{\lambda=5} = N(A - \lambda I) = \left\{ \begin{bmatrix} v_1 \\ v_2 \end{bmatrix} = t \begin{bmatrix} \frac{1}{2} \\ 1 \end{bmatrix}, t \in \mathbb{R} \right\},$$

in which $E_{\lambda=5}$ is the eigenspace for the second eigenvector:

$$\begin{bmatrix} \frac{1}{2} \\ 1 \end{bmatrix}.$$

We recall this is only a brief introduction on eigenvectors and eigenvalues, and suggest the reader to proceed with the following book [8].

## 4.3  Using Basic Algebra to Build a Classification Algorithm

We start by introducing a kernel-based classification algorithm $\mathbb{A}$ that receives a training set $\mathscr{D} = \{(x_1, y_1), (x_2, y_2), \ldots, (x_n, y_n)\}$, in which $x_i \in \mathbb{R}^2$ are examples in the input space, and $y_i \in \{+1, -1\}$ are their corresponding classes [7]. Let every $x_i$ be composed of the variables temperature and humidity for some world region, collected over years.[3] Let the class define whether a person plays ($+1$) or does not play ($-1$) soccer under such weather conditions. Figure 4.11a illustrates a given instance for this classification task.

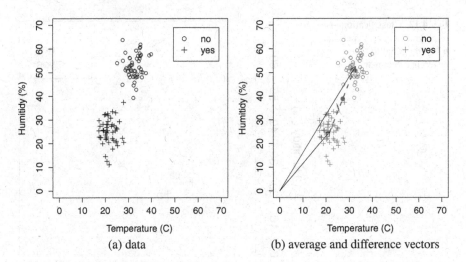

(a) data                                    (b) average and difference vectors

**Fig. 4.11** Classification task of playing soccer: (**a**) dataset (**b**) average vectors for each class, and difference vector **w** indicated with a dashed line

---

[3]No temporal relation is here assumed.

Let we compute the average values as defined in Eqs. (4.5) and (4.6), in which $m_+$ and $m_-$ correspond to the number of examples labeled as $+1$ and $-1$, respectively; $\mathbf{x}_i$ is the vectorial form[4] of example $x_i$; and, finally, $\mathbf{c}_+$ and $\mathbf{c}_-$ are the average vectors serving as class prototypes (as illustrated in Fig. 4.11b).

$$\mathbf{c}_+ = \frac{1}{m_+} \sum_{i|y_i=+1} \mathbf{x}_i \tag{4.5}$$

$$\mathbf{c}_- = \frac{1}{m_-} \sum_{i|y_i=-1} \mathbf{x}_i \tag{4.6}$$

Let vector $\mathbf{w} = \mathbf{c}_+ - \mathbf{c}_-$ be the difference between averages, whose central point is $\mathbf{c} = \frac{\mathbf{c}_+ + \mathbf{c}_-}{2}$. Then, let a linear hyperplane $\mathbf{h}$, orthogonal to $\mathbf{w}$, crossing it at $\mathbf{c}$ (Fig. 4.12). Observe $\mathbf{h}$ divides the input space into two regions (a.k.a. two half spaces), one associated with positive examples and another with the negatives.

As next step, consider $\mathbb{A}$ is used to predict the output label for an unseen example, represented by vector $\mathbf{x}$ (see Fig. 4.13).

**Fig. 4.12** Classification task of playing soccer: hyperplane **h** is orthogonal to the difference vector **w**

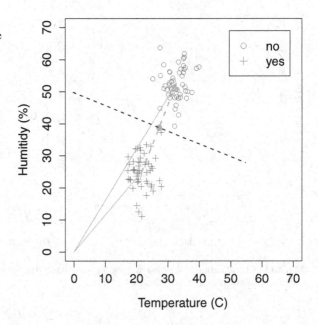

---

[4]Our notation considers $x_i$ to be an identification of some example, without precisely defining its representation (e.g. it might be in a Topological, Hausdorff, Normed, or any other space). However, $\mathbf{x}_i$ is its vectorial form in some Hilbert space.

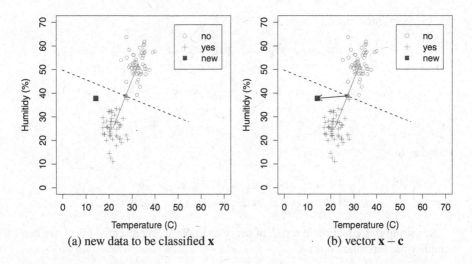

Fig. 4.13 Classification task of playing soccer: adding an unseen example into the input space.
(a) New example to be classified, (b) distance vector between the new example and the midpoint
of the hyperplane

Our algorithm estimates the output class by computing on which side of **h** the
vector **x** lies on. For that, the dot product of vector **x** − **c** with **w** provides the result
(Fig. 4.13b). In particular, by computing $d = < \mathbf{x} - \mathbf{c}, \mathbf{w} >$, it is possible to interpret
the results in $\mathbb{R}^2$ as follows:

1. vectors **x** − **c** and **w** "pull" to the same side, then $d > 0$;
2. vectors **x** − **c** and **w** "pull" to opposite sides, then $d < 0$;
3. otherwise, they are orthogonal to each other when $d = 0$.

Term "pull" can be used in this two-dimensional space to study the angle between
vectors **w** and **x** − **c**. We say vectors "pull" to the same side when they form an
acute angle; "pulling" to opposite sides mean their angle is obtuse; and finally,
when orthogonal, the angle is $\frac{\pi}{2}$. The classification result is obtained from the sign
function, i.e. $y = \text{sign}(< \mathbf{x} - \mathbf{c}, \mathbf{w} >)$, because the magnitude of the dot product
does not affect the result. However, note that when $y = 0$, we have a tie, meaning
the example lies exactly on the hyperplane, and therefore it might belong to any
class.

Next, we detail the dot product to obtain a closed form that facilitates the
implementation of the algorithm:

$$y = \text{sign}(< \mathbf{x} - \mathbf{c}, \mathbf{w} >)$$

$$= \text{sign}(< \mathbf{x} - \mathbf{c}, \mathbf{c}_+ - \mathbf{c}_- >)$$

$$= \text{sign}(< \mathbf{x} - \frac{1}{2}(\mathbf{c}_+ + \mathbf{c}_-), \mathbf{c}_+ - \mathbf{c}_- >)$$

$$= \text{sign}(< \mathbf{x}, \mathbf{c}_+ > - < \mathbf{x}, \mathbf{c}_- >$$

$$- \frac{1}{2}(< \mathbf{c}_+, \mathbf{c}_+ > - < \mathbf{c}_+, \mathbf{c}_- > + < \mathbf{c}_-, \mathbf{c}_+ > - < \mathbf{c}_-, \mathbf{c}_- >))$$

$$= \text{sign}(< \mathbf{x}, \mathbf{c}_+ > - < \mathbf{x}, \mathbf{c}_- > + \frac{1}{2}(< \mathbf{c}_-, \mathbf{c}_- > - < \mathbf{c}_+, \mathbf{c}_+ >)),$$

resulting in:

$$\begin{cases} y = \text{sign}(< \mathbf{x}, \mathbf{c}_+ > - < \mathbf{x}, \mathbf{c}_- > + b) \\ b = \frac{1}{2}(< \mathbf{c}_-, \mathbf{c}_- > - < \mathbf{c}_+, \mathbf{c}_+ >). \end{cases}$$

Rewriting this equation in terms of every example $x_i$, represented by $\mathbf{x}_i$, we can simplify the formulation for $y$:

$$y = \text{sign}(< \mathbf{x}, \mathbf{c}_+ > - < \mathbf{x}, \mathbf{c}_- > + b)$$

$$= \text{sign}\left( \frac{1}{m_+} \sum_{i|y_i=+1} < \mathbf{x}, \mathbf{x}_i > - \frac{1}{m_-} \sum_{i|y_i=-1} < \mathbf{x}, \mathbf{x}_i > + b \right),$$

in which $b$ is also rewritten as:

$$b = \frac{1}{2}(< \mathbf{c}_-, \mathbf{c}_- > - < \mathbf{c}_+, \mathbf{c}_+ >)$$

$$= \frac{1}{2}\left( \left\langle \frac{1}{m_-} \sum_{i|y_i=-1} < \mathbf{x}, \mathbf{x}_i >, \frac{1}{m_-} \sum_{i|y_i=-1} < \mathbf{x}, \mathbf{x}_i > \right\rangle \right.$$

$$\left. - \left\langle \frac{1}{m_+} \sum_{i|y_i=+1} < \mathbf{x}, \mathbf{x}_i >, \frac{1}{m_+} \sum_{i|y_i=+1} < \mathbf{x}, \mathbf{x}_i > \right\rangle \right)$$

$$= \frac{1}{2}\left( \frac{1}{m_-^2} \sum_{(i,j)|y_i=y_j=-1} < \mathbf{x}_i, \mathbf{x}_j > - \frac{1}{m_+^2} \sum_{(i,j)|y_i=y_j=+1} < \mathbf{x}_i, \mathbf{x}_j > \right).$$

Listing 4.1 details the classification algorithm described above. Function *first.classification.algorithm()* runs the training and test stages. A complete example of usage of this algorithm is found in *test.first()*, which employs the Normal probability distribution to produce synthetic positive and negative examples (see Fig. 4.14). Positive examples are significantly far from the negative ones, allowing a perfect linear separation between classes, that is why this algorithm works properly in this scenario.

**Fig. 4.14** Positive and negative examples produced by function *test.first()* (Code 4.1)

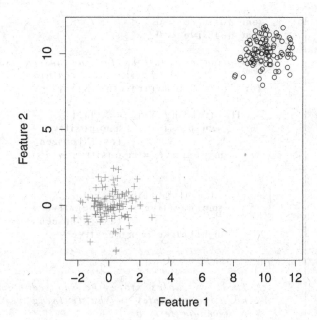

**Listing 4.1** Our first classification algorithm

```
first.classification.algorithm <- function(training.set,
                                                test.set) {
  # Defining the column id representing the expected class
  classAttributeId = ncol(training.set)

  # Setting X and Y for training
  training.X = training.set[,1:(classAttributeId-1)]
  training.Y = training.set[,classAttributeId]

  # Setting X and Y for testing
  test.X = test.set[,1:(classAttributeId-1)]
  test.Y = test.set[,classAttributeId]

  # The final results are saved in this variable
  results = NULL

  cat("#_Outcome\tExpected_class\n")
  # For every unseen example in the test set
  for (unseen in 1:nrow(test.X)) {

    # These variables count the number of positive
    # and negative examples in the training set
    m_positive = 0
    m_negative = 0

    # To sum up the dot product of the unseen example
    # against every other example contained in the
    # positive and the negative classes
```

```
sum_positive = 0
sum_negative = 0

# Apply the equations for the unseen example
# test.X[unseen,] given the training set
for (i in 1:nrow(training.X)) {

    if (training.Y[i] == +1) {
       sum_positive = sum_positive +
                      test.X[unseen,] %*% training.X[i,]
       m_positive = m_positive + 1
    }

    if (training.Y[i] == -1) {
       sum_negative = sum_negative +
                      test.X[unseen,] %*% training.X[i,]
       m_negative = m_negative + 1
    }
}

# These variables store the squared number of positive and
# negative examples in the training set. They are required
# to compute term b
m_squared_positive = 0
m_squared_negative = 0

# To sum up the dot product of the unseen example against
# every example contained in the positive and negative
# classes. They are used to compute term b
sum_b_positive = 0
sum_b_negative = 0

# Starting the computation of term b
for (i in 1:nrow(training.X)) {
    for (j in 1:nrow(training.X)) {

        if (training.Y[i] == -1 && training.Y[j] == -1 ) {
        sum_b_negative = sum_b_negative +
                         training.X[i,] %*% training.X[j
                         ,]
        m_squared_negative = m_squared_negative + 1
    }

        if (training.Y[i] == +1 && training.Y[j] == +1 ) {
        sum_b_positive = sum_b_positive + \
                         training.X[i,] %*% training.X[j
                         ,]
        m_squared_positive = m_squared_positive + 1

    }
  }
}

# Finally, we have term b.
```

```r
        # We do not square variables m_squared_negative and
        # m_squared_positive because they were already squared
        # due to the double loops used above
        b = 1/2 * ( 1/m_squared_negative * sum_b_negative
                    - 1/m_squared_positive * sum_b_positive)

        # Now term y is computed to classify the unseen
        # example, assigning either a positive or negative label
        y = sign( 1/m_positive * sum_positive
                  - 1/m_negative * sum_negative + b)

        # Saving the output and expected labels, respectively
        results = rbind(results, cbind(y, test.Y[unseen]))

        # Printing out the results
        cat(y, "", test.Y[unseen], "\n")
    }

  return (results)
}

test.first <- function() {

    # Generating the positive examples
    dataset = cbind(rnorm(mean=0, sd=1, n=100),
                    rnorm(mean=0, sd=1, n=100), rep(1, 100))

    # Generating the negative examples
    dataset = rbind(dataset, cbind(rnorm(mean=10, sd=1, n=100),
                    rnorm(mean=10, sd=1, n=100), rep(-1, 100)))

    # Plotting the dataset
    plot(dataset[,1:2], col=dataset[,3]+2)
    cat("Click_on_the_chart_to_continue...\n")
    locator(1)

    # Setting the training set size
    train.size = round(nrow(dataset)/2)

    # Sampling half of this dataset for training
    id = sample(1:nrow(dataset), size=train.size)

    # Building up the training set
    train.set = dataset[id,]

    # Building up the test set
    test.set = dataset[-id,]

    # Calling our classification algorithm to check the results
    results= first.classification.algorithm(train.set, test.set)

  return (results)
}
```

**Fig. 4.15** A more complex
input space to apply the
classification algorithm

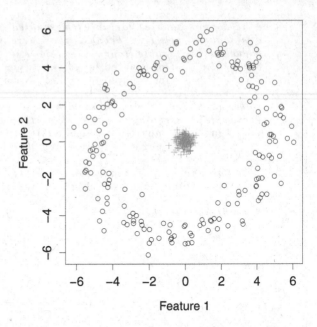

Feature 1

Poor results are obtained if we apply this classification algorithm in a more
complex input space, such as in Fig. 4.15, once it attempts to shatter a non-linearly
separable problem using a single hyperplane (Code 4.2). Such scenario motivates
the use of kernels to modify the input space, making positive and negative classes
linearly separable.

**Listing 4.2** Applying our classification algorithm on a more complex task

```
source("first-classification-algorithm.r")

test.complex <- function() {

    # Generating the positive examples
    dataset = cbind(rnorm(mean=0, sd=0.25, n=200),
        rnorm(mean=0, sd=0.25, n=200), rep(1, 200))

    # Generating the negative examples
    negative.set = 5*sin(2*pi*seq(0,9,len=200)) +
        rnorm(mean=0, sd=0.5, n=200)
    dataset = rbind( dataset,
        cbind(negative.set[1:(length(negative.set)-5)],
            negative.set[6:length(negative.set)],
            rep(-1, length(negative.set)-5)) )

    # Plotting the dataset
    plot(dataset[,1:2], col=dataset[,3]+2)
    cat("Click_on_the_chart_to_continue...\n")
    locator(1)
```

```
# Setting the training set size
train.size = round(nrow(dataset)/2)

# Sampling half of this dataset for training
id = sample(1:nrow(dataset), size=train.size)

# Building up the training set
train.set = dataset[id,]

# Building up the test set
test.set = dataset[-id,]

# Calling our classification algorithm to check results
results= first.classification.algorithm(train.set, test.
    set)

return (results)
}
```

A kernel function $k(x_i, x_j)$ maps examples $x_i$ and $x_j$ into another space, also referred to as features space, so that the algorithm computes their dot product. When kernelizing the algorithm, a slightly different formulation is obtained:

$$y = \text{sign}\left(\frac{1}{m_+}\sum_{i|y_i=+1} k(x, x_i) - \frac{1}{m_-}\sum_{i|y_i=-1} k(x, x_i) + b\right),$$

in which term $b$ also needs to be redefined:

$$b = \frac{1}{2}\left(\frac{1}{m_-^2}\sum_{(i,j)|y_i=y_j=-1} k(x_i, x_j) - \frac{1}{m_+^2}\sum_{(i,j)|y_i=y_j=+1} k(x_i, x_j)\right).$$

In this situation, every example $x_i$ is mapped from some input space $\mathbb{R}^m$ to a features space $\mathbb{R}^p$, for $p > m$, so that the classification algorithm computes the dot product in such target space. Let us consider the second-order polynomial kernel working on vectors in $\mathbb{R}^2$:

$$k(x_i, x_j) = <\mathbf{x}_i, \mathbf{x}_j>^2,$$

what is the same as:

$$k(x_i, x_j) = (\mathbf{x}_{i,1}\mathbf{x}_{j,1} + \mathbf{x}_{i,2}\mathbf{x}_{j,2})^2,$$

given:

$$\mathbf{x}_i = \begin{bmatrix} \mathbf{x}_{i,1} \\ \mathbf{x}_{i,2,} \end{bmatrix}, \ \mathbf{x}_j = \begin{bmatrix} \mathbf{x}_{j,1} \\ \mathbf{x}_{j,2.} \end{bmatrix}$$

In open form, this kernel function results in:

$$k(x_i, x_j) = \mathbf{x}_{i,1}^2 \mathbf{x}_{j,1}^2 + 2\mathbf{x}_{i,1}\mathbf{x}_{j,1}\mathbf{x}_{i,2}\mathbf{x}_{j,2} + \mathbf{x}_{i,2}^2 \mathbf{x}_{j,2}^2,$$

which is equivalent to the following space transformation:

$$\Phi(x_i) = \Phi\left(\begin{bmatrix} \mathbf{x}_{i,1} \\ \mathbf{x}_{i,2} \end{bmatrix}\right) = \begin{bmatrix} \mathbf{x}_{i,1}^2 \\ \sqrt{2}\mathbf{x}_{i,1}\mathbf{x}_{i,2} \\ \mathbf{x}_{i,2}^2 \end{bmatrix},$$

and, then, the dot product is computed on the features space as follows:

$$k(x_i, x_j) = \mathbf{x}_{i,1}^2 \mathbf{x}_{j,1}^2 + 2\mathbf{x}_{i,1}\mathbf{x}_{j,1}\mathbf{x}_{i,2}\mathbf{x}_{j,2} + \mathbf{x}_{i,2}^2 \mathbf{x}_{j,2}^2$$

$$= \begin{bmatrix} \mathbf{x}_{i,1}^2 \\ \sqrt{2}\mathbf{x}_{i,1}\mathbf{x}_{i,2} \\ \mathbf{x}_{i,2}^2 \end{bmatrix} \cdot \begin{bmatrix} \mathbf{x}_{j,1}^2 \\ \sqrt{2}\mathbf{x}_{j,1}\mathbf{x}_{j,2} \\ \mathbf{x}_{j,2}^2 \end{bmatrix}.$$

Now it is clear that every $\mathbf{x} \in \mathbb{R}^2$ is transformed into a 3-dimensional vector, as follows:

$$T(\mathbf{x}) = \begin{bmatrix} \mathbf{x}_1^2 \\ \sqrt{2}\mathbf{x}_1\mathbf{x}_2 \\ \mathbf{x}_2^2 \end{bmatrix},$$

so that the classification algorithm applies the dot product over those new vectors. Figure 4.16 illustrates the kernelized vectors.

The features space is linearly separable, drastically improving the algorithm accuracy. Listing 4.3 details such implementation.

**Listing 4.3** Applying the second-order polynomial kernel to proceed with the classification

```
source("first−classification−complex.r")

require(rgl)

Transformation <− function(vec) {
    class = vec[3]
    # Observe the expected class will be the same
    return (c(vec[1]^2, sqrt(2)*vec[1]*vec[2], vec[2]^2,
        class))
}

test.complex.kernel <− function() {

    # Generating the positive examples
    dataset = cbind(rnorm(mean=0, sd=0.25, n=200),
```

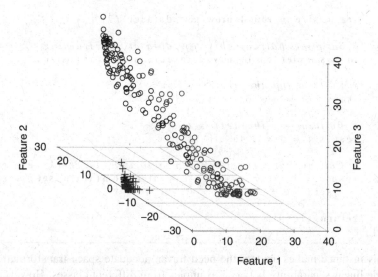

**Fig. 4.16** The three-dimensional features space obtained after applying the second-order polynomial kernel on the input space illustrated in Fig. 4.15

```
                        rnorm(mean=0, sd=0.25, n=200), rep(1, 200))

# Generating the negative examples
negative.set = 5*sin(2*pi*seq(0,9,len=200)) +
                        rnorm(mean=0, sd=0.5, n=200)
dataset = rbind(dataset,
                cbind(negative.set[1:(length(negative.set)
                -5)],
                negative.set[6:length(negative.set)],
                rep(-1, length(negative.set)-5)))

# Plotting the original dataset
plot(dataset[,1:2], col=dataset[,3]+2)
cat("Click on the chart to continue ...\n")
locator(1)

# Applying the kernel function to map every example
# into the features space
new.dataset = NULL
for (i in 1:nrow(dataset)) {
   new.dataset = rbind(new.dataset,
                        Transformation(dataset[i,]))
}

print(new.dataset)
# Plotting the transformed dataset
plot3d(new.dataset[,1:3], col=new.dataset[,4]+2)

# Setting the training set size
```

```
train.size = round(nrow(new.dataset)/2)

# Sampling half of this new.dataset for training
id = sample(1:nrow(new.dataset), size=train.size)

# Building up the training set
train.set = new.dataset[id,]

# Building up the test set
test.set = new.dataset[-id,]

# Calling our classification algorithm to check results
results= first.classification.algorithm(train.set, test.
    set)

return (results)
}
```

This instance makes evident the need for an adequate space transformation to provide linear separability between examples from different classes. However, the kernel design is not a trivial task, motivating researchers to algebraically study the input space. In fact, the best kernel function to make an input space linearly separable depends on the target problem. As consequence, studying the input space is more important than designing new classification algorithms.

## 4.4   Hyperplane-Based Classification: An Intuitive View

In this section, we start the design of Support Vector Machines from an intuitive point of view. Let a dataset labeled according to classes $\{-1, +1\}$, in which examples are generated using two Normal probability distributions that allow their linear separation (see Fig. 4.17).

In this scenario, our intention is to build up a linear hyperplane with the maximal margin, i.e., the hyperplane has the same distance to both closest positive and negative examples, as shown in Fig. 4.20. This is the best hyperplane given it provides the smallest as possible class overlapping, considering distribution deviations from means.

Figure 4.18 illustrates the practical impact of maximal-margin hyperplanes. Consider one-dimensional examples and the probability distributions responsible for generating them.[5] Observe $h_1$ or $h_2$ might be obtained after running any regular classification algorithm (e.g. Perceptron), while $h_{best}$ corresponds to the maximal-margin hyperplane (e.g. SVM). Although all of them produce the same classification results and, therefore, the same empirical risks $R_{emp}(h)$, $h_{best}$ provides the greatest

---

[5]Distributions are used for illustration purposes. In fact, we remind the reader that the Statistical Learning Theory assumes they are unknown at the time of training.

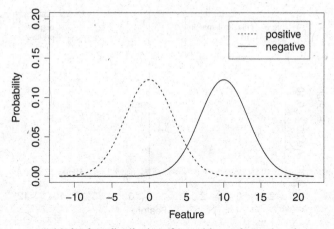

(a) the data distributions for positive and negative classes

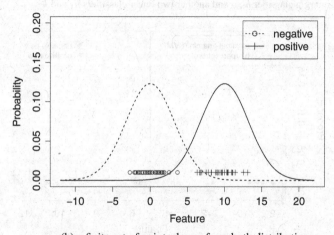

(b) a finite set of points drawn from both distributions

**Fig. 4.17** Input space of examples produced by two different Normal probability distributions (**a**) the distribution densities, (**b**) densities and points drawn from the distributions to be used as training examples

learning generalization given both class distributions are shattered in the best as possible place. Figures permit us to conclude that $h_1$ will mostly misclassify the negative class, while $h_2$ the positive class. The reader may also refer to the type-I and type-II errors from Statistical Inference to interpret the results [6].

In Fig. 4.18 illustration, suppose the hyperplane $h_2$ is estimated by the Perceptron algorithm. Its hyperplane is skewed towards one of the data distributions (the positive examples), so that new examples drawn and eventually close to such decision boundary will tend to be misclassified, as shown in Fig. 4.19.

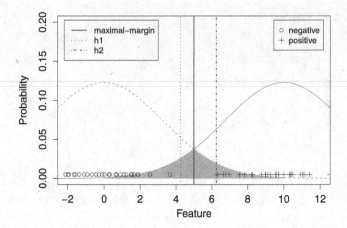

**Fig. 4.18** Assessing the maximal-margin hyperplane in an one-dimensional problem, comparing the maximal-margin classifier $h_{\mathrm{best}}$ and another two linear classifiers $h_1$ and $h_2$

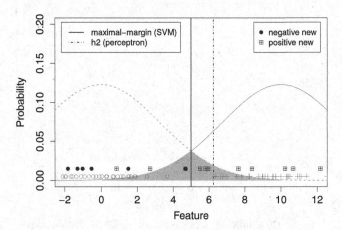

**Fig. 4.19** Illustration of the effect of non-maximal margin hyperplane on unseen (new) data examples drawn from the class distributions. The vertical solid line is the optimal classifier $h_1$, while the vertical dashed line represents a suboptimal classifier such as $h_2$

Support Vector Machines (SVMs) were designed to find this maximal margin hyperplane. As discussed in Chap. 2, if the best as possible space is provided, then SVM outperforms any other classification algorithm.

Consider a linearly separable problem with two classes, as shown in Fig. 4.20. Intuitively, SVM builds up the best as possible hyperplane, solving the following problem for the closest positive and negative examples:

$$|< \mathbf{w}, \mathbf{x}_+ > + b| = 1,$$

$$|< \mathbf{w}, \mathbf{x}_- > + b| = 1,$$

**Fig. 4.20** Input space to illustrate how SVM works

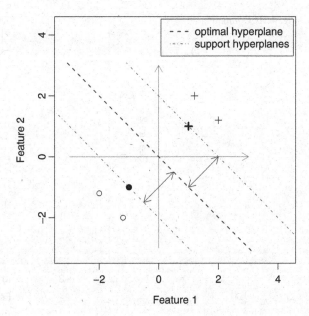

in which vector $\mathbf{w}$ corresponds to weights, $\mathbf{x}_+$ and $\mathbf{x}_-$ are the examples nearby the hyperplane (a.k.a. support vectors), and finally $b$ is a bias term.[6] Such equality constraints must produce an absolute value equals to one for $\mathbf{x}_+$ and $\mathbf{x}_-$, what is associated to the relative distance they have to the decision boundary. When $\mathbf{x}_i$ is far from such boundary, those constraints should be adapted to produce:

$$|<\mathbf{w}, \mathbf{x}_i> +b| > 1,$$

meaning it is $|<\mathbf{w}, \mathbf{x}_i> +b| - 1$ units more distant from the hyperplane than the support vectors. In summary, the SVM optimization problem intends to find $\mathbf{w}$ and $b$ while respecting such constraints, so the best as possible linear separation can be used to classify further unseen examples. But how could one estimate those variables? As first step, vector $\mathbf{w}$ must be found:

$$\mathbf{w} = \mathbf{x}_+ - \mathbf{x}_-,$$

which is given by the difference between the positive and the negative support vectors. Figure 4.21 illustrates vector $\mathbf{w}$ in a simple classification scenario:

$$\mathbf{w} = \mathbf{x}_+ - \mathbf{x}_-$$

---

[6]This work as an intercept term, such as $\theta$ for the Perceptron and the Multilayer Perceptron algorithms (see Chap. 1).

**Fig. 4.21** Computing
vector **w**

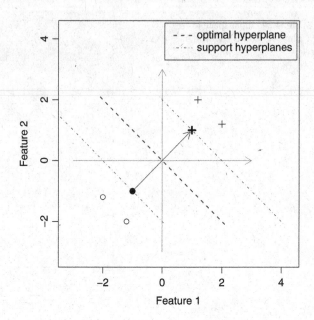

$$= \begin{bmatrix} 1 \\ 1 \end{bmatrix} - \begin{bmatrix} -1 \\ -1 \end{bmatrix} = \begin{bmatrix} 2 \\ 2 \end{bmatrix}.$$

Then, we solve the equality constraints using the support vectors:

$$|< \mathbf{w}, \mathbf{x}_+ > +b| = 1$$
$$|< \mathbf{w}, \mathbf{x}_- > +b| = 1,$$

in form:

$$\left\langle \begin{bmatrix} 2 \\ 2 \end{bmatrix}, \mathbf{x}_+ \right\rangle + b = +1$$

$$\left\langle \begin{bmatrix} 2 \\ 2 \end{bmatrix}, \mathbf{x}_- \right\rangle + b = -1.$$

Afterwards, the bias term $b$ is computed:

$$\begin{bmatrix} 2 \\ 2 \end{bmatrix} \begin{bmatrix} 1 & 1 \end{bmatrix} + b = +1$$

$$\begin{bmatrix} 2 \\ 2 \end{bmatrix} \begin{bmatrix} -1 & -1 \end{bmatrix} + b = -1,$$

to obtain:

$$2 + 2 + b = +1$$

$$-2 + (-2) + b = -1$$
$$b = \pm 3.$$

Observe term $b$ is not unique, consequently it does not provide a single solution satisfying this problem. In order to obtain unicity, we multiply both sides of equality constraints by their corresponding classes ($y_+$ and $y_-$):

$$y_+(< \mathbf{w}, \mathbf{x}_+ > +b) = +1y_+$$
$$y_-(< \mathbf{w}, \mathbf{x}_- > +b) = -1y_-,$$

obtaining:

$$y_+(< \mathbf{w}, \mathbf{x}_+ > +b) = 1$$
$$y_-(< \mathbf{w}, \mathbf{x}_- > +b) = 1,$$

thus, ensuring the constraints only for the support vectors. We then extend constraints to consider any other example by using inequalities:

$$y_+(< \mathbf{w}, \mathbf{x}_+ > +b) \geq 1$$
$$y_-(< \mathbf{w}, \mathbf{x}_- > +b) \geq 1,$$

which are simplified to the general case (for any label), as follows:

$$y_i(< \mathbf{w}, \mathbf{x}_i > +b) \geq 1.$$

Going back to our previous example, we check if this last inequality holds for both values of term $b$, e.g., $+3$ and $-3$. For $b = +3$ (question mark means assessment):

$$+1\left(\begin{bmatrix} 2 \\ 2 \end{bmatrix} [1\ 1] + 3\right) \overset{?}{\geq} 1,$$

yes, it holds since $7 \geq 1$.

$$-1\left(\begin{bmatrix} 2 \\ 2 \end{bmatrix} [-1\ -1] + 3\right) \overset{?}{\geq} 1,$$

yes, given $1 \geq 1$.
Assessing for $b = -3$:

$$+1\left(\begin{bmatrix} 2 \\ 2 \end{bmatrix} [1\ 1] - 3\right) \overset{?}{\geq} 1,$$

**Fig. 4.22** Two hyperplanes
for both values of the bias
term $b = \pm 3$

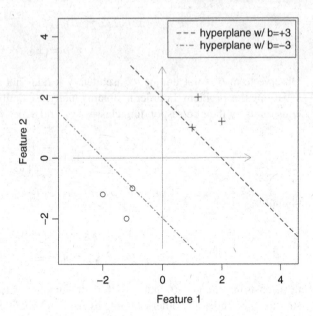

yes, it holds since $1 \geq 1$.

$$-1\left(\begin{bmatrix}2\\2\end{bmatrix}[-1\ -1] - 3\right) \overset{?}{\geq} 1,$$

yes, given $7 \geq 1$. As consequence, we conclude both values satisfy our general
inequality taking us to the following conclusions:

1. Different hyperplanes are found for $b = +3$ and $b = -3$ (see Fig. 4.22);
2. The bias term $b$ is seen as a correction term to produce positive and negative
   values on each side of the decision boundary;
3. Vector **w** is orthogonal to the hyperplane (see Fig. 4.23).

In order to empirically explore the SVM problem, we now study the effect of
increasing the norm (the length) of vector **w**. For example, let:

$$\mathbf{w} = \begin{bmatrix}10\\10\end{bmatrix},$$

what would require a bias term $b \pm 19$. Two solutions were again obtained, but
with scaled correction terms $b$. In fact, the greater the norm of **w**, the greater the
difference in inequality results as seen next:

$$+1\left(\begin{bmatrix}10\\10\end{bmatrix}[1\ 1] + 19\right) \geq 1 \Rightarrow 39 \geq 1$$

**Fig. 4.23** Illustrating the orthogonality of vector **w**, drawn as a gray arrow, with respect to the optimal hyperplane

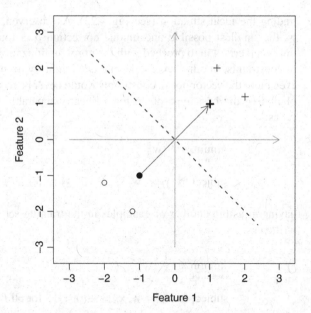

$$-1\left(\begin{bmatrix}10\\10\end{bmatrix}[-1\ -1]+19\right)\geq 1 \Rightarrow 1\geq 1,$$

and:

$$+1\left(\begin{bmatrix}10\\10\end{bmatrix}[1\ 1]-19\right)\geq 1 \Rightarrow 1\geq 1$$

$$-1\left(\begin{bmatrix}10\\10\end{bmatrix}[-1\ -1]-19\right)\geq 1 \Rightarrow 39\geq 1.$$

To contrapose, we now assess the best as possible scenario in which vector **w** is:

$$\mathbf{w}=\begin{bmatrix}0.5\\0.5\end{bmatrix},$$

allowing to obtain unicity for the bias term $b = 0$, as required. Applying such solution on the inequality, we have:

$$+1\left(\begin{bmatrix}0.5\\0.5\end{bmatrix}[1\ 1]+0\right)\geq 1 \Rightarrow 1\geq 1$$

$$-1\left(\begin{bmatrix}0.5\\0.5\end{bmatrix}[-1\ -1]+0\right)\geq 1 \Rightarrow 1\geq 1,$$

finding the ideal situation (see Fig. 4.23). As observed, by reducing the norm of **w**, the smallest possible and unique correction $b$ is found. However, there is an important criterion to proceed with the norm minimization, which is the satisfaction of constraints. In other words, despite one might try to improve results by reducing even more the vector norm, constraints would never be respected. Thus, this problem of finding the best hyperplane for a linearly separable task could be, intuitively, set as:

$$\underset{\mathbf{w},b}{\text{minimize}} \quad \|\mathbf{w}\|$$

$$\text{subject to} \quad y_i(<\mathbf{w}, \mathbf{x}_i> +b) \geq 1, \text{ for all } i = 1, \ldots, m,$$

having $m$ as the number of examples in the training set. This problem is typically written as:

$$\underset{\mathbf{w}\in\mathbb{H},b\in\mathbb{R}}{\text{minimize}} \quad \frac{1}{2}\|\mathbf{w}\|^2$$

$$\text{subject to} \quad y_i(<\mathbf{w}, \mathbf{x}_i> +b) \geq 1, \text{ for all } i = 1, \ldots, m,$$

in which $\mathbb{H}$ defines the space the hyperplane is built in (the Hilbert space), $\mathbb{R}$ ensures the bias term $b$ lies on the real line, and fraction $\frac{1}{2}$ is only used as a mathematical convenience for solving the error derivative. It is worth to mention that the norm is powered to two to transform the objective function into a convex surface, facilitating the usage of mathematical tools to drive the optimization process (e.g. gradient descent method). Chapter 5 details this formulation and why it is the most used in the literature.

## 4.5  Hyperplane-Based Classification: An Algebraic View

After an intuitive approach for the SVM optimization problem, this section introduces its step-by-step algebraic formulation. Consider examples lying on $\mathbb{R}^2$ and let two linear hyperplanes (a.k.a. support hyperplanes) define the boundaries for the positive and negative labels such as the "sidewalks" for a street, as in Fig. 4.24a. Those support hyperplanes are parallel to the maximal-margin hyperplane as well as to each other. It is important to notice that they lie on the tip of the support vectors, defining the frontiers of each class. The best as possible hyperplane divides the region between support hyperplanes in half (Fig. 4.24b).

In that scenario, if we take the difference vector $\mathbf{x}_+ - \mathbf{x}_-$ and project it over the unitary vector **v**, we obtain the "street" width (see Fig. 4.25). This is only possible because **v** is orthogonal to both support hyperplanes. By measuring such width, it is possible to compute the gap between classes and, consequently, find the maximal margin as half of such value.

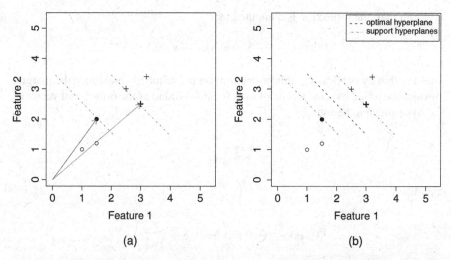

(a)                                    (b)

**Fig. 4.24** Support hyperplanes define the "sidewalks", lying on the tip of the highlighted support vectors (**a**). The best as possible hyperplane is in between those support hyperplanes (**b**)

**Fig. 4.25** Computing the difference vector $\mathbf{x}_+ - \mathbf{x}_-$ to be projected on the unitary vector $\mathbf{v}$ and obtain the "street" width

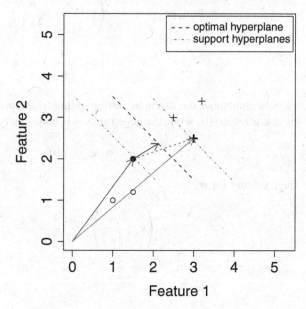

From that, the projection is computed as:

$$\text{Proj}_\mathbf{v}(\mathbf{x}_+ - \mathbf{x}_-) = \langle \mathbf{x}_+ - \mathbf{x}_-, \mathbf{v} \rangle .$$

Observe that by maximizing this projection, we consequently maximize the margin. Besides the unitary vector $\mathbf{v}$ is unknown, $\mathbf{w}$ can be found as the orthogonal vector of the hyperplane to be found:

$$\mathbf{v} = \frac{\mathbf{w}}{\|\mathbf{w}\|},$$

and, then:

$$\text{Proj}_\mathbf{v}(\mathbf{x}_+ - \mathbf{x}_-) = \left\langle \mathbf{x}_+ - \mathbf{x}_-, \frac{\mathbf{w}}{\|\mathbf{w}\|} \right\rangle,$$

thus, after some manipulation:

$$\text{Proj}_\mathbf{v}(\mathbf{x}_+ - \mathbf{x}_-) = \left\langle \mathbf{x}_+ - \mathbf{x}_-, \frac{\mathbf{w}}{\|\mathbf{w}\|} \right\rangle$$

$$= \left\langle \mathbf{x}_+, \frac{\mathbf{w}}{\|\mathbf{w}\|} \right\rangle - \left\langle \mathbf{x}_-, \frac{\mathbf{w}}{\|\mathbf{w}\|} \right\rangle.$$

Now remember the example corresponding to the positive support vector must be classified as $+1$, while the negative as $-1$, both satisfying:

$$y_i(< \mathbf{w}, \mathbf{x}_i > +b) = 1,$$

then, solving for $\mathbf{w}$:

$$\left\langle \mathbf{x}_+, \frac{\mathbf{w}}{\|\mathbf{w}\|} \right\rangle = \frac{< \mathbf{x}_+, \mathbf{w} >}{\|\mathbf{w}\|},$$

and, thus:

$$< \mathbf{w}, \mathbf{x}_i > +b = y_i$$

$$< \mathbf{w}, \mathbf{x}_i > = y_i - b,$$

so, setting $y_i = +1$:

$$\left\langle \mathbf{x}_+, \frac{\mathbf{w}}{\|\mathbf{w}\|} \right\rangle = \frac{1 - b}{\|\mathbf{w}\|},$$

and, setting $y_i = -1$:

$$\left\langle \mathbf{x}_-, \frac{\mathbf{w}}{\|\mathbf{w}\|} \right\rangle = \frac{-1-b}{\|\mathbf{w}\|}.$$

Finally, the projection is given by:

$$\mathrm{Proj}_\mathbf{v}(\mathbf{x}_+ - \mathbf{x}_-) = \frac{1-b}{\|\mathbf{w}\|} - \frac{-1-b}{\|\mathbf{w}\|} = \frac{2}{\|\mathbf{w}\|}.$$

Therefore, by maximizing the projection, we look for the maximal margin:

maximize $\mathrm{Proj}_\mathbf{v}(\mathbf{x}_+ - \mathbf{x}_-)$

subject to $y_i(<\mathbf{w}, \mathbf{x}_i> +b) \geq 1$, for all $i = 1, \ldots, m$,

what is the same as:

maximize $\dfrac{2}{\|\mathbf{w}\|}$

subject to $y_i(<\mathbf{w}, \mathbf{x}_i> +b) \geq 1$, for all $i = 1, \ldots, m$.

By inverting the objective function, we can reformulate it as a minimization problem, as usually found in literature:

minimize $\dfrac{1}{2}\|\mathbf{w}\|^2$

subject to $y_i(<\mathbf{w}, \mathbf{x}_i> +b) \geq 1$, for all $i = 1, \ldots, m$,

having the norm of $\mathbf{w}$ powered to two in order to simplify error derivatives as for a convex function, so it converges to a minimum (see Chap. 5 for more details). Defining the sets for variables $\mathbf{w}$ and the bias term $b$, we finally have:

$$\underset{\mathbf{w} \in \mathbb{H}, b \in \mathbb{R}}{\text{minimize}} \frac{1}{2}\|\mathbf{w}\|^2$$

subject to $y_i(<\mathbf{w}, \mathbf{x}_i> +b) \geq 1$, for all $i = 1, \ldots, m$,

noting $\mathbf{w}$ is in a Hilbert space $\mathbb{H}$ (see Chap. 2), and $b$ is a real number.

## 4.5.1  *Lagrange Multipliers*

Lagrange multipliers is a mathematical tool to find local maxima or minima subject to equality constraints. To formalize, consider the following optimization problem:

$$\text{maximize } f(x, y)$$

$$\text{subject to } g(x, y) = c,$$

in which $f(x, y)$ defines the objective function, while $g(x, y)$ ensures equality constraints. According to Lagrange multipliers, there is one or more solutions only if $f$ and $g$ are differentiable, i.e.:

$$\nabla f = -\lambda \nabla g,$$

given some point $(x, y)$, $\lambda \neq 0$, and $\nabla$ meaning the gradient vector. Therefore, this optimization problem has a solution only if the gradient vectors of $f$ and $g$ are equal (a.k.a. parallel) given some constant $\lambda \neq 0$.

As an example, consider the following optimization problem:

$$\text{maximize } f(x, y) = x + y$$

$$\text{subject to } x^2 + y^2 = 1.$$

To solve this problem using the Lagrange multipliers, the following must hold:

1. All constraints must be defined in terms of equalities;
2. Functions $f$ and $g$ must be differentiable, so the gradient vectors exist;
3. Gradient vectors for $f$ and $g$ must be different from zero.

Thus, the following system of equations must be solved:

$$\begin{cases} \nabla f = -\lambda \nabla g \\ g(x, y) - c = 0 \end{cases},$$

to find:

$$\begin{cases} \begin{bmatrix} 1 \\ 1 \end{bmatrix} = -\lambda \begin{bmatrix} 2x \\ 2y \end{bmatrix}, \\ x^2 + y^2 - 1 = 0 \end{cases}$$

what results in:

$$\begin{cases} 1 = -\lambda 2x \\ 1 = -\lambda 2y \\ x^2 + y^2 - 1 = 0 \end{cases}$$

Finally, solving such system:

$$\begin{cases} 1 + \lambda 2x = 0 & \Rightarrow x = -\frac{1}{2\lambda} \\ 1 + \lambda 2y = 0 & \Rightarrow y = -\frac{1}{2\lambda} \\ x^2 + y^2 - 1 = 0 \Rightarrow \left(-\frac{1}{2\lambda}\right)^2 + \left(-\frac{1}{2\lambda}\right)^2 - 1 = 0, & \text{for } \lambda \neq 0, \end{cases}$$

we find:

$$\lambda = \pm \frac{1}{\sqrt{2}},$$

which is then used to calculate, by substitution, $x$ and $y$. By setting $\lambda = -\frac{1}{\sqrt{2}}$, the following solution is obtained:

$$x = y = -\frac{1}{2\lambda} = \left(-\frac{1}{2}\right)\left(-\frac{\sqrt{2}}{1}\right) = \frac{\sqrt{2}}{2},$$

and, setting $\lambda = \frac{1}{\sqrt{2}}$, there is another solution:

$$x = y = -\frac{1}{2\lambda} = \left(-\frac{1}{2}\right)\left(\frac{\sqrt{2}}{1}\right) = -\frac{\sqrt{2}}{2}.$$

As a consequence, there are two coordinates $(x, y)$ satisfying the Lagrange multipliers:

$$\left(\frac{\sqrt{2}}{2}, \frac{\sqrt{2}}{2}\right), \left(-\frac{\sqrt{2}}{2}, -\frac{\sqrt{2}}{2}\right).$$

The objective value is only found by evaluating them in terms of $f$:

$$f\left(\frac{\sqrt{2}}{2}, \frac{\sqrt{2}}{2}\right) = \sqrt{2},$$

which is the global maximum, and:

$$f\left(-\frac{\sqrt{2}}{2}, -\frac{\sqrt{2}}{2}\right) = -\sqrt{2},$$

which is the global minimum.

Figure 4.26 illustrates this optimization problem, having the vertical axis associated with the output values of $f$, while the other two axes are related to variables

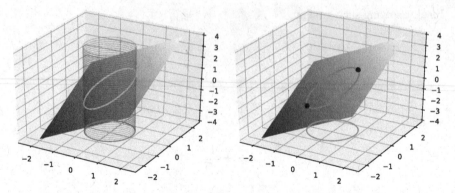

**Fig. 4.26** Illustration of the optimization problem and its solutions (global maximum and minimum) found according to Lagrange multipliers

**Fig. 4.27** Studying the same optimization problem from Fig. 4.26, but using contour lines that help visualizing the gradient vectors

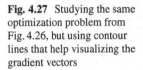

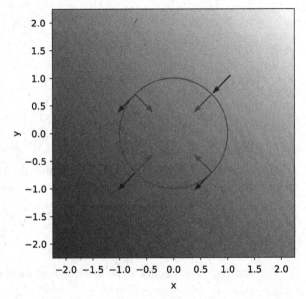

$x$ and $y$. Observe constraint $x^2 + y^2 = 1$ is plotted into the space formed by axes $x$ and $y$, which is then projected into the surface formed by $f$. Notice one of the points corresponds to the maximum, while the other to the minimum value of $f$, considering this equality constraint.

Figure 4.27 shows contour lines for the same problem, in which arrows denote the gradient vectors. Notice there are two points in which the gradient vectors of the constraint and the objective function are parallel to each other (independently of their directions). Those points are found through Lagrange multipliers and both lie on the projection of the constraint into the surface produced by the objective function. This idea motivated Joseph Louis Lagrange to find the solution for any optimization problem, given the objective function $f$ and one or more equality

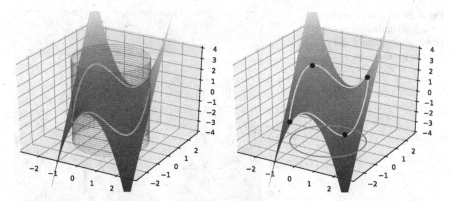

**Fig. 4.28** Illustration of the second optimization problem and its solutions according to Lagrange multipliers

constraints $g_i$, in form:

$$\begin{cases} \nabla f = -\lambda \nabla g_i \\ g_i(x, y) - c_i = 0 \end{cases},$$

having $\lambda \neq 0$ as the factor to correct the magnitudes and directions of gradient vectors, so they become equal.

Lagrange multipliers does not provide any solution within the region defined by constraint $x^2 + y^2 = 1$, i.e., for $x, y$ such that $x^2 + y^2 < 1$, nor even outside, i.e., for $x, y$ such that $x^2 + y^2 > 1$. It is a mathematical tool to find solutions that touch the projection of constraints into the objective function.

As a second example, let:

$$\text{maximize} \ \ f(x, y) = x^2 y$$

$$\text{subject to} \ \ x^2 + y^2 = 3,$$

as illustrated in Fig. 4.28. The single constraint function is defined in terms of $x$ and $y$, and the vertical axis refers to the objective function outputs. To complement, Fig. 4.29 shows the contour lines, from which the solution points are easily seen.

Thus, using Lagrange multipliers, the following problem must be solved:

$$\begin{cases} \begin{bmatrix} 2xy \\ x^2 \end{bmatrix} = -\lambda \begin{bmatrix} 2x \\ 2y \end{bmatrix}, \\ x^2 + y^2 - 3 = 0 \end{cases}$$

from which six critical points are obtained:

$$(\sqrt{2}, 1), (-\sqrt{2}, 1), (\sqrt{2}, -1), (-\sqrt{2}, -1), (0, \sqrt{3}), (0, -\sqrt{3}),$$

**Fig. 4.29** Contour lines for the second optimization problem. Note that $f$ and $g$ become parallel only when constraints are respected

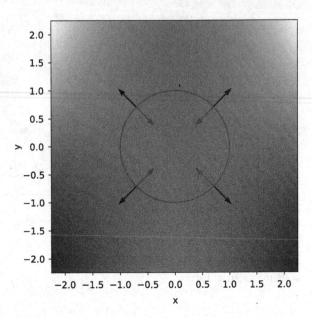

which must be evaluated by the objective function $f$:

$$f(\pm\sqrt{2}, 1) = 2, \; f(\pm\sqrt{2}, -1) = -2, \; f(0, \pm\sqrt{3}) = 0,$$

so that the global maxima occurs at two coordinates $(\sqrt{2}, 1)$ and $(-\sqrt{2}, 1)$, while the global minima are at $(\sqrt{2}, -1)$ and $(-\sqrt{2}, -1)$.

### 4.5.2   Karush-Kuhn-Tucker Conditions

Although the Lagrange multipliers provides an important optimization tool, it can only be employed under equality constraints, while the SVM minimization problem is restricted by inequalities. In order to tackle the maximal-margin problem, we must introduce the Karush-Kuhn-Tucker (KKT) conditions which complement Lagrange multipliers for inequality constraints.

$$\underset{\mathbf{w}\in\mathbb{H}, b\in\mathbb{R}}{\text{minimize}} \quad \frac{1}{2} \left\| \mathbf{w} \right\|^2$$

$$\text{subject to} \quad y_i(<\mathbf{w}, \mathbf{x}_i> +b) \geq 1, \; \text{for all } i = 1, \ldots, m,$$

as defined in Sect. 4.5. The Karush-Kuhn-Tucker (KKT) conditions extend Lagrange multipliers to solve problems in the following standard forms:

$$\text{maximize } f(x)$$

$$\text{subject to } g_i(x) \geq 0, h_j(x) = 0,$$

and:

$$\text{minimize } f(x)$$

$$\text{subject to } g_i(x) \geq 0, h_j(x) = 0,$$

given $f$ is the objective function, $g_i$ defines a set of inequality constraints for $i = 1, \ldots, n$, and $h_j$ corresponds to the set of equality constraints for $j = 1, \ldots, m$.

KKT conditions determine the circumstances in which this type of optimization problem has solution:

1. Primal feasibility—it must exist an optimal solution $x^*$, for which all constraints are satisfied, i.e., $g_i(x^*) \geq 0$ and $h_j(x^*) = 0$, for all $i, j$;
2. Stationarity—there is at least one (stationary) point for which the gradient of $f$ is parallel to the gradient of constraints $g_i$ and $h_j$, such as in Lagrange multipliers, i.e.:

$$\nabla f(x^*) = \sum_{i=1}^{n} \mu_i^* \nabla g_i(x^*) + \sum_{j=1}^{m} \lambda_j^* \nabla h_j(x^*),$$

in which $\mu_i^*$ and $\lambda_j^*$ are known as the KKT multipliers (instead of Lagrange multipliers), and stars mean the optimal value for each variable;
3. Complementary slackness—KKT multipliers times inequality constraints must be equal to zero, i.e., $\mu_i^* g_i(x^*) = 0$ for all $i = 1, \ldots, m$;
4. Dual feasibility—All KKT multipliers must be positive for inequality constraints, i.e., $\mu_i^* \geq 0$ for all $i = 1, \ldots, m$.

Later in Chap. 5, we detail how those conditions are found. For now, we simply exemplify their usefulness.

Consider the following optimization problem (from [2, 3]):

$$\text{minimize } f(x_1, x_2) = 4x_1^2 + 2x_2^2$$

$$\text{subject to } 3x_1 + x_2 = 8$$

$$2x_1 + 4x_2 \leq 15,$$

and consider someone has informed us that the second constraint in non-binding. A **binding constraint** is the one whose solution depends on, so if we change it, the optimal solution also changes. On the other hand, a **non-binding constraint** does not affect the optimal solution, so it can be discarded without any loss.

However, before dropping the second constraint, the problem is rewritten in the standard form:

$$\text{minimize} \quad f(x_1, x_2) = 4x_1^2 + 2x_2^2$$
$$\text{subject to} \quad 3x_1 + x_2 = 8$$
$$- 2x_1 - 4x_2 \geq -15,$$

and, then, we need to check whether the KKT conditions are held:

1. Primal feasibility—all constraints have to be satisfied, which, in this case, is:

$$3x_1 + x_2 = 8;$$

2. Stationarity—we must find the stationary point:

$$\nabla f(x^*) = \sum_{i=1}^{n} \mu_i^* \nabla g_i(x^*) + \sum_{j=1}^{m} \lambda_j^* \nabla h_j(x^*)$$

$$\begin{bmatrix} 8x_1 \\ 4x_2 \end{bmatrix} - \lambda_1 \begin{bmatrix} 3 \\ 1 \end{bmatrix} - \mu_1 \begin{bmatrix} -2 \\ -4 \end{bmatrix} = 0$$

but, given the second constraint is non-binding, then $\mu_1 = 0$:

$$\nabla f(x^*) = \sum_{i=1}^{n} \mu_i^* \nabla g_i(x^*) + \sum_{j=1}^{m} \lambda_j^* \nabla h_j(x^*)$$

$$\begin{bmatrix} 8x_1 \\ 4x_2 \end{bmatrix} - \lambda_1 \begin{bmatrix} 3 \\ 1 \end{bmatrix} = 0,$$

therefore, the linear system of equations is solved in form $A\mathbf{x} = B$:

$$\begin{bmatrix} 3 & 1 & 0 \\ 8 & 0 & -3 \\ 0 & 4 & -1 \end{bmatrix} \begin{bmatrix} x_1 \\ x_2 \\ \lambda_1 \end{bmatrix} = \begin{bmatrix} 8 \\ 0 \\ 0 \end{bmatrix},$$

to obtain $\mathbf{x} = A^{-1}B$:

$$x = \begin{bmatrix} 2.182 \\ 1.455 \\ 5.818 \end{bmatrix};$$

3. The further conditions (Complementary slackness and Dual feasibility) are only necessary for inequality constraints, what is not the case in this example.

As a consequence, we found a feasible solution $x_1 = 2.182, x_2 = 1.455$, $\lambda_1 = 5.818$ and $\mu_1 = 0$ (given the inequality constraint is non-binding) for this optimization problem, considering all KKT conditions. Observe that a solver may be used instead of the inverse of $A$.

Now consider another problem with two inequality constraints:

$$\text{minimize} \quad f(x_1, x_2, x_3) = x_1^2 + 2x_2^2 + 3x_3^2$$
$$\text{subject to} \quad g_1(x_1, x_2, x_3) = -5x_1 + x_2 + 3x_3 \leq -3$$
$$g_2(x_1, x_2, x_3) = 2x_1 + x_2 + 2x_3 \geq 6,$$

which must be rewritten using the standard form:

$$\text{minimize} \quad f(x_1, x_2, x_3) = x_1^2 + 2x_2^2 + 3x_3^2$$
$$\text{subject to} \quad g_1(x_1, x_2, x_3) = 5x_1 - x_2 - 3x_3 \geq 3$$
$$g_2(x_1, x_2, x_3) = 2x_1 + x_2 + 2x_3 \geq 6,$$

in order to assess the KKT conditions:

1. Primal feasibility—initially, every constraint is assumed to have a solution. Later on, that will be checked;
2. Stationarity—finding the stationary points:

$$\nabla f(x^*) = \sum_{i=1}^{n} \mu_i^* \nabla g_i(x^*) + \sum_{j=1}^{m} \lambda_j^* \nabla h_j(x^*),$$

to obtain:

$$\begin{cases} \begin{bmatrix} 2 \\ 4 \\ 6 \end{bmatrix} - \mu_1 \begin{bmatrix} 5 \\ -1 \\ -3 \end{bmatrix} - \mu_2 \begin{bmatrix} 2 \\ 1 \\ 2 \end{bmatrix} = \begin{bmatrix} 0 \\ 0 \\ 0 \end{bmatrix}, \\ 5x_1 - x_2 - 3x_3 = 3 \\ 2x_1 + x_2 + 2x_3 = 6 \end{cases}$$

from which constraints are solved by using equalities instead. Then, solving the linear system $Ax = B$:

$$\begin{bmatrix} 2 & 0 & 0 & -5 & -2 \\ 0 & 4 & 0 & +1 & -1 \\ 0 & 0 & 6 & +3 & -2 \\ 5 & -1 & -3 & 0 & 0 \\ 2 & 1 & 2 & 0 & 0 \end{bmatrix} \begin{bmatrix} x_1 \\ x_2 \\ x_3 \\ \mu_1 \\ \mu_2 \end{bmatrix} = B$$

$$\begin{bmatrix} 2 & 0 & 0 & -5 & -2 \\ 0 & 4 & 0 & +1 & -1 \\ 0 & 0 & 6 & +3 & -2 \\ 5 & -1 & -3 & 0 & 0 \\ 2 & 1 & 2 & 0 & 0 \end{bmatrix} x = \begin{bmatrix} 0 \\ 0 \\ 0 \\ 3 \\ 6 \end{bmatrix},$$

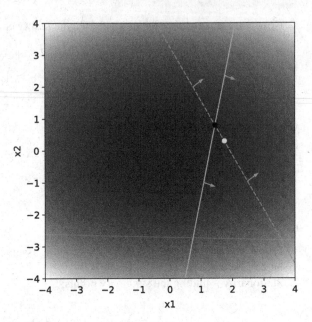

**Fig. 4.30** Contour lines in terms of variables $x_1$ and $x_2$. This supports the analysis of gradient vectors for the objective and constraint functions. The continuous and dashed lines represent the constraints, the black dot is the solution found considering both constraints, and the white dot is the optimal solution, obtained by considering the first constraint non-binding

the following solutions are found:

$$x_1 = 1.450, \, x_2 = 0.8, \, x_3 = 1.150, \, \mu_1 = -0.5, \, \mu_2 = 2.70;$$

3. Complementary slackness—we now verify if $\mu_i g_i(x^*) = 0$ for $i = 1, 2$:

$$\mu_1 g_1(x^*) = -0.5(5(1.450) - (0.8) - 3(1.150) - 3) \approx 0$$

$$\mu_2 g_2(x^*) = 2.70(2(1.450) + (0.8) + 2(1.150) - 6) = 0,$$

what ensures this third condition.

4. Dual feasibility—this fourth condition is violated due to $\mu_1 = -0.5 < 0$, meaning this is not an optimal solution.

Consider the contour lines in Fig. 4.30 from which we can visually inspect the solution found. Observe constraint $g_1(x^*)$ is non-binding as indicated by $\mu_1 = -0.5 < 0$. Consequently, to solve this problem, one must set $\mu_1 = 0$ to cancel $g_1(x^*)$ in form:

$$\begin{cases} \begin{bmatrix} 2 \\ 4 \\ 6 \end{bmatrix} - \mu_2 \begin{bmatrix} 2 \\ 1 \\ 2 \end{bmatrix} = \begin{bmatrix} 0 \\ 0 \\ 0 \end{bmatrix}, \\ 2x_1 + x_2 + 2x_3 = 6 \end{cases}$$

providing the system $A\mathbf{x} = B$:

$$\begin{bmatrix} 2 & 0 & 0 & 0 & -2 \\ 0 & 4 & 0 & 0 & -1 \\ 0 & 0 & 6 & 0 & -2 \\ 2 & 1 & 2 & 0 & 0 \end{bmatrix} \begin{bmatrix} x_1 \\ x_2 \\ x_3 \\ \mu_2 \end{bmatrix} = B$$

$$\begin{bmatrix} 2 & 0 & 0 & 0 & -2 \\ 0 & 4 & 0 & 0 & -1 \\ 0 & 0 & 6 & 0 & -2 \\ 2 & 1 & 2 & 0 & 0 \end{bmatrix} \mathbf{x} = \begin{bmatrix} 0 \\ 0 \\ 0 \\ 3 \\ 6 \end{bmatrix},$$

in order to obtain the final solution $x_1 = 2.0571429$, $x_2 = 0.5142857$, $x_3 = 0.6857143$, $\mu_1 = 0$, and $\mu_2 = 2.0571429$, which indeed respects the KKT conditions.

For the sake of curiosity, the Karush-Kuhn-Tucker conditions were originally named after Kuhn and Tucker [5]. Later on, other researchers discovered that those conditions were already stated by William Karush in his master's thesis [4].

## 4.6 Formulating the Hard-Margin SVM Optimization Problem

After building up the concepts on Lagrange multipliers and KKT conditions, we are finally able to formulate the optimal linear hyperplane in terms of an optimization problem:

$$\underset{\mathbf{w} \in \mathbb{H}, b \in \mathbb{R}}{\text{minimize}} \quad \frac{1}{2} \|\mathbf{w}\|^2$$

subject to $y_i(< \mathbf{w}, \mathbf{x}_i > +b) \geq 1$, for all $i = 1, \ldots, m$.

The KKT conditions for this optimization problem are:

1. Primal feasibility—we assume there is an optimal solution for this problem to be verified later;
2. Stationarity—we formulate the Lagrangian as follows:

$$\Lambda(\mathbf{w}, b, \alpha) = \frac{1}{2} \|\mathbf{w}\|^2 - \sum_{i=1}^{m} \alpha_i (y_i (< \mathbf{x}_i, \mathbf{w} > +b) - 1),$$

in which $\alpha_i$ is used instead of $\mu_i$ to represent the KKT multipliers, a notation commonly found in the SLT literature;

**Fig. 4.31** Simple
classification task, using
positive and negative
examples

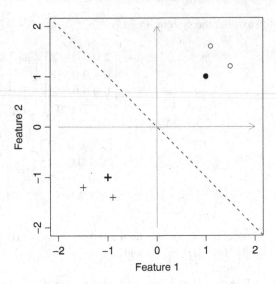

3. Complementary slackness—KKT multipliers times inequality constraints must
   be equal to zero: either because the multiplier makes the constraint equals to
   zero or the constraint is non-binding. To illustrate, consider Fig. 4.31 whose
   support vectors provide enough constraints, thus all other points might be simply
   disconsidered once their constraints are weaker (thus, $\alpha_j = 0$ for every $j$
   associated to those examples);
4. Dual feasibility—All KKT multipliers must be positive for inequality constraints,
   i.e., $\mu_i^* \geq 0$ for all $i = 1, \ldots, m$.

Thus, if all KKT conditions are held, a solution exists. To solve this problem,
the gradient vectors for the objective $f$ and the constraint functions $g$ must be
computed:

$$\nabla f = \alpha \nabla g$$
$$\nabla f - \alpha \nabla g = 0,$$

given:

$$f = \frac{1}{2} \left\| \mathbf{w} \right\|^2,$$

and:

$$g_i = y_i (< \mathbf{x}_i, \mathbf{w} > + b) - 1,$$

for every $i = 1, \ldots, m$. As a consequence of this formulation, we derive our
optimization problem in terms of free variables, which are the ones we can adapt:

$$\frac{\partial \Lambda}{\partial \mathbf{w}} = \mathbf{w} - \sum_{i=1}^{m} \alpha_i y_i \mathbf{x}_i = 0 \Rightarrow \mathbf{w} = \sum_{i=1}^{m} \alpha_i y_i \mathbf{x}_i$$

$$\frac{\partial \Lambda}{\partial b} = -\sum_{i=1}^{m} \alpha_i y_i = 0.$$

Returning to the Lagrangian:

$$\Lambda(\mathbf{w}, b, \alpha) = \frac{1}{2} \|\mathbf{w}\|^2 - \sum_{i=1}^{m} \alpha_i y_i < \mathbf{x}_i, \mathbf{w} > -b \sum_{i=1}^{m} \alpha_i y_i + \sum_{i=1}^{m} \alpha_i,$$

and, then, substituting the terms found through derivatives in order to ensure equality for the gradient vectors (see Sect. 4.5.1):

$$-\sum_{i=1}^{m} \alpha_i y_i = 0,$$

we have:

$$\Lambda(\mathbf{w}, b, \alpha) = \frac{1}{2} \|\mathbf{w}\|^2 - \sum_{i=1}^{m} \alpha_i y_i < \mathbf{x}_i, \mathbf{w} > + \sum_{i=1}^{m} \alpha_i,$$

and plugging term:

$$\mathbf{w} = \sum_{i=1}^{m} \alpha_i y_i \mathbf{x}_i,$$

we find:

$$\Lambda(\mathbf{w}, b, \alpha) = \frac{1}{2} < \sum_{i=1}^{m} \alpha_i y_i \mathbf{x}_i, \sum_{j=1}^{m} \alpha_j y_j \mathbf{x}_j > - \sum_{i=1}^{m} \alpha_i y_i < \mathbf{x}_i, \sum_{j=1}^{m} \alpha_j y_j \mathbf{x}_j >$$

$$+ \sum_{i=1}^{m} \alpha_i.$$

By simplifying the formulation, we finally obtain the Dual form for this optimization problem:

$$\Lambda(\mathbf{w}, b, \alpha) = \frac{1}{2} \sum_{i=1}^{m} \sum_{j=1}^{m} \alpha_i \alpha_j y_i y_j < \mathbf{x}_i, \mathbf{x}_j > . - \sum_{i=1}^{m} \sum_{j=1}^{m} \alpha_i \alpha_j y_i y_j < \mathbf{x}_i, y_j \mathbf{x}_j >$$

$$+ \sum_{i=1}^{m} \alpha_i,$$

$$\Lambda(\mathbf{w}, b, \alpha) = -\frac{1}{2} \sum_{i=1}^{m} \sum_{j=1}^{m} \alpha_i \alpha_j y_i y_j < \mathbf{x}_i, \mathbf{x}_j > + \sum_{i=1}^{m} \alpha_i,$$

addressing the same problem but using different variables. In this situation, observe **w** and $b$ are now represented in terms of $\alpha$. This helps us to deal with only one vectorial variable at a single scale, while **w** and $b$ had different ones. This is only possible because the derivation preserves the optimization problem characteristics. Note that after finding the gradient vectors, we can use every possible Lagrangian simplification to obtain an alternative form to express a target problem. The original optimization problem is referred to as Primal, while this second is called the Dual form. There are particular situations in which the Dual solution does not match the Primal, as discussed in more details in Chap. 5. For now, consider we were told the solution will be approximately the same, what is sufficient to address the SVM optimization.

The Dual form still requires some particular constraints obtained from the Lagrangian derivation. Those are necessary to ensure parallel gradient vectors (see Sect. 4.5.1). So, constraint $-\sum_{i=1}^{m} \alpha_i y_i = 0$ is added, which was cancelled of the objective function while formulating the Dual. By including it, the gradients are as parallel as in the Primal form.

While the original problem is formulated in terms of a minimization, the Dual necessarily employs the opposite goal. We still need to ensure all KKT conditions to make this Dual problem complete. Again, the first comes from the assumption that there is at least one solution for the primal constraints (primal feasibility), the second comes from the Lagrangian derivation (stationarity), the third is guaranteed by adding constraint $-\sum_{i=1}^{m} \alpha_i y_i = 0$. Finally, the fourth KKT condition (Dual feasibility) must be respected, implying $\alpha_i \geq 0$ for all $i = 1, \ldots, m$, so that the complete Dual form is:

$$\underset{\alpha}{\text{maximize}} \ \ W(\alpha) = -\frac{1}{2} \sum_{i=1}^{m} \sum_{j=1}^{m} \alpha_i \alpha_j y_i y_j < \mathbf{x}_i, \mathbf{x}_j > + \sum_{i=1}^{m} \alpha_i$$

$$\text{subject to } \alpha_i \geq 0, \ \text{ for all } i = 1, \ldots, m$$

$$\sum_{i=1}^{m} \alpha_i y_i = 0.$$

It is worth to mention that whenever some Lagrangian derivation implies a substitution in an objective function, the problem is kept the same. However, when the same implies canceling out some term, that derivative must be taken as an additional constraint. In our example, one of the derivatives resultant from the Lagrangian, i.e., $\mathbf{w} = \sum_{i=1}^{m} \alpha_i y_i \mathbf{x}_i$, was not added as constraint given it was substituted and not canceled such as $-\sum_{i=1}^{m} \alpha_i y_i = 0$. Without any loss of generality, also notice this latter was multiplied by $-1$. We also highlight that the objective function was renamed to $W(\alpha)$, given it is a different function and only $\alpha$ is adapted during the optimization process.

Once the Dual form is solved, we can now find $\mathbf{w}$ and $b$, in form:

$$f(\mathbf{x}_+) = <\mathbf{w}, \mathbf{x}_+> +b \geq 1$$
$$f(\mathbf{x}_-) = <\mathbf{w}, \mathbf{x}_-> +b \leq 1,$$

having $\mathbf{x}_+$ and $\mathbf{x}_-$ as positive and negative training examples, respectively. From that, the sign function provides labels:

$$f(\mathbf{x}) = \text{sign}(<\mathbf{w}, \mathbf{x}> +b),$$

for any unseen example $\mathbf{x}$. Substituting vector $\mathbf{w}$ by the term found after differentiating the Primal form:

$$f(\mathbf{x}) = \text{sign}\left( \sum_{i=1}^{m} \alpha_i y_i <\mathbf{x}_i, \mathbf{x}> +b \right),$$

and term $b$ found by solving the following system of equations given support vectors $\mathbf{x}_+$ and $\mathbf{x}_-$:

$$\begin{cases} <\mathbf{w}, \mathbf{x}_+> +b = +1 \\ <\mathbf{w}, \mathbf{x}_-> +b = -1 \end{cases}.$$

Adding both equations:

$$<\mathbf{w}, \mathbf{x}_+> +b+ <\mathbf{w}, \mathbf{x}_-> +b = +1 - 1$$

$$<\mathbf{w}, \mathbf{x}_+ + \mathbf{x}_-> +2b = 0$$

$$b = -\frac{1}{2} <\mathbf{w}, \mathbf{x}_+ + \mathbf{x}_->,$$

and substituting $\mathbf{w}$, we finally have:

$$b = -\frac{1}{2} \sum_{i=1}^{m} <\mathbf{x}_i, \mathbf{x}_+ + \mathbf{x}_->.$$

In order to illustrate this Dual form, consider a simple scenario as shown in Fig. 4.32. Trying different values for $\alpha_i$ (two examples, so there are two constraints) such as $\{0.1, 0.25, 0.3, 0.7\}$, one can analyze the outcomes of $W(\alpha)$. Listing 4.4 supports the user to attempt other values for $\alpha_i$.

**Fig. 4.32** Simple
classification task

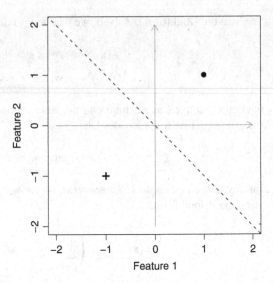

**Listing 4.4** Assessing the objective function $W(\alpha)$

```
find_w_b <- function(alphas, x, y) {

        # Computing vector w
        w = 0
        for (i in 1:length(alphas)) {
                w = w + alphas[i] * y[i] * x[i,]
        }

        # Computing term b
        positive_support_vector = x[y == +1,]
        negative_support_vector = x[y == -1,]
        b = 1/2* ((positive_support_vector
                + negative_support_vector) %*% w)

        cat("Vector_w\n")
        print(w)

        cat("Real_number_b\n")
        print(b)

        ret = list()
        ret$w = w
        ret$b = b

        ret
}

classify <- function(w, b, newX) {
        ret = w %*% newX + b
        ret
}
```

```
x  =  NULL
x  =  rbind(c(+1,  +1))
x  =  rbind(x,  c(-1,  -1))

y  =  c(+1,  -1)

alphas  =  c(0.25,  0.25)

W_of_alpha  =  0
for  (i  in  1:2)  {
        for  (j  in  1:2)  {
                W_of_alpha  =  W_of_alpha
                        + alphas[i]*alphas[j]*y[i]*y[j]*x[i,]%*%x[j
                        ,]
        }
}

result  =  -1/2  *  W_of_alpha  +  sum(alphas)

cat("This_is_the_result_for_W(alpha),_which_we_want_to_maximize!
    \n")
print(result)

find_w_b(alphas,  x,  y)
```

For instance, using $\alpha_1 = \alpha_2 = 0.1$, the objective function $W(\alpha) = 0.16$, vector $\mathbf{w} = \begin{bmatrix} 0.2\ 0.2 \end{bmatrix}^T$, and $b = 0$. Other examples are:

$$\alpha_1 = \alpha_2 = 0.25 \Rightarrow W(\alpha) = 0.25, \mathbf{w} = \begin{bmatrix} 0.5\ 0.5 \end{bmatrix}^T, b = 0$$

$$\alpha_1 = \alpha_2 = 0.3 \Rightarrow W(\alpha) = 0.24, \mathbf{w} = \begin{bmatrix} 0.6\ 0.6 \end{bmatrix}^T, b = 0$$

$$\alpha_1 = \alpha_2 = 0.7 \Rightarrow W(\alpha) = -0.56, \mathbf{w} = \begin{bmatrix} 1.4\ 1.4 \end{bmatrix}^T, b = 0,$$

consequently $\alpha_1 = \alpha_2 = 0.25$ are the best parameters, once they provide the maximum value when compared to others. How can one maximize the objective function $W(\alpha)$, for any $\alpha_i \geq 0$, and still respect all constraints? This answer requires a special optimization algorithm, as discussed in Chap. 5. For now, we suggest the reader to call function *classify()* from Listing 4.4 to test the classification performance for such a simple scenario.

To illustrate, consider the following instances:

**Listing 4.5** Running function classify() from Listing 4.4

```
print(classify(w=c(0.5,  0.5), b=0, newX=c(-5,-5)))
print(classify(w=c(0.5,  0.5), b=0, newX=c(-2,-2)))
print(classify(w=c(0.5,  0.5), b=0, newX=c(-0.1,-0.5)))
print(classify(w=c(0.5,  0.5), b=0, newX=c(2,2)))
print(classify(w=c(0.5,  0.5), b=0, newX=c(5,5)))
print(classify(w=c(0.5,  0.5), b=0, newX=c(0.1,0.5)))
```

whose outputs are, respectively:

**Listing 4.6** Text output produced by Listing 4.5

```
-5
-2
-0.3
2
5
0.3
```

confirming the positive and negative signs indicate the label, i.e., the point position relative to the hyperplane.

It is worth to mention that a kernel can be plugged into the Dual form:

$$\text{maximize}_{\alpha} \ W(\alpha) = -\frac{1}{2} \sum_{i=1}^{m} \sum_{j=1}^{m} \alpha_i \alpha_j y_i y_j k(x_i, x_j) + \sum_{i=1}^{m} \alpha_i$$

$$\text{subject to } \alpha_i \geq 0, \text{ for all } i = 1, \ldots, m$$

$$\sum_{i=1}^{m} \alpha_i y_i = 0,$$

so that the classification result would be:

$$f(x) = \text{sign}\left(\sum_{i=1}^{m} \alpha_i y_i k(x_i, x) + b\right).$$

Observe the kernel function is applied to transform examples $x_i$ and $x_j$ from an input space to some features space, while all the SVM formulation remains the same. There are additional conditions to maintain the same SVM formulation which are discussed in Chap. 5.

## 4.7 Formulating the Soft-Margin SVM Optimization Problem

All formulation provided in the previous section is known as the hard-margin SVM problem, in which training examples are assumed to be linearly separable. However, real-world tasks typically present some degree of class overlap so that a more flexible SVM formulation is required. The soft-margin problem introduces slack variables to relax the assumption of perfect linear separability.

Let the slack variables $\xi_i \geq 0$, and the new problem constraints in form:

$$y_i(< \mathbf{w}, \mathbf{x}_i > +b) \geq 1 - \xi_i,$$

for all $i = 1, \ldots, m$. This new formulation allows constraints to produce values smaller than one, required by overlapping examples that appear close to the hyperplane or even on the opposite side. In other words, slack variables permit positive points to lay on the negative side of the hyperplane and vice-versa. Figure 4.33 illustrates a problem requiring such relaxation.

The greater slack variables are, the greater is the uncertainty region associated with the hyperplane. This is the same as having a wider hyperplane in which every point lying inside it does not have to respect constraints. Figure 4.34 illustrate the impact of such slackness on the soft-margin hyperplane.

The hard-margin SVM does not work for the task illustrated in Fig. 4.34, because constraints cannot be held. In such scenario, the only solution is provided by the soft-margin formulation, having slack variables $\xi_i$ as relaxations for every constraint $i$. Consider the positive point located at the negative side of the hyperplane. The soft-margin output is:

**Fig. 4.33** Classification task involving the class overlapping

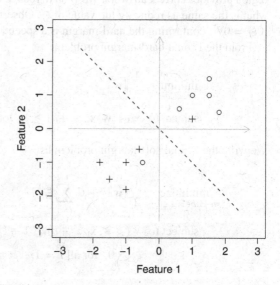

**Fig. 4.34** Classification task
employing the Soft-Margin
Support Vector Machine

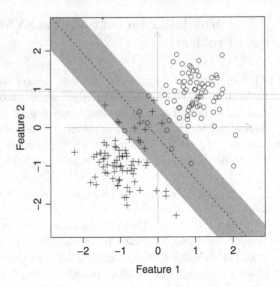

$$y_i(< \mathbf{w}, \mathbf{x}_i > +b) \geq 1 - \xi_i$$

$$+1(< \mathbf{w}, \mathbf{x}_i > +b) \geq 1 - \xi_i.$$

Note the hyperplane should produce a negative value for $< \mathbf{w}, \mathbf{x}_i > +b$, but this
constraint would only be valid for $\xi_i > 1$ with sufficient magnitude to make it true.
The same happens for the negative point at the positive side. This reinforces the idea
that $\xi_i$ works by increasing the thickness of the hyperplane, which is associated to
a region of uncertain classification. Every training example out of that uncertainty
region provides correct answers. We wish to reduce uncertainty as much as possible,
what is the same as reducing the value of $\xi_i$. Observe we have a hard-margin SVM
if $\xi_i = 0 \forall i$, confirming the hard-margin is a special case of the soft-margin SVM.

From the Primal hard-margin problem:

$$\underset{\mathbf{w} \in \mathbb{H}, b \in \mathbb{R}}{\text{minimize}} \quad \frac{1}{2} \|\mathbf{w}\|^2$$

subject to $y_i(< \mathbf{w}, \mathbf{x}_i > +b) \geq 1$, for all $i = 1, \ldots, m$.

We write the Primal soft-margin problem as:

$$\underset{\mathbf{w} \in \mathbb{H}, b \in \mathbb{R}, \xi \in \mathbb{R}_+^m}{\text{minimize}} \quad \frac{1}{2} \|\mathbf{w}\|^2 + C \sum_{i=1}^{m} \xi_i$$

subject to $y_i(< \mathbf{w}, \mathbf{x}_i > +b) \geq 1 - \xi_i$, for all $i = 1, \ldots, m$

$$\xi_i \geq 0, \text{ for all } i = 1, \ldots, m,$$

in which constant $C > 0$ defines the trade-off between margin maximization and the constraints relaxation. This constant is hyper-parameter set by the end-user, which is referred to as *cost* in several libraries [1]. The additional constraint $\xi_i \geq 0$ is necessary to ensure feasibility for the other constraints.

Applying the Karush-Kuhn-Tucker conditions to obtain the Dual form:

1. Primal feasibility—we assume there is an optimal solution for which all constraints are satisfied, which is verified later;
2. Stationarity—there is at least one stationary point for which the gradient of $f$ is parallel to the gradient of constraints $g_i$ and $h_j$:

$$\nabla f(x^*) = \sum_{i=1}^{n} \mu_i^* \nabla g_i(x^*) + \sum_{j=1}^{m} \lambda_j^* \nabla h_j(x^*),$$

in this case, it is written in form:

$$\Lambda(\mathbf{w}, b, \boldsymbol{\xi}, \boldsymbol{\alpha}, \boldsymbol{\beta}, \boldsymbol{\lambda}) = \frac{1}{2} \|\mathbf{w}\|^2 + C \sum_{i=1}^{m} \xi_i - \sum_{i=1}^{m} \alpha_i (y_i (< \mathbf{x}_i, \mathbf{w} > +b) - 1 + \xi_i)$$

$$- \sum_{i=1}^{m} \beta_i \xi_i,$$

having $\sum_{i=1}^{m} \lambda_i \xi_i$ as additional constraint to ensure slack variables are greater than or equal to zero in order to satisfy the main classification constraints, $\alpha_i$ and $\beta_i$ are the KKT multipliers;
3. Complementary slackness—KKT multipliers times inequality constraints must be equal to zero, what must be evaluated later;
4. Dual feasibility—all KKT multipliers must be positive for inequality constraints, i.e., $\alpha_i \geq 0$ and $\beta_i$ for all $i = 1, \ldots, m$. This is also assessed later.

Solving the Lagrangian:

$$\Lambda(\mathbf{w}, b, \boldsymbol{\xi}, \boldsymbol{\alpha}, \boldsymbol{\beta}, \boldsymbol{\lambda}) = \frac{1}{2} \|\mathbf{w}\|^2 + C \sum_{i=1}^{m} \xi_i - \sum_{i=1}^{m} \alpha_i (y_i (< \mathbf{x}_i, \mathbf{w} > +b) - 1 + \xi_i)$$

$$- \sum_{i=1}^{m} \beta_i \xi_i,$$

the derivatives in terms of free variables $\mathbf{w}, b, \boldsymbol{\xi}$ are:

$$\frac{\partial \Lambda}{\partial \mathbf{w}} = \mathbf{w} - \sum_{i=1}^{m} \alpha_i y_i \mathbf{x}_i = 0$$

$$\mathbf{w} = \sum_{i=1}^{m} \alpha_i \, y_i \mathbf{x}_i,$$

then:

$$\frac{\partial \Lambda}{\partial b} = \sum_{i=1}^{m} \alpha_i \, y_i = 0,$$

and, finally:

$$\frac{\partial \Lambda}{\partial \xi_i} = C - \alpha_i - \beta_i = 0, \text{ for all } i = 1, \dots, m.$$

From this last equation, we have:

$$\alpha_i = C - \beta_i, \text{ for all } i = 1, \dots, m.$$

Observe $\alpha_i$ is constrained by $C$ and $\beta_i$. Due to the fourth KKT condition, $\alpha_i \geq 0$ and $\beta_i \geq 0$, so that:

$$0 \leq \alpha_i \leq C - \beta_i,$$

and, as $\beta_i$ can assume zero:

$$0 \leq \alpha_i \leq C, \text{ for all } i = 1, \dots, m.$$

As in the hard-margin SVM problem, we first distribute the Lagrangian terms:

$$\Lambda(\mathbf{w}, b, \boldsymbol{\xi}, \boldsymbol{\alpha}, \boldsymbol{\lambda}) = \frac{1}{2} \|\mathbf{w}\|^2 + C \sum_{i=1}^{m} \xi_i - \sum_{i=1}^{m} \alpha_i (y_i (< \mathbf{x}_i, \mathbf{w} > +b) - 1 + \xi_i)$$

$$- \sum_{i=1}^{m} \beta_i \xi_i,$$

to obtain:

$$\Lambda(\mathbf{w}, b, \boldsymbol{\xi}, \boldsymbol{\alpha}, \boldsymbol{\lambda}) = \frac{1}{2} \|\mathbf{w}\|^2 + C \sum_{i=1}^{m} \xi_i - \sum_{i=1}^{m} \alpha_i y_i < \mathbf{x}_i, \mathbf{w} > - \sum_{i=1}^{m} \alpha_i y_i b$$

$$- \sum_{i=1}^{m} \alpha_i (-1) - \sum_{i=1}^{m} \alpha_i \xi_i - \sum_{i=1}^{m} \beta_i \xi_i,$$

and, then, plugging $\mathbf{w} = \sum_{i=1}^{m} \alpha_i y_i \mathbf{x}_i$:

$$\Lambda(\mathbf{w}, b, \boldsymbol{\xi}, \boldsymbol{\alpha}, \boldsymbol{\lambda}) = \frac{1}{2} \sum_{i=1}^{m} \sum_{j=1}^{m} \alpha_i \alpha_j y_i y_j < \mathbf{x}_i, \mathbf{x}_j > + C \sum_{i=1}^{m} \xi_i$$

$$- \sum_{i=1}^{m} \sum_{j=1}^{m} \alpha_i \alpha_j y_i y_j < \mathbf{x}_i, \mathbf{x}_j >$$

$$- \sum_{i=1}^{m} \alpha_i y_i b - \sum_{i=1}^{m} \alpha_i(-1) - \sum_{i=1}^{m} \alpha_i \xi_i - \sum_{i=1}^{m} \beta_i \xi_i,$$

next, substituting $\sum_{i=1}^{m} \alpha_i y_i = 0$:

$$\Lambda(\mathbf{w}, b, \boldsymbol{\xi}, \boldsymbol{\alpha}, \boldsymbol{\lambda}) = \frac{1}{2} \sum_{i=1}^{m} \sum_{j=1}^{m} \alpha_i \alpha_j y_i y_j < \mathbf{x}_i, \mathbf{x}_j > + C \sum_{i=1}^{m} \xi_i$$

$$- \sum_{i=1}^{m} \sum_{j=1}^{m} \alpha_i \alpha_j y_i y_j < \mathbf{x}_i, \mathbf{x}_j >$$

$$- \sum_{i=1}^{m} \alpha_i(-1) - \sum_{i=1}^{m} \alpha_i \xi_i - \sum_{i=1}^{m} \beta_i \xi_i,$$

and using:

$$\alpha_i = C - \beta_i$$
$$\beta_i = C - \alpha_i,$$

term $\beta_i$ is no longer necessary:

$$\Lambda(\mathbf{w}, b, \boldsymbol{\xi}, \boldsymbol{\alpha}, \boldsymbol{\lambda}) = \frac{1}{2} \sum_{i=1}^{m} \sum_{j=1}^{m} \alpha_i \alpha_j y_i y_j < \mathbf{x}_i, \mathbf{x}_j > + C \sum_{i=1}^{m} \xi_i$$

$$- \sum_{i=1}^{m} \sum_{j=1}^{m} \alpha_i \alpha_j y_i y_j < \mathbf{x}_i, \mathbf{x}_j >$$

$$- \sum_{i=1}^{m} \alpha_i(-1) - \sum_{i=1}^{m} \alpha_i \xi_i - \sum_{i=1}^{m} (C - \alpha_i) \xi_i.$$

We can now manipulate three summation terms of that formulation, as follows:

$$+C\sum_{i=1}^{m}\xi_i - \sum_{i=1}^{m}\alpha_i\xi_i - \sum_{i=1}^{m}(C-\alpha_i)\xi_i =$$

$$+\sum_{i=1}^{m}C\xi_i - \sum_{i=1}^{m}\alpha_i\xi_i - \sum_{i=1}^{m}(C-\alpha_i)\xi_i =$$

$$+\sum_{i=1}^{m}(C-\alpha_i)\xi_i - \sum_{i=1}^{m}(C-\alpha_i)\xi_i = 0,$$

allowing us to reduce it to:

$$\Lambda(\mathbf{w},b,\boldsymbol{\xi},\boldsymbol{\alpha},\boldsymbol{\lambda}) = \frac{1}{2}\sum_{i=1}^{m}\sum_{j=1}^{m}\alpha_i\alpha_j y_i y_j <\mathbf{x}_i,\mathbf{x}_j>$$

$$-\sum_{i=1}^{m}\sum_{j=1}^{m}\alpha_i\alpha_j y_i y_j <\mathbf{x}_i,\mathbf{x}_j> - \sum_{i=1}^{m}\alpha_i(-1),$$

to finally obtain:

$$W(\boldsymbol{\alpha}) = -\frac{1}{2}\sum_{i=1}^{m}\sum_{j=1}^{m}\alpha_i\alpha_j y_i y_j <\mathbf{x}_i,\mathbf{x}_j> + \sum_{i=1}^{m}\alpha_i,$$

which was renamed as $W(\boldsymbol{\alpha})$, given it depends only on $\alpha_i$, for $i = 1,\ldots,m$. This can be seen as a vector $\boldsymbol{\alpha}$ of variables to be found instead of $\mathbf{w}, b, \boldsymbol{\xi}, \alpha$ and $\boldsymbol{\beta}$. This is again the effect of applying the derivation as a way to find the Dual form for the original soft-margin optimization problem.

From all those steps, we write the Dual form by adding constraints:

$$\underset{\alpha}{\text{maximize}}\;\; W(\boldsymbol{\alpha}) = -\frac{1}{2}\sum_{i=1}^{m}\sum_{j=1}^{m}\alpha_i\alpha_j y_i y_j <\mathbf{x}_i,\mathbf{x}_j> + \sum_{i=1}^{m}\alpha_i$$

subject to $0 \le \alpha_i \le C$, for all $i = 1,\ldots,m$,

$$\sum_{i=1}^{m}\alpha_i y_i = 0,$$

in which the constraint $\sum_{i=1}^{m}\alpha_i y_i = 0$ was added because it was cancelled out in the Lagrangian derivation. To remind the reader, the classification process is:

$$f(\mathbf{x}) = \text{sign}\,(<\mathbf{w},\mathbf{x}> +b)$$

$$f(\mathbf{x}) = \text{sign}\left(\sum_{i=1}^{m} \alpha_i \, y_i < \mathbf{x}_i, \mathbf{x} > +b\right),$$

given some unseen example $\mathbf{x}$, in which $b$ is found as:

$$\begin{cases} < \mathbf{w}, \mathbf{x}_+ > +b = +1 - \xi_+ \\ < \mathbf{w}, \mathbf{x}_- > +b = -1 - \xi_- \end{cases},$$

by summing up equations:

$$< \mathbf{w}, \mathbf{x}_+ + \mathbf{x}_- > +2b = -\xi_+ - \xi_-,$$

and, thus:

$$b = -\frac{1}{2}(\xi_+ + \xi_-) - \frac{1}{2} < \mathbf{w}, \mathbf{x}_+ + \mathbf{x}_- >.$$

After all those steps, we have the complete soft-margin SVM optimization problem, which is the most employed in practical scenarios.

## 4.8   Concluding Remarks

In this chapter, we first introduced a simple classification algorithm based on Linear Algebra concepts, supporting the reader to understand why SVM employs the dot product as similarity measurement. Next, we presented both an intuitive and an algebraic view of the SVM problem. Afterwards, Lagrange multipliers and Karush-Kuhn-Tucker conditions were detailed in order to support the formulation of both maximal-margin problems. Next chapter addresses the minimal requirements on optimization concepts, tools and implementations.

## 4.9   List of Exercises

1. Using Lagrange multipliers, solve the following optimization problems:

   (a)

   $$\text{maximize } (x, y) = x^2 y$$

   $$\text{subject to } x + y = 7$$

(b)

$$\text{maximize } f(x, y) = x^2 - y^2$$

$$\text{subject to } y - x^2 = 1$$

(c) Let the price of coffee beans be given by the seed cost $x$ and the labor $y$, as follows: $f(x, y) = x + 3y$. Find the maximum and minimum prices, while respecting constraint $x^2 + y^2 = 1500$.

2. Using the KKT conditions, solve the following problem [2]:

   (a)

$$\text{minimize } f(x, y) = 2x^2 + 3y$$

$$\text{subject to } g_1(x, y) = x^2 + y^2 - 7 \le 0$$

$$g_2(x, y) = x + y - 1 \le 0$$

which must be written in the standard form (do not forget it) before being solved, i.e.:

$$\text{minimize } f(x, y) = 2x^2 + 3y$$

$$\text{subject to } g_1(x, y) = -x^2 - y^2 + 7 \ge 0$$

$$g_2(x, y) = -x - y + 1 \ge 0.$$

# References

1. C.-C. Chang, C.-J. Lin, Libsvm: a library for support vector machines. ACM Trans. Intell. Syst. Technol. **2**(3), 27:1–27:27 (2011)
2. J.D. Hedengren, A.R. Parkinson, KKt conditions with inequality constraints (2013). https://youtu.be/JTTiELgMyuM
3. J.D. Hedengren, R.A. Shishavan, K.M. Powell, T.F. Edgar, Nonlinear modeling, estimation and predictive control in APMonitor. Comput. Chem. Eng. **70**(Manfred Morari Special Issue), 133–148 (2014)
4. W. Karush, Minima of functions of several variables with inequalities as side conditions, Master's thesis, Department of Mathematics, University of Chicago, Chicago, IL, 1939
5. H.W. Kuhn, A.W. Tucker, Nonlinear programming, in *Proceedings of the Second Berkeley Symposium on Mathematical Statistics and Probability, Berkeley, CA* (University of California Press, Berkeley, 1951), pp. 481–492
6. W.C. Schefler, *Statistics: Concepts and Applications* (Benjamin/Cummings Publishing Company, San Francisco, 1988)
7. B. Scholkopf, A.J. Smola, *Learning with Kernels: Support Vector Machines, Regularization, Optimization, and Beyond* (MIT Press, Cambridge, 2001)
8. G. Strang, *Introduction to Linear Algebra* (Wellesley-Cambridge Press, Wellesley, 2009)

# Chapter 5
# In Search for the Optimization Algorithm

## 5.1  Motivation

In this chapter, we provide the necessary foundation for completely design and implement SVM optimization algorithm. The concepts are described so that those can be broadly applied to general-purpose optimization problems.

## 5.2  Introducing Optimization Problems

There are several real-world problems that we wish to minimize cost or maximize profit while respecting certain constraints. For instance, Formula One designers attempt to find the best car aerodynamics to maximize speed constrained to fuel consumption, aircraft manufacturers intend to minimize drag while respecting turbulence thresholds, agricultural producers must decide how to divide farms into different crops in order to maximize profit or reduce costs constrained to the total investment, etc. In all those scenarios, an optimization problem has the following form:

$$\text{minimize/maximize} \quad f_0(\mathbf{x})$$

$$\text{subject to} \quad f_i(x) \leq b_i, \text{ for all } i = 1, \ldots, m,$$

in which:

1. vector $\mathbf{x} = \begin{bmatrix} x_1 \ldots x_n \end{bmatrix}^T$ contains the optimization variables, i.e., the ones that are allowed to be modified in order to provide a solution for the problem;
2. function $f_0 : \mathbb{R}^n \to \mathbb{R}$ is referred to as the objective function;
3. functions $f_i : \mathbb{R}^n \to \mathbb{R}$ are the constraint functions (or simply constraints);
4. constants $b_1, \ldots, b_m$ are the constraint bounds.

© Springer International Publishing AG, part of Springer Nature 2018
R. Fernandes de Mello, M. Antonelli Ponti, *Machine Learning*,
https://doi.org/10.1007/978-3-319-94989-5_5

Given this problem, vector $\mathbf{x}^*$ is referred to as the *optimal* or the *problem solution* if it produces the minimal (or maximal when that is the case) value of the objective function, given all constraints are satisfied. Thus, for a minimization problem:

$$f_0(\mathbf{x}^*) \leq f_0(\mathbf{x}),$$

and, for a maximization problem:

$$f_0(\mathbf{x}^*) \geq f_0(\mathbf{x}),$$

given any feasible value for $\mathbf{x}$. In summary, feasibility means all constraints are satisfied:

$$f_1(\mathbf{x}) \leq b_1, \ldots, f_m(\mathbf{x}) \leq b_m.$$

## 5.3  Main Types of Optimization Problems

Optimization problems are typically organized in:

1. Linear optimization problems;
2. Nonlinear optimization problems:

   (a) Convex problems;
   (b) Quasi-convex problems;
   (c) Non-convex problems.

In Linear optimization problems, every function $f_i$, for $i = 0, \ldots, m$ is linear. This means the objective function $f_0(.)$, and all constraints $f_1, \ldots, f_m$ are linear. For instance, Fig. 5.1 illustrates a linear problem in which constraints are defined in terms of vector $\mathbf{x} = \begin{bmatrix} x_1 & x_2 \end{bmatrix}^T$. This means constraint functions are in $\mathbb{R}^2$ while the objective is mapped in a third dimension, thus the whole problem is in $\mathbb{R}^3$.

In this scenario, we have four linear constraints which are seen as the four straight lines in the plane formed by combinations of variables $x_1$ and $x_2$. For now, we will not detail how constraints were defined in order to take the internal region of this 4-sided polygon as the feasible set. By feasible, we mean that $x_1$ and $x_2$ can only assume values inside this polygon. It is common to illustrate an optimization problem by projecting the constraints into the objective function (whenever possible, i.e., if the problem is in $\mathbb{R}^2$ or $\mathbb{R}^3$) as shown in Fig. 5.2.

Such projection helps us to solve linear optimization problems in a simpler way, only by pointing out the minimum or the maximum value for the objective function. Of course, there are many scenarios in which projection is impossible or too complex to provide any visual inspection, therefore algorithmic tools help us to tackle problems, as seen in Sect. 5.4.

**Fig. 5.1** Example of a linear optimization problem. Constraints depend on variables $x_1$ and $x_2$, therefore such functions are plotted in terms of the space $\mathbb{R}^2$ formed by those two variables. Observe the outputs provided by the objective function are plotted using an additional axis

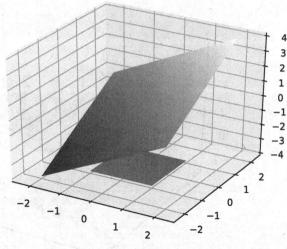

**Fig. 5.2** Projection of the constraints on the objective function in order to support the interpretation of the linear optimization problem from Fig. 5.1

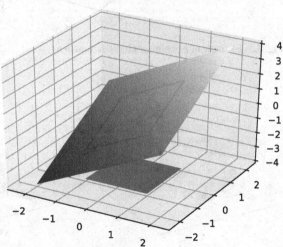

A Nonlinear optimization problem has either the objective function or any of the constraints as nonlinear. Figures 5.3 and 5.4 illustrate a nonlinear objective function (convex), and examples of nonlinear constraints, respectively. A single nonlinear function, either objective or constraint is enough to make it a nonlinear problem.

There is a particular class of nonlinear problems called Convex optimization problems for which there are solutions. In such class, either the objective function is convex or the constraints form a convex set, or both. Figure 5.5 illustrates the situation in which constraints form a convex set for $x_1$ and $x_2$, and the objective function is linear. In such scenario, constraints are nonlinear but convex, simplifying the design of optimization algorithms.

In summary, a feasible region, defined by constraints, is referred to as a convex set when every possible line segment (or affine function) connecting two points

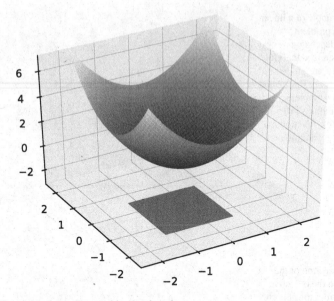

**Fig. 5.3** Example of a nonlinear objective function

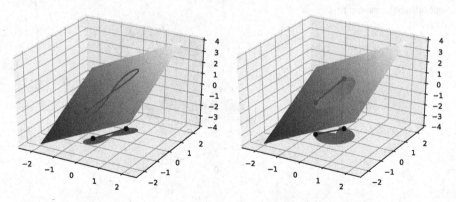

**Fig. 5.4** Two examples of linear optimization problems with nonlinear constraint functions

$A$ and $B$ inside that region produces points also laying down inside the same set in form $(1-\alpha)A+\alpha B$, for $0 \leq \alpha \leq 1$. Figure 5.5 illustrates a line segment connecting two points of the feasible region. Observe that all line points are inside the same set.

However, to solve this class of problems, we need complementary methods such as the Lagrange multipliers and the Karush-Kuhn-Tucker conditions [1, 2, 13, 14]. In addition, when the objective function is convex, we need interior point methods. All those are discussed in Sect. 5.5.

On the other hand, we could have a Nonlinear and Nonconvex optimization problem, as depicted in Fig. 5.4. Besides the objective function is linear, notice the constraints form a nonconvex set for the feasible region. To confirm the

**Fig. 5.5** Example of a
convex optimization problem.
While the objective function
is linear, the constraints form
a convex set

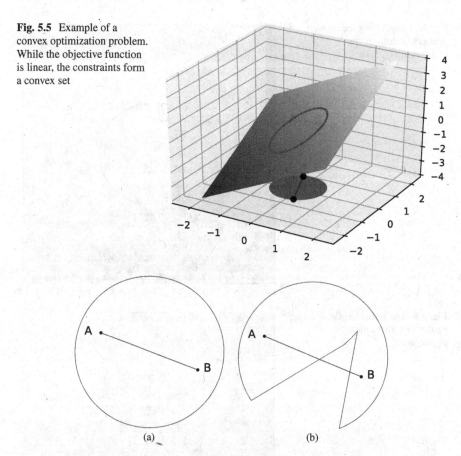

**Fig. 5.6** Illustration of a convex versus a nonconvex set. (**a**) Convex, (**b**) non-convex

nonconvexity of the feasible set, we suggest the reader to trace a line segment
connecting two points $A$ and $B$ as shown. If at least one of those line segments
has at least one point outside the region, then it is nonconvex. To better illustrate
convexity and nonconvexity of sets, see Fig. 5.6.

Nonlinear and Nonconvex optimization problems are more difficult to tackle.
Attempt to picture an algorithm that "walks" on the feasible region considering the
gradient of the linear objective function shown in Fig. 5.4. Figure 5.7 shows the
gradients, but using vectors in a bidimensional plane to simplify understanding.
Consider the starting point $A$ and that one wishes to maximize such objective
function, in that situation the algorithm should "walk" according to the gradient
vectors and cross the boundary for the feasible set, what is undesirable once it would
assume invalid solutions (they do not satisfy constraints). A few possible methods
can be used in such scenarios.

There are some methods that avoid crossing the boundaries of the feasible region
by considering something similar to a physical barrier (Fig. 5.8). For instance,

**Fig. 5.7** Illustrating the
gradient vectors for the
problem in Fig. 5.4b

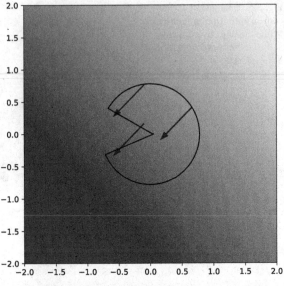

**Fig. 5.8** The impact of the
barrier in optimization
problems

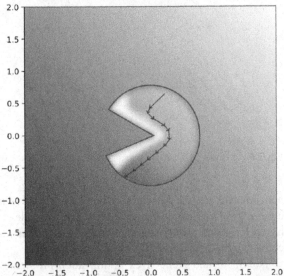

consider the algorithm starts at point *A* but, while moving according to the gradient
vectors, it finds an intransponible barrier (such as the Everest, at least for most
people), then "walks" around it until finding a way to follow the gradient vectors, in
order to finally obtain the optimal solution.

As an alternative, one can also simplify nonconvex sets by creating convex
approximations for the feasible region, as illustrated in Fig. 5.9, making optimization much simpler.

**Fig. 5.9** Relaxing the nonconvex optimization problem by creating some convex approximation

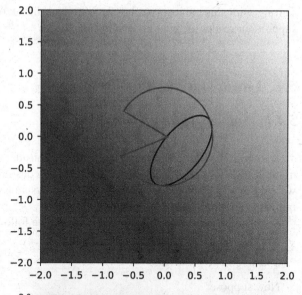

**Fig. 5.10** Example of a grid-based approach to solve a nonconvex optimization problem

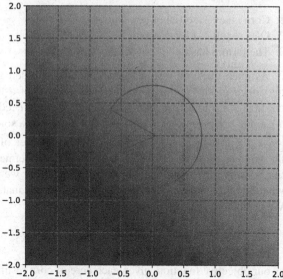

We can also have nonconvex objective functions while having convex feasible sets (see Fig. 5.10). In such situation, we may consider algorithms that attempt to apply cover as much as possible the feasible region and evaluate several candidate points. Then, after finding the best as possible maximum or minimum, it could use gradient vectors to converge to a fair enough candidate solution. Would such solution be the optimal? Not sure.

In this book, we are interested in discussing Convex optimization, because Support Vector Machines correspond to this class of problems. However, before starting with this subject, we briefly introduce linear optimization in attempt to support our reader to better understand all details.

## 5.4   Linear Optimization Problems

### 5.4.1   Solving Through Graphing

To start with the linear optimization problems, consider either a maximization or a minimization problem given linear objective and constraint functions. As first step, we employ the graphical interpretation and analysis of vertex points to obtain the optimal solution.

For example, consider 240 acres of land and the following possibilities [17]:

1. $40/acre if we grow corn, and;
2. $30/acre if we grow oats.

   Now suppose:

1. Corn takes 2 h of labor per acre;
2. Oats requires 1 h per acre;
3. The farmer has only 320 h of labor available.

As main question: "How many acres of each cereal should the farmer grow to maximize profit?". This is a practical example of a linear optimization problem, but first is has to be properly formulated. We first need to define the free variables, a.k.a. the optimization variables. In this scenario, a solution is given by vector $\mathbf{x} = \begin{bmatrix} x_1 & x_2 \end{bmatrix}^T$, in which $x_1$ is associated to the number of acres of corn, and $x_2$ with the number of acres of oats. Next, we need to define the objective function. For that, observe the main question. In this situation, we wish to maximize profit, so what equation (objective function) would be adequate/enough to quantify profit? We decided to use the following:

$$f_0(\mathbf{x}) = 40x_1 + 30x_2,$$

as our objective function, given each acre of corn results in $40 and oats in $30.

As next step, we need to define the constraints, otherwise they could grow up to an infinite number. In fact, variables $x_1$ and $x_2$ obviously accept only positive numbers, because a negative amount would be impossible, thus:

$$x_1 \geq 0, \ x_2 \geq 0,$$

from this we have the first two constraints for our problem. In addition, we cannot grow more than 240 acres, once that is the farm size, so:

$$x_1 + x_2 \leq 240.$$

Finally, the farmer has 320 h of labor available:

$$2x_1 + x_2 \leq 320,$$

given every acre of corn requires 2 h of labor, while oats just one.

The full problem formulation is as follows:

$$\text{maximize} \quad f_0(\mathbf{x}) = 40x_1 + 30x_2$$

$$\text{subject to} \quad x_1 + x_2 \leq 240$$

$$2x_1 + x_2 \leq 320$$

$$x_1 \geq 0, \ x_2 \geq 0,$$

which can be graphed because there are only two optimization variables (two axes) plus a third axis for the objective function, so that problem can be represented in a three-dimensional space. At first, we graph constraint $x_1 + x_2 \leq 240$, as shown in Fig. 5.11. We found this linear function by considering $x_2 = 0$, i.e. without growing oats, providing $x_1 + 0 \geq 240$, so that $(x_1, x_2) = (240, 0) = 240$, and by considering $x_1 = 0$, the coordinate $(x_1, x_2) = (0, 240) = 240$ is found.

After obtaining this linear function, we need to define which of its sides holds the constraint. We basically analyze pairs of values $(x_1, x_2)$ to satisfy it. For instance

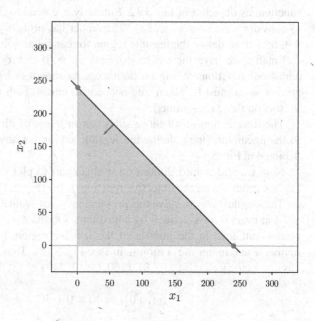

**Fig. 5.11** Graphing constraint $x_1 + x_2 \leq 240$. The gray shaded area represents the feasible region for this constraint. The arrow is a vector indicating the correct linear function side that satisfies the constraint

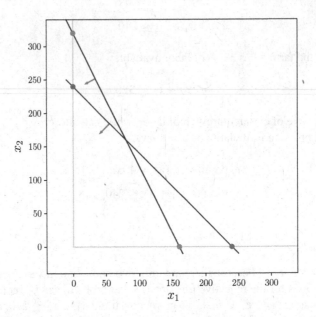

**Fig. 5.12** Graphing constraint $2x_1 + x_2 \leq 320$ with the previous one

$(5, 5) = 10 \leq 240$, what holds it, while $(130, 140) = 270$ does not. We can trace a vector to inform us the correct linear function side, depicted as an arrow in Fig. 5.11.

Now we go for the next constraint $2x_1 + x_2 \leq 320$ and find the second linear function, as depicted in Fig. 5.12. Similarly, we need to determine the correct side to consider values for $x_1$ and $x_2$. The reader has probably already noticed the side will help us to define the feasible region for candidate solutions **x**.

Finally, we have the two constraints $x_1 \geq 0$ and $x_2 \geq 0$, what implies two additional functions laying on both axes, as shown in Fig. 5.13. It is obvious the correct sides must be taken into account to ensure both constraints (see direction vectors on those constraints).

The intersection of all those sides, set in terms of direction vectors associated to the linear functions, define the feasible region for any candidate solution **x**, as depicted in Fig. 5.14.

Now we add a third dimension in this chart to plot the results every pair $\mathbf{x} = \begin{bmatrix} x_1 & x_2 \end{bmatrix}$ produces on the objective function $f_0(\mathbf{x})$ (Fig. 5.15).

The simplest way to solve this problem involves evaluating the objective function $f_0(.)$ at every vertex defined by constraints. Observe there is no need of assessing points contained in the interior of the feasible region, because they will provide neither a maximum nor a minimum (see Fig. 5.15). Thus, we have to assess $f_0(\mathbf{x})$ having vector **x** equals to $(0, 0)$, $(160, 0)$, $(0, 240)$ and $(80, 160)$:

$$f_0(\begin{bmatrix} 0 & 0 \end{bmatrix}) = 40 \times 0 + 30 \times 0 = 0$$

**Fig. 5.13** All constraints plotted for the linear optimization problem involving the farmer's decision, including the $x_1 \geq 0$ and $x_2 \geq 0$, which are indicated by the vectors at the origin

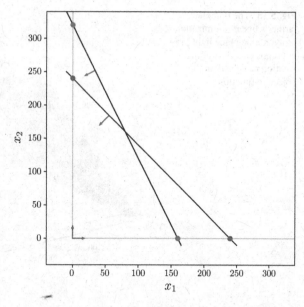

**Fig. 5.14** Feasible region for the farmer's decision

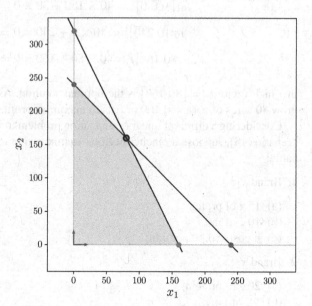

**Fig. 5.15** The complete
farmer's linear optimization
problem: the plane light gray
dot represents the optimal
solution projected on the
objective function

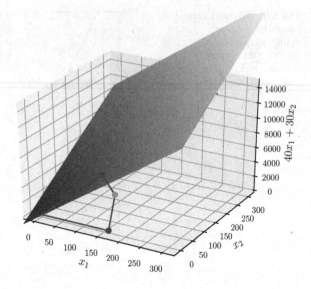

$$f_0([160\ 0]) = 40 \times 160 + 30 \times 0 = 6400$$

$$f_0([0\ 240]) = 40 \times 0 + 240 \times 0 = 7200$$

$$f_0([80\ 160]) = 40 \times 80 + 160 \times 0 = 8000,$$

in which vector $\mathbf{x} = [80\ 160]$ is the optimal solution. As consequence, we should
grow 80 acres of corn and 160 of oats to maximize profit.

Considering a different linear optimization problem to illustrate the minimization
scenario [18], suppose a rancher needing to mix two brands of food to feed his/her
cattle:

1. Brand $x_1$:

   (a) 15 g of protein;
   (b) 10 g of fat;
   (c) it costs $0.80 per unit;

2. Brand $x_2$:

   (a) 20 g of protein;
   (b) 5 g of fat;
   (c) it costs $0.50 per unit.

Consider each serving is required to have at least:

1. 60 g of protein;
2. 30 g of fat.

The main question is: "How much of each brand should he/she use to minimize the total feeding cost?". Observe again the need for formulating the problem. As we already defined, the optimization variables will be $x_1$ and $x_2$, the first associated to the amount of Brand $x_1$, and the second to Brand $x_2$. Then, the objective function must be set:

$$f_0(\mathbf{x}) = 0.80x_1 + 0.50x_2,$$

which is inherently obtained from the cost per unit. The first constraints will certainly be $x_1 \geq 0$ and $x_2 \geq 0$, given negative values are not acceptable to compose the serving. Then, constraints must ensure the necessary amount of protein and fat per serving:

$$15x_1 + 20x_2 \geq 60,$$

and:

$$10x_1 + 5x_2 \geq 30,$$

respectively. Finally, the complete linear optimization problem is:

$$\begin{aligned}
\text{minimize} \quad & f_0(\mathbf{x}) = 0.80x_1 + 0.50x_2 \\
\text{subject to} \quad & 15x_1 + 20x_2 \geq 60 \\
& 10x_1 + 5x_2 \geq 30 \\
& x_1 \geq 0, \; x_2 \geq 0,
\end{aligned}$$

which requires a three-dimensional space representation, having two axes for variables $x_1$ and $x_2$, and an additional for the objective function $f_0(.)$. We then start graphing the constraints in terms of variables $x_1$ and $x_2$ to next add the third dimension. Figure 5.16 illustrates all four constraints.

The feasible region for vector $\mathbf{x} = \begin{bmatrix} x_1 & x_2 \end{bmatrix}$ is depicted in Fig. 5.17. Then, we add a third dimension to represent the values every vector $\mathbf{x}$ produces for the objective function $f_0(\mathbf{x})$ (Fig. 5.18).

Observe the gradient vectors in Fig. 5.18. It is easy to notice that smaller costs are associated to small values for $x_1$ and $x_2$, that is why the analysis of vertices $(4, 0)$, $(0, 6)$ and $(2.4, 1.2)$ allows to find the minimum (optimum). By assessing the objective function at those points, we have:

$$f_0(\begin{bmatrix} 4 & 0 \end{bmatrix}) = 0.80 \times 4 + 0.50 \times 0 = 3.2$$

$$f_0(\begin{bmatrix} 0 & 6 \end{bmatrix}) = 0.80 \times 0 + 0.50 \times 6 = 3$$

$$f_0(\begin{bmatrix} 2.4 & 1.2 \end{bmatrix}) = 0.80 \times 2.4 + 0.50 \times 1.2 = 2.52,$$

**Fig. 5.16** Graphing all four
constraints for the rancher's
decision

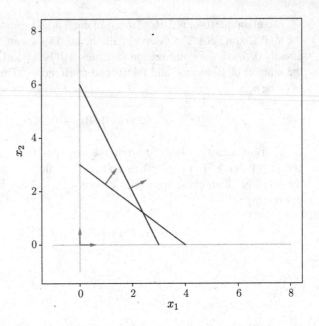

**Fig. 5.17** Feasible region for
the rancher's decision

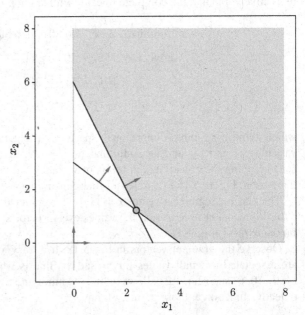

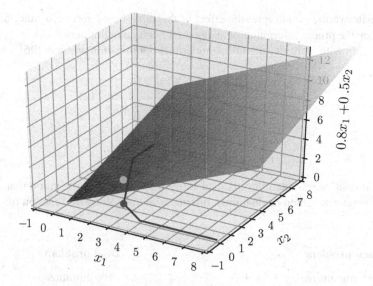

**Fig. 5.18** The complete rancher's linear optimization problem

therefore the optimal solution is given by vector $\mathbf{x} = \begin{bmatrix} 2.4 & 1.2 \end{bmatrix}$, according to which this rancher will mix 2.4 units of Brand $x_1$ with 1.2 unit of Brand $x_2$ to get the minimal cost while satisfying all constraints.

## 5.4.2 Primal and Dual Forms of Linear Problems

We now address how to translate those problems into equivalent forms in attempt to improve modeling and devise simpler algorithms. Here, the original optimization problem is referred to as the *primal* form (or the *primal* problem), which is then translated into the *dual* form (or the *dual* problem). What would it be the benefit of rewriting the original problem into an equivalent form? That usually helps us to simplify the free variables and constraints. Next sections introduce the tools to write the dual forms from the original problems.

### 5.4.2.1 Using the Table and Rules

In this section, a practical approach of building up dual forms is to use a table mapping relationships between the primal and dual problems. This provides a systematic approach to rewrite problems without requiring any additional knowledge. Afterwards, we interpret the effect of this table-based approach[1] and, finally, introduce the proper reformulation using the Lagrange multipliers.

To begin with, consider the following primal optimization problem [26]:

$$\text{maximize } 3x + 6y + 2z$$
$$\text{subject to } 3x + 4y + z \leq 2$$
$$x + 2y + 3z = 10$$
$$y \geq 0,$$

which should be rewritten into its dual form. It is important to mention that if the primal maximizes, then its dual must minimize another objective function (or vice-versa):

**Primal problem**                                                    **Dual problem**

$\quad$ maximize $3x + 6y + 2z$                                    minimize

$\quad$ subject to $3x + 4y + z \leq 2$

$\qquad\qquad x + 2y + 3z = 10$

$\qquad\qquad y \geq 0,$

Then, the upper bounds for all constraints become quantities of the new objective function to be minimized:

**Primal problem**                                                    **Dual problem**

$\quad$ maximize $3x + 6y + 2z$                                    minimize  2      10

$\quad$ subject to $3x + 4y + z \leq 2$

$\qquad\qquad x + 2y + 3z = 10$

$\qquad\qquad y \geq 0,$

Those quantities are now associated with free variables:

**Primal problem**                                                    **Dual problem**

$\quad$ maximize $3x + 6y + 2z$                                    minimize  $2\lambda_1 + 10\lambda_2$

$\quad$ subject to $3x + 4y + z \leq 2$

$\qquad\qquad x + 2y + 3z = 10$

$\qquad\qquad y \geq 0,$

---

[1] More details about this table-based method can be found in [1].

Then, from the primal constraints:

$$3x + 4y + z \leq 2$$
$$x + 2y + 3z = 10,$$

we build up a matrix $A$:

$$\begin{bmatrix} 3 & 4 & 1 \\ 1 & 2 & 3 \end{bmatrix},$$

which, once transposed, provide the quantities for the constraints of the dual form:

**Primal problem**

maximize $3x + 6y + 2z$

subject to $3x + 4y + z \leq 2$

$x + 2y + 3z = 10$

$y \geq 0,$

**Dual problem**

minimize $2\lambda_1 + 10\lambda_2$

subject to $3 \quad 1$

$4 \quad 2$

$1 \quad 3$

in which the column order is associated with each new variable of the dual form:

**Primal problem**

maximize $3x + 6y + 2z$

subject to $3x + 4y + z \leq 2$

$x + 2y + 3z = 10$

$y \geq 0,$

**Dual problem**

minimize $2\lambda_1 + 10\lambda_2$

subject to $3\lambda_1 \quad 1\lambda_2$

$4\lambda_1 \quad 2\lambda_2$

$1\lambda_1 \quad 3\lambda_2$

Next, all constraint terms are summed and bounded by the quantities of the primal objective function:

**Primal problem**

maximize $3x + 6y + 2z$

subject to $3x + 4y + z \leq 2$

$x + 2y + 3z = 10$

$y \geq 0,$

**Dual problem**

minimize $2\lambda_1 + 10\lambda_2$

subject to $3\lambda_1 + 1\lambda_2 \quad 3$

$4\lambda_1 + 2\lambda_2 \quad 6$

$1\lambda_1 + 3\lambda_2 \quad 2$

This dual formulation still requires the following: (1) defining bounds for $\lambda_1$ and $\lambda_2$; and (2) setting the relational operators for the new constraints. Having two constraints in the primal, two variables are needed in the dual form: $\lambda_1$ and $\lambda_2$. The relational operator associated with the first primal constraint $\leq$ implies a bound for $\lambda_1$ to be greater or equal to zero. Consequently, the operator $=$ in the second primal constraint implies a bound for $\lambda_2$, as seen in Table 5.1.

**Table 5.1** Conversion between the primal and dual forms

| Variables | Minimization problem | Maximization problem | Constraints |
|---|---|---|---|
|  | $\geq 0$ | $\leq$ |  |
|  | $\leq 0$ | $\geq$ |  |
|  | Unrestricted | $=$ |  |
| Constraints | Minimization problem | Maximization problem | Variables |
|  | $\geq$ | $\geq 0$ |  |
|  | $\leq$ | $\leq 0$ |  |
|  | $=$ | Unrestricted |  |

Analyzing Table 5.1, when the primal is in maximization form and the constraint uses a relational operator $\leq$, then the corresponding dual variable must be bounded as $\geq 0$. While variable $\lambda_2$ will be unbounded, given the constraint in the maximization form uses the relational operator $=$:

**Primal problem**

$$\text{maximize } 3x + 6y + 2z$$
$$\text{subject to } 3x + 4y + z \leq 2$$
$$x + 2y + 3z = 10$$
$$y \geq 0,$$

**Dual problem**

$$\text{minimize } 2\lambda_1 + 10\lambda_2$$
$$\text{subject to } 3\lambda_1 + 1\lambda_2 \quad 3$$
$$4\lambda_1 + 2\lambda_2 \quad 6$$
$$1\lambda_1 + 3\lambda_2 \quad 2$$
$$\lambda_1 \geq 0$$

Now we need to define the relational operators for the dual constraints. Given the constraints in the primal form affect the dual variables, the variables of the primal affect the dual constraints. Observe variables $x$ and $z$ are unbounded, that is why there is no additional constraint for them. By looking at Table 5.1, observe that unrestricted variables in maximization problems correspond to the relational operator $=$ for dual constraints. Then, the first and third constraints must use operator $=$, once the first is associated with $x$, and the third with $z$:

**Primal problem**

$$\text{maximize } 3x + 6y + 2z$$
$$\text{subject to } 3x + 4y + z \leq 2$$
$$x + 2y + 3z = 10$$
$$y \geq 0,$$

**Dual problem**

$$\text{minimize } 2\lambda_1 + 10\lambda_2$$
$$\text{subject to } 3\lambda_1 + 1\lambda_2 = 3$$
$$4\lambda_1 + 2\lambda_2 \quad 6$$
$$1\lambda_1 + 3\lambda_2 = 2$$
$$\lambda_1 \geq 0$$

Let us analyze the impacts of variable $y$ from the primal form. It is bounded as $y \geq 0$ in the maximization problem, affecting the associated dual constraint to use the relational operator $\geq$ (Table 5.1):

**Primal problem**

maximize $3x + 6y + 2z$

subject to $3x + 4y + z \leq 2$

$x + 2y + 3z = 10$

$y \geq 0,$

**Dual problem**

minimize $2\lambda_1 + 10\lambda_2$

subject to $3\lambda_1 + 1\lambda_2 = 3$

$4\lambda_1 + 2\lambda_2 \geq 6$

$1\lambda_1 + 3\lambda_2 = 2$

$\lambda_1 \geq 0$

Then we finally have the dual form for the original primal problem. It is interesting to note that, by solving any of these problems, the same solution is found. This is guaranteed for linear problems, but not necessarily for other problem types. In this case, it is worth to obtain the dual form, because it considers only two variables ($\lambda_1$ and $\lambda_2$) instead of three. This makes the problem simpler and, in addition, allows to solve it by graphing. In the next section, we find again the dual form for another problem and, then, analyze the solutions found using both forms.

### 5.4.2.2 Graphical Interpretation of Primal and Dual Forms

We start formulating the dual problem for another linear optimization problem and, then, provide a graphical interpretation for the solutions in both forms. Thus, consider the following primal problem:

$$\text{maximize } 6x + 4y$$

$$\text{subject to } x + y \leq 2$$

$$2x - y \leq 2$$

$$x \geq 0, y \geq 0,$$

The dual form will be a minimization problem, and the bounds for the primal constraints will define the quantities associated with the dual variables:

**Primal problem**

maximize $6x + 4y$

subject to $x + y \leq 2$

$2x - y \leq 2$

$x \geq 0, y \geq 0,$

**Dual problem**

minimize $2 \quad 2$

Now we define the two new variables $\lambda_1$ and $\lambda_2$, and sum both quantities to compose the dual objective function:

**Primal problem**                                    **Dual problem**

maximize $6x + 4y$                                     minimize $2\lambda_1 + 2\lambda_2$

subject to $x + y \leq 2$

$2x - y \leq 2$

$x \geq 0, y \geq 0,$

Next, quantities associated with the primal constraints, i.e.:

$$x + y \leq 2$$
$$2x - y \leq 2,$$

are used to build up a matrix $A$:

$$\begin{bmatrix} 1 & 1 \\ 2 & -1 \end{bmatrix}.$$

and, by transposing $A$, we find the quantities for the dual constraints:

**Primal problem**                                    **Dual problem**

maximize $6x + 4y$                                     minimize $2\lambda_1 + 2\lambda_2$

subject to $x + y \leq 2$                               subject to $1 \quad 2$

$2x - y \leq 2$                                            $1 \quad -1$

$x \geq 0, y \geq 0,$

Afterwards, the bounds for dual variables are defined:

**Primal problem**                                    **Dual problem**

maximize $6x + 4y$                                     minimize $2\lambda_1 + 2\lambda_2$                   and

subject to $x + y \leq 2$                               subject to $1\lambda_1 + 2\lambda_2$

$2x - y \leq 2$                                            $1\lambda_1 - 1\lambda_2,$

$x \geq 0, y \geq 0,$

the quantities of the primal objective function are used to set the bounds for the constraints:

**Primal problem**

maximize $6x + 4y$

subject to $x + y \leq 2$

$2x - y \leq 2$

$x \geq 0, y \geq 0,$

**Dual problem**

minimize $2\lambda_1 + 2\lambda_2$

subject to $1\lambda_1 + 2\lambda_2 \quad 6$

$1\lambda_1 - 1\lambda_2 \quad 4.$

Next, we look at Table 5.1 to define constraints for variables $\lambda_1$ and $\lambda_2$. The first primal constraint is responsible for restricting $\lambda_1$, and the second for $\lambda_2$. Observe both primal constraints use the relational operator $\leq$, thus $\lambda_1 \geq 0$ and $\lambda_2 \geq 0$:

**Primal problem**

maximize $6x + 4y$

subject to $x + y \leq 2$

$2x - y \leq 2$

$x \geq 0, y \geq 0,$

**Dual problem**

minimize $2\lambda_1 + 2\lambda_2$

subject to $1\lambda_1 + 2\lambda_2 \quad 6$

$1\lambda_1 - 1\lambda_2 \quad 4$

$\lambda_1 \geq 0, \lambda_2 \geq 0.$

Finally, we analyze bounds for $x$ and $y$ to define the relational operators for the dual constraints. Both $x$ and $y$ must be greater or equal to zero, thus from Table 5.1, the relational operator $\geq$ is then set for both constraints:

**Primal problem**

maximize $6x + 4y$

subject to $x + y \leq 2$

$2x - y \leq 2$

$x \geq 0, y \geq 0,$

**Dual problem**

minimize $2\lambda_1 + 2\lambda_2$

subject to $1\lambda_1 + 2\lambda_2 \geq 6$

$1\lambda_1 - 1\lambda_2 \geq 4$

$\lambda_1 \geq 0, \lambda_2 \geq 0,$

which is the final dual problem formulation.

Let us analyze both forms by graphing solutions. Figure 5.19 illustrates the constraints in a space with $x$ and $y$ as defined in the primal problem.

The corners of the feasible region provide the solution:

$$\text{Given}(0, 0) \rightarrow 6x + 4y = 6(0) + 4(0) = 0$$

$$\text{Given}(1, 0) \rightarrow 6x + 4y = 6(1) + 4(0) = 6$$

$$\text{Given}(0, 2) \rightarrow 6x + 4y = 6(0) + 4(2) = 8$$

$$\text{Given}\left(\frac{4}{3}, \frac{2}{3}\right) \rightarrow 6x + 4y = 6\left(\frac{4}{3}\right) + 4\left(\frac{2}{3}\right) = \frac{32}{3} \approx 10.66,$$

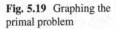
**Fig. 5.19** Graphing the primal problem

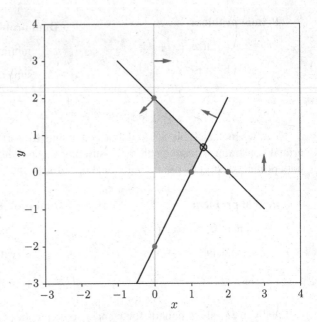

consequently, the solution for this maximization primal problem is at $x = \frac{4}{3}$ and $y = \frac{2}{3}$, for which the objective function produces $\frac{32}{3} \approx 10.66$.

Figure 5.20 illustrates the dual problem, in which two corners must be evaluated:

$$\text{Given}(6, 0) \rightarrow 2(6) + 2(0) = 12$$

$$\text{Given}\left(\frac{14}{3}, \frac{2}{3}\right) \rightarrow 2\left(\frac{14}{3}\right) + 2\left(\frac{2}{3}\right) = \frac{32}{3} \approx 10.66,$$

allowing us to conclude that both primal and dual forms provide the same solution.

While the primal problem is a maximization under a constrained space spanned by $x$ and $y$, the dual minimizes under a *different* space defined by $\lambda_1$ and $\lambda_2$. The outcome provided by both objective functions is the same, however the optimization process adapts different free variables while looking for the solution.

In the particular scenario of linear optimization problems, the primal and dual objective functions always provide the same result, what is referred to as *Strong Duality*. On the other hand, there is a class of nonlinear problems for which the outcomes of the objective functions are different, meaning we have a *Weak Duality*. When the duality is weak for a primal maximization (minimization) problem, the objective function will provide an outcome that is less (greater) than the one produced by its dual form. As a consequence, when solving the primal and the dual forms simultaneously, we can analyze whether the final outcomes are equal or not. When equal, the problems are completely equivalent. When different, it is possible to measure the *gap* between them (how distant one is from the other).

**Fig. 5.20** Graphing the dual problem

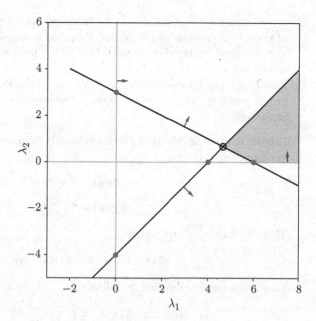

### 5.4.2.3   Using Lagrange Multipliers

The previous section illustrated a table-based method to formulate dual problems. Such a table is build up using Lagrange multipliers. Now Lagrange multipliers are employed to analytically formulate the dual problem without needing a mapping table.

For example, consider the following primal problem:

$$\text{maximize }\ F(x)$$
$$\text{subject to }\ g(x) \le b,$$

from which the dual is found by solving the following Lagrangian:

$$\text{minimize }\ L(x, \lambda) = F(x) - \lambda\,[g(x) - b]$$
$$\text{subject to }\ \frac{\partial L}{\partial x} = 0$$
$$\lambda \ge 0$$
$$x \in D,$$

whose terms can be described as:

1. first constraint $\frac{\partial L}{\partial x} = 0$ is required to make the gradient vector of $F(x)$ parallel to the gradient vector of constraint $g(x) - b$ (this is detailed in Sect. 4.5.1);

2. second constraint $\lambda \geq 0$ is due to the relational operator $\leq$ used in the constraint of the primal problem as previously discussed in last section (see Table 5.1). A more detailed explanation is given later in the text, when discussing the KKT conditions;
3. third constraint informs us that $x \in D$, i.e. in the feasible region $D$, for this primal problem. Any $x$ outside $D$ would be invalid by disrespecting the primal constraints.

In order to exercise this formal framework to build up the dual form, consider the following primal problem:

$$\text{maximize } \mathbf{c}^T \mathbf{x}$$

$$\text{subject to } A\mathbf{x} \leq \mathbf{b}.$$

Then, we build the Lagrangian:

$$L(\mathbf{x}, \lambda) = \mathbf{c}^T \mathbf{x} - \lambda (A\mathbf{x} - \mathbf{b}),$$

and use it to formulate the dual, as follows:

$$\text{minimize } L(\mathbf{x}, \lambda) = \mathbf{c}^T \mathbf{x} - \lambda (A\mathbf{x} - \mathbf{b})$$

$$\text{subject to } \frac{\partial L}{\partial \mathbf{x}} = 0$$

$$\lambda \geq 0$$

$$x \in D.$$

We now solve constraint $\frac{\partial L}{\partial \mathbf{x}} = 0$:

$$\frac{\partial L}{\partial \mathbf{x}} = \frac{\partial}{\partial \mathbf{x}} \mathbf{c}^T \mathbf{x} - \lambda (A\mathbf{x} - \mathbf{b})$$

$$= \mathbf{c} - A^T \lambda = 0$$

$$A^T \lambda = \mathbf{c}.$$

Since this formulation respects the parallel gradients, any substitution resultant of this last derivative provides the same stationary solutions. Therefore, we can take advantage of using this equation to make substitutions in the original Lagrangian. So, from:

$$L(\mathbf{x}, \lambda) = \mathbf{c}^T \mathbf{x} - \lambda (A\mathbf{x} - \mathbf{b}),$$

we obtain:

$$L(\mathbf{x}, \lambda) = \mathbf{c}^T \mathbf{x} - A^T \lambda \mathbf{x} + \lambda \mathbf{b},$$

and given that $A^T\lambda = \mathbf{c}$, then:

$$L(\mathbf{x}, \lambda) = \mathbf{c}^T\mathbf{x} - \mathbf{c}^T\mathbf{x} + \lambda\mathbf{b}$$
$$= \lambda\mathbf{b},$$

finally, the dual problem will be:

$$\text{minimize } L(\mathbf{x}, \lambda) = \lambda\mathbf{b}$$
$$\text{subject to } A^T\lambda = \mathbf{c}$$
$$\lambda \geq 0,$$

then observe we disconsidered $x \in D$ because the whole problem is no longer represented in terms of $x$. In fact, after solving the dual problem, we can also find $x$.

To complement, we suggest the reader to observe that $\mathbf{c}$, associated with the primal objective function, defines the upper bounds for the dual constraints, justifying part of the table-based method discussed in the previous section. Matrix $A$, that defines the quantities associated with primal constraints, is now transposed in the dual. Moreover, pay attention in the new dual variable $\lambda$, which corresponds to the Lagrange multipliers, ensuring gradient vectors are parallel. Finally, $\lambda$ is required to be greater than or equal to zero ($\geq 0$), due to the relational operator $\leq$ found in the constraint of the primal problem.[2]

Let the primal form be:

$$\text{minimize } \lambda\mathbf{b}$$
$$\text{subject to } A^T\lambda = \mathbf{c}$$
$$\lambda \geq 0.$$

Say the corresponding dual form is:

$$\text{maximize } \mathbf{c}^T\mathbf{x}$$
$$\text{subject to } A\mathbf{x} \leq \mathbf{b}.$$

In order to confirm this is indeed the correct formulation, we must first build up the Lagrangian for the primal problem:

$$L(\lambda, \mathbf{x}) = \lambda\mathbf{b} - \mathbf{x}\left(A^T\lambda - \mathbf{c}\right),$$

---

[2]We will not detail this information in here, but discuss it later in a convex optimization scenario. We suggest [1] as a more detailed introduction to linear optimization problems.

in which the Lagrange multiplier is $\mathbf{x}$, only because that will help us to obtain the same problem form as before.

Solving the Lagrangian to ensure parallel gradient vectors:

$$\frac{\partial L}{\partial \lambda} = \frac{\partial}{\partial \lambda} \lambda \mathbf{b} - \mathbf{x} \left( A^T \lambda - \mathbf{c} \right)$$

$$= \mathbf{b} - A\mathbf{x} = 0$$

$$= A\mathbf{x} = b,$$

thus, substituting this term in the Lagrangian:

$$L(\lambda, \mathbf{x}) = \lambda \mathbf{b} - \mathbf{x} \left( A^T \lambda - \mathbf{c} \right)$$

$$= \lambda \mathbf{b} - A\mathbf{x}\lambda + \mathbf{c}^T \mathbf{x}$$

$$= \lambda \mathbf{b} - \lambda \mathbf{b} + \mathbf{c}^T \mathbf{x}$$

$$= \mathbf{c}^T \mathbf{x}.$$

Then, the dual form is:

$$\text{maximize } \mathbf{c}^T \mathbf{x},$$

and the constraints are obtained from $A\mathbf{x} = b$. By considering the same concepts of the previous section (see Table 5.1 for more information), minimization problems whose variables are bounded in form $\geq 0$ require the relational operator $\leq$ in the dual constraints:

$$\text{maximize } \mathbf{c}^T \mathbf{x}$$

$$\text{subject to } A\mathbf{x} \leq \mathbf{b}.$$

We should also analyze the eventual constraints that $\mathbf{x}$ will take in the dual form. From Table 5.1, we see that constraints in the primal form of a minimization problem using the relational operator $=$ define an unbounded variable in the dual. Therefore, there is no need to add any other constraint for $\mathbf{x}$. This finally confirms the dual problem is found after applying the Lagrangian on the primal. The reader might not be completely satisfied with the usage of Table 5.1 to represent relational operators, however that will be better explained analyzing the KKT conditions for convex optimization problems.

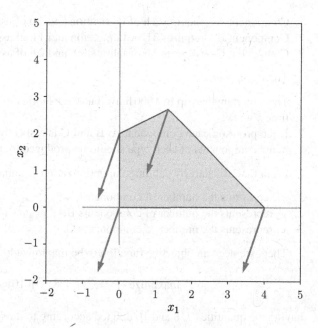

**Fig. 5.21** Illustrating the Simplex algorithm

### 5.4.3 Using an Algorithmic Approach to Solve Linear Problems

We now provide an algorithmical view on how to solve linear problems using the Simplex algorithm. This section does not intend to explain all Simplex forms and details, but solely an introduction[3] to solve linear problems in maximization form given constraints are defined using $\leq$, or in minimization form provided constraints use $\geq$.

Intuitively, the Simplex algorithm is an approach to visit the corners of the feasible region in order to solve the linear optimization problem. At every corner, it computes the objective function and decides which vertex provides the solution.

To illustrate, consider Fig. 5.21 that defines a linear optimization problem with four constraints. The feasible region is the set for which all constraints are held. The gradient of the objective function is traced using directional vectors, supporting the visual inspection of the maximal and the minimal corners. Simplex starts at any corner and "walks" in the direction of the gradient vectors to reach another candidate solution. More specifically, it moves to another corner while the gradient vector is increased given a maximization problem, or decreased when a minimization is considered.

For example, suppose a company manufacturing different car parts [19]:

---

[3]More details in [1].

1. Component A—requires 2 h of fabrication and 1 h of assembly;
2. Component B—requires 3 h of fabrication and 1 h of assembly;
3. Component C—requires 2 h of fabrication and 2 h of assembly.

Then, consider:

1. This company has up to 1000 h available for fabrication, and 800 h of assembly time a week;
2. If the profit on each component A, B and C is respectively $7, $8 and $10, how many components of each type should be produced to maximize profit?

From that, we start by defining the optimization variables as:

1. $x_1$ represents the number of components A;
2. $x_2$ represents the number of components B;
3. $x_3$ represents the number of components C.

Then, we need an objective function to be maximized:

$$\text{maximize}\quad P = 7x_1 + 8x_2 + 10x_3,$$

having the quantities 7, 8 and 10 defined according to the individual profit provided by each component A, B and C, respectively.

Next, we need to set the problem constraints. There are three associated with the variables:

$$\text{maximize}\, P = 7x_1 + 8x_2 + 10x_3$$

$$\text{subject to}\quad x_1, x_2, x_3 \geq 0,$$

due to one cannot manufacture a negative number of components.

Another constraint is necessary to bound the fabrication time. Remind the company has up to 1000 h for fabrication, so:

$$2x_1 + 3x_2 + 2x_3 \leq 1000,$$

what is a consequence of the hours to fabricate components A, B and C, respectively.

Finally, the next constraint bounds the available assembly time, as follows:

$$x_1 + x_2 + 2x_3 \leq 800,$$

which is resultant of the assembly time spent on each component A, B and C, respectively.

Putting all pieces together, the linear optimization problem is:

$$\text{maximize}\quad P = 7x_1 + 8x_2 + 10x_3$$

$$\text{subject to}\ 2x_1 + 3x_2 + 2x_3 \leq 1000$$

$$x_1 + x_2 + 2x_3 \leq 800$$

$$x_1, x_2, x_3 \geq 0.$$

To apply the Simplex algorithm, we must transform our formulation to the standard form. We start by taking the objective function:

$$P = 7x_1 + 8x_2 + 10x_3,$$

and making it equal to zero:

$$-7x_1 - 8x_2 - 10x_3 + P = 0.$$

Then, the *slack variables* $s_1$ and $s_2$ are introduced to convert the inequality constraints to equalities:

$$2x_1 + 3x_2 + 2x_3 + s_1 = 1000$$

$$x_1 + x_2 + 2x_3 + s_2 = 800,$$

and observe $s_1, s_2 \geq 0$ to ensure such inequalities, i.e., they must sum up in order to be equal to 1000 and 800. This standard transformation makes the whole problem easier to be addressed [1].

We now have the following system of equations:

$$-7x_1 - 8x_2 - 10x_3 + P = 0$$

$$2x_1 + 3x_2 + 2x_3 + s_1 = 1000$$

$$x_1 + x_2 + 2x_3 + s_2 = 800,$$

which is used to build up the Simplex Tableau. Then, labeling our variables as columns:

| $x_1$ | $x_2$ | $x_3$ | $s_1$ | $s_2$ | $P$ | Results |
|-------|-------|-------|-------|-------|-----|---------|

and adding the quantities associated to the new equality constraints:

| $x_1$ | $x_2$ | $x_3$ | $s_1$ | $s_2$ | $P$ | Results |
|-------|-------|-------|-------|-------|-----|---------|
| 2 | 3 | 2 | 1 | 0 | 0 | 1000 |
| 1 | 1 | 2 | 0 | 1 | 0 | 800 |

and, finally, the objective function is set as equal to zero (after a solid line to make it more evident):

| $x_1$ | $x_2$ | $x_3$ | $s_1$ | $s_2$ | $P$ | Results |
|---|---|---|---|---|---|---|
| 2 | 3 | 2 | 1 | 0 | 0 | 1000 |
| 1 | 1 | 2 | 0 | 1 | 0 | 800 |
| $-7$ | $-8$ | $-10$ | 0 | 0 | 1 | 0 |

From that, we start executing the Simplex steps. Initially, we label the two first Tableau rows with the slack variables, and the last with variable $P$:

| Basis | $x_1$ | $x_2$ | $x_3$ | $s_1$ | $s_2$ | $P$ | Results |
|---|---|---|---|---|---|---|---|
| $s_1$ | 2 | 3 | 2 | 1 | 0 | 0 | 1000 |
| $s_2$ | 1 | 1 | 2 | 0 | 1 | 0 | 800 |
| $P$ | $-7$ | $-8$ | $-10$ | 0 | 0 | 1 | 0 |

what is the same as defining the *basis* for a given space in Linear Algebra. This means only a vector composed of $s_1$ and $s_2$ is currently considered to solve this problem. This is the same as taking only the submatrix with the rows and columns corresponding to the slack variables (current basis):

| Basis | $x_1$ | $x_2$ | $x_3$ | $s_1$ | $s_2$ | $P$ | Results |
|---|---|---|---|---|---|---|---|
| $s_1$ | 2 | 3 | 2 | **1** | **0** | 0 | 1000 |
| $s_2$ | 1 | 1 | 2 | **0** | **1** | 0 | 800 |
| $P$ | $-7$ | $-8$ | $-10$ | 0 | 0 | 1 | 0 |

Being the same as:

$$B = \begin{bmatrix} 1 & 0 \\ 0 & 1 \end{bmatrix},$$

which currently defines the solution basis in terms of Linear Algebra.[4]

Given basis $B$, we must solve the following system of equations:

$$Bx = b,$$

in which:

$$b = \begin{bmatrix} 1000 \\ 800 \end{bmatrix},$$

---

[4]More details in [23].

to obtain $\mathbf{x}$:

$$Bx = b$$

$$x = B^{-1}b$$

$$x = \begin{bmatrix} s_1 \\ s_2 \end{bmatrix} = \begin{bmatrix} 1 & 0 \\ 0 & 1 \end{bmatrix} \begin{bmatrix} 1000 \\ 800 \end{bmatrix} = \begin{bmatrix} 1000 \\ 800 \end{bmatrix}.$$

Considering this basis $B$, slack variables are $s_1 = 1000$ and $s_2 = 800$. Consequently, no component is fabricated $x_1 = x_2 = x_3 = 0$, yet the constraints are held, what is not a reasonable solution. Observe the Simplex must find the best basis $B$ that maximizes the profit by adapting the Tableau and bringing other variables to form a next candidate basis. Thus, as two terms compose our current basis, only two components will be fabricated. If another constraint is included, it would be possible to define three variables.

As in Linear Algebra, the best basis is found by using the *row-reduced echelon form*, what depends on a pivot term. The most relevant term to be corrected is made evident at the last row, which is associated with the objective function. The variable that better supports the optimization criterion is the one having the most negative value (in bold):

| Basis | $x_1$ | $x_2$ | $x_3$ | $s_1$ | $s_2$ | $P$ | Results |
|---|---|---|---|---|---|---|---|
| $s_1$ | 2 | 3 | 2 | 1 | 0 | 0 | 1000 |
| $s_2$ | 1 | 1 | 2 | 0 | 1 | 0 | 800 |
| $P$ | $-7$ | $-8$ | $\mathbf{-10}$ | 0 | 0 | 1 | 0 |

defining the *pivot column*. This column informs us that variable $x_3$ must be considered at this point as part of our next basis $B$. However, one of the slack variables must provide room for $x_3$ (see row labels). In order to take such decision, column "Results" is divided by the corresponding values (values for $x_3$) in the pivot column:

| Basis | $x_1$ | $x_2$ | $x_3$ | $s_1$ | $s_2$ | $P$ | Results |
|---|---|---|---|---|---|---|---|
| $s_1$ | 2 | 3 | **2** | 1 | 0 | 0 | **1000** |
| $s_2$ | 1 | 1 | **2** | 0 | 1 | 0 | **800** |
| $P$ | $-7$ | $-8$ | $\mathbf{-10}$ | 0 | 0 | 1 | 0 |

thus:

$$\frac{1000}{2} = 500$$

$$\frac{800}{2} = 400,$$

allowing to find the *pivot row* as the one producing the smallest value, i.e. 400, which is associated with the smallest as possible gradient correction, however still respecting constraints. Therefore, the row associated to variable $s_2$ is the *pivot row*:

| Basis | $x_1$ | $x_2$ | $x_3$ | $s_1$ | $s_2$ | $P$ | Results |
|-------|-------|-------|-------|-------|-------|-----|---------|
| $s_1$ | 2 | 3 | 2 | 1 | 0 | 0 | 1000 |
| $s_2$ | 1 | 1 | 2 | 0 | 1 | 0 | 800 |
| $P$ | −7 | −8 | −10 | 0 | 0 | 1 | 0 |

Having both the pivot row and column, the **pivot term** is (in bold):

| Basis | $x_1$ | $x_2$ | $x_3$ | $s_1$ | $s_2$ | $P$ | Results |
|-------|-------|-------|-------|-------|-------|-----|---------|
| $s_1$ | 2 | 3 | 2 | 1 | 0 | 0 | 1000 |
| $s_2$ | 1 | 1 | **2** | 0 | 1 | 0 | 800 |
| $P$ | −7 | −8 | −10 | 0 | 0 | 1 | 0 |

Next, the row-reduced echelon form must be used to make the pivot term equals to one, requiring row operations. Firstly, we multiply the second row by $\frac{1}{2}$ (using the following notation: $R_2 = \frac{1}{2}R_2$) to obtain (such as when solving a linear system):

| Basis | $x_1$ | $x_2$ | $x_3$ | $s_1$ | $s_2$ | $P$ | Results |
|-------|-------|-------|-------|-------|-------|-----|---------|
| $s_1$ | 2 | 3 | 2 | 1 | 0 | 0 | 1000 |
| $s_2$ | $\frac{1}{2}$ | $\frac{1}{2}$ | 1 | 0 | $\frac{1}{2}$ | 0 | 400 |
| $P$ | −7 | −8 | −10 | 0 | 0 | 1 | 0 |

Secondly, row operations are used to make the other column terms equal to zero, therefore we operate on the first and third rows as follows ($R$ means row and the index corresponds to the row number):

$$R_1 = R_1 - 2R_2$$

$$R_3 = R_3 + 10R_2$$

to obtain:

| Basis | $x_1$ | $x_2$ | $x_3$ | $s_1$ | $s_2$ | $P$ | Results |
|-------|-------|-------|-------|-------|-------|-----|---------|
| $s_1$ | 1 | 2 | 0 | 1 | −1 | 0 | 200 |
| $s_2$ | $\frac{1}{2}$ | $\frac{1}{2}$ | 1 | 0 | $\frac{1}{2}$ | 0 | 400 |
| $P$ | −2 | −3 | 0 | 0 | 5 | 1 | 4000 |

Thirdly, we substitute the row variable $R_2$ ($s_2$) by the pivot column variable $x_3$ to compose the next basis:

| Basis | $x_1$ | $x_2$ | $x_3$ | $s_1$ | $s_2$ | $P$ | Results |
|-------|-------|-------|-------|-------|-------|-----|---------|
| $s_1$ | 1 | 2 | 0 | 1 | $-1$ | 0 | 200 |
| $x_3$ | $\frac{1}{2}$ | $\frac{1}{2}$ | 1 | 0 | $\frac{1}{2}$ | 0 | 400 |
| $P$ | $-2$ | $-3$ | 0 | 0 | 5 | 1 | 4000 |

Now there is another basis to represent the solution for our linear optimization problem, defined by variables $s_1$ and $x_3$ whose values are found by computing[5]:

$$B\mathbf{x} = \mathbf{b}$$

$$\mathbf{x} = B^{-1}\mathbf{b}$$

$$\mathbf{x} = \begin{bmatrix} s_1 \\ x_3 \end{bmatrix} = \begin{bmatrix} 1 & 0 \\ 0 & 1 \end{bmatrix}\begin{bmatrix} 200 \\ 400 \end{bmatrix} = \begin{bmatrix} 200 \\ 400 \end{bmatrix}.$$

which confirms the correction $s_1 = 200$ is necessary to hold the constraints for this optimization problem, given $x_3 = 400$. As matter of fact, all slack variables should be equal to zero, allowing to find a solution with no relaxation. That is always possible for linear optimization problems.

Iterations repeat all previous steps. Thus, in this example, we must analyze the last row $R_3$ looking for the current most negative value, which is now at the column associated with $x_2$. Then, we proceed in the same way as before by dividing the results (last column) by the quantities at column $x_2$:

$$\frac{200}{2} = 100$$

$$\frac{400}{\frac{1}{2}} = 800,$$

and the pivot row has the smallest value after such operation, i.e. $R_1$. Therefore, the pivot term is 2 (in bold):

| Basis | $x_1$ | $x_2$ | $x_3$ | $s_1$ | $s_2$ | $P$ | Results |
|-------|-------|-------|-------|-------|-------|-----|---------|
| $s_1$ | 1 | **2** | 0 | 1 | $-1$ | 0 | 200 |
| $x_3$ | $\frac{1}{2}$ | $\frac{1}{2}$ | 1 | 0 | $\frac{1}{2}$ | 0 | 400 |
| $P$ | $-2$ | $-3$ | 0 | 0 | 5 | 1 | 4000 |

---

[5]Observe we build matrix $B$ using the columns associated to those variables indexing rows.

and now the next basis must be found for this Tableau by applying the row-reduced echelon form again. So, we operate on rows in the following order (the order is relevant):

$$R_1 = \frac{1}{2}R_1$$

$$R_2 = R_2 - \frac{1}{2}R_1$$

$$R_3 = R_3 + 3R_1,$$

to obtain:

| Basis | $x_1$ | $x_2$ | $x_3$ | $s_1$ | $s_2$ | $P$ | Results |
|-------|-------|-------|-------|-------|-------|-----|---------|
| $s_1$ | $\frac{1}{2}$ | 1 | 0 | $\frac{1}{2}$ | $-\frac{1}{2}$ | 0 | 100 |
| $x_3$ | $\frac{1}{4}$ | 0 | 1 | $-\frac{1}{4}$ | $\frac{3}{4}$ | 0 | 350 |
| $P$ | $-\frac{1}{2}$ | 0 | 0 | $\frac{3}{2}$ | $\frac{7}{2}$ | 1 | 4300 |

and the row variable $s_1$ (pivot row) is exchanged with the column variable $x_2$ (pivot column):

| Basis | $x_1$ | $x_2$ | $x_3$ | $s_1$ | $s_2$ | $P$ | Results |
|-------|-------|-------|-------|-------|-------|-----|---------|
| $x_2$ | $\frac{1}{2}$ | 1 | 0 | $\frac{1}{2}$ | $-\frac{1}{2}$ | 0 | 100 |
| $x_3$ | $\frac{1}{4}$ | 0 | 1 | $-\frac{1}{4}$ | $\frac{3}{4}$ | 0 | 350 |
| $P$ | $-\frac{1}{2}$ | 0 | 0 | $\frac{3}{2}$ | $\frac{7}{2}$ | 1 | 4300 |

Using this new basis:

$$B\mathbf{x} = \mathbf{b}$$

$$\mathbf{x} = B^{-1}\mathbf{b}$$

$$\mathbf{x} = \begin{bmatrix} x_2 \\ x_3 \end{bmatrix} = \begin{bmatrix} 1 & 0 \\ 0 & 1 \end{bmatrix}\begin{bmatrix} 100 \\ 350 \end{bmatrix} = \begin{bmatrix} 100 \\ 350 \end{bmatrix},$$

given there is no slack variable labeling rows, so we may conclude this is a candidate solution. Thus, let us plug those values for $x_2$ and $x_3$ into our original problem:

$$\text{maximize} \quad P = 7x_1 + 8x_2 + 10x_3$$

$$\text{subject to} \quad 2x_1 + 3x_2 + 2x_3 \leq 1000$$

$$x_1 + x_2 + 2x_3 \leq 800$$

$$x_1, x_2, x_3 \geq 0,$$

to obtain ($x_1 = 0$ as this variable does not compose the basis):

$$\text{maximize } P = 7(0) + 8(100) + 10(350)$$
$$\text{subject to } 2(0) + 3(100) + 2(350) \leq 1000$$
$$(0) + (100) + 2(350) \leq 800$$
$$x_1, x_2, x_3 \geq 0,$$

what is the same as:

$$\text{maximize } P = 7(0) + 8(100) + 10(350) = 4300$$
$$\text{subject to } 1000 \leq 1000$$
$$800 \leq 800$$
$$x_1, x_2, x_3 \geq 0,$$

therefore, this is indeed a candidate solution, provided all constraints were held. Observe the objective function produced 4300 as output, which is the same value at the last column and row of the Tableau (see Table 5.2). In fact, such a Tableau cell always provides the objective function output, therefore there is no need of computing it using the original problem form. However, the last row of this Tableau still contains a negative number (variable $x_1$), indicating that there is still room for improvements.

As consequence, we must proceed with another iteration to find the next pivot term which is at the first column. So, dividing the results by the quantities at the first column:

$$\frac{100}{\frac{1}{2}} = 200$$

$$\frac{350}{\frac{1}{4}} = 1400,$$

then, the first is the pivot row, and the pivot term is (in bold—see Table 5.2):

**Table 5.2** Tableau representing the first candidate solution

| Basis | $x_1$ | $x_2$ | $x_3$ | $s_1$ | $s_2$ | $P$ | Results |
|-------|-------|-------|-------|-------|-------|-----|---------|
| $x_2$ | $\frac{1}{2}$ | 1 | 0 | $\frac{1}{2}$ | $-\frac{1}{2}$ | 0 | 100 |
| $x_3$ | $\frac{1}{4}$ | 0 | 1 | $-\frac{1}{4}$ | $\frac{3}{4}$ | 0 | 350 |
| $P$ | $-\frac{1}{2}$ | 0 | 0 | $\frac{3}{2}$ | $\frac{7}{2}$ | 1 | 4300 |

Operating on rows to make the pivot term equals to 1 and any other term equals to zero:

$$R_1 = 2R_1$$

$$R_2 = R_2 - \frac{1}{4}R_1$$

$$R_3 = R_3 + \frac{1}{2}R_1,$$

we obtain:

| Basis | $x_1$ | $x_2$ | $x_3$ | $s_1$ | $s_2$ | $P$ | Results |
|-------|-------|-------|-------|-------|-------|-----|---------|
| $x_2$ | 1 | 2 | 0 | 1 | $-1$ | 0 | 200 |
| $x_3$ | 0 | $-\frac{1}{2}$ | 1 | $-\frac{1}{2}$ | 1 | 0 | 300 |
| $P$ | 0 | 1 | 0 | 2 | 3 | 1 | 4400 |

and, then, exchanging the variable associated with the pivot row with $x_1$:

Finally, observe there is no negative value at the last row, meaning the Simplex algorithm has reached its stop criterion. Our last basis is:

$$B\mathbf{x} = \mathbf{b}$$

$$\mathbf{x} = B^{-1}\mathbf{b}$$

$$\mathbf{x} = \begin{bmatrix} x_1 \\ x_3 \end{bmatrix} = \begin{bmatrix} 1 & 0 \\ 0 & 1 \end{bmatrix} \begin{bmatrix} 200 \\ 300 \end{bmatrix} = \begin{bmatrix} 200 \\ 300 \end{bmatrix},$$

given the objective function produces 4400. Such output is confirmed by plugging $x_1 = 200$ and $x_3 = 300$ ($x_2 = 0$ as it is not part of the basis):

$$\text{maximize} \quad P = 7(200) + 8(0) + 10(300) = 4400$$

$$\text{subject to} \quad 2(200) + 3(0) + 2(300) \leq 1000$$

$$(200) + (0) + 2(300) \leq 800$$

$$x_1, x_2, x_3 \geq 0.$$

Notice Simplex has improved the previous candidate solution. Consequently, the maximum profit is obtained for:

1. $x_1 = 200$ components of type A;
2. $x_2 = 0$ component of type B;
3. $x_3 = 300$ components of type C;

4. No slackness is necessary to hold equality constraints,[6] thus $s_1 = 0$ and $s_2 = 0$.

An interesting observation is that the algorithm actually found a transformation basis. Now that we know the basis is formed by $x_1$ and $x_3$, it is easy to notice from the first Tableau:

| $x_1$ | $x_2$ | $x_3$ | $s_1$ | $s_2$ | $P$ | Results |
|------|------|------|------|------|-----|---------|
| 2 | 3 | 2 | 1 | 0 | 0 | 1000 |
| 1 | 1 | 2 | 0 | 1 | 0 | 800 |
| $-7$ | $-8$ | $-10$ | 0 | 0 | 1 | 0 |

that we originally had a basis represented in matrix form as:

$$B = \begin{bmatrix} 2 & 2 \\ 1 & 2 \end{bmatrix},$$

and then finding its inverse, $B^{-1}$:

$$B^1 = \begin{bmatrix} 1 & -1 \\ -\frac{1}{2} & 1 \end{bmatrix},$$

which is exactly the values obtained at the last Tableau (see Table 5.3) for the columns associated with the slack variables $s_1$ and $s_2$. Thus, observe the initial solution provides full relevance to slack variables and, via step by step reductions, the final basis is found for this linear optimization problem. In addition, the row-reduced echelon form provided us the basis and its inverse.

### 5.4.4  On the KKT Conditions for Linear Problems

The Karush-Kuhn-Tucker conditions define necessary properties so the primal and the dual forms of some optimization problem provide close enough results [1, 2,

**Table 5.3** Tableau representing the final solution

| Basis | $x_1$ | $x_2$ | $x_3$ | $s_1$ | $s_2$ | $P$ | Results |
|-------|------|-------|-------|-------|-------|-----|---------|
| $x_1$ | 1 | 2 | 0 | 1 | $-1$ | 0 | 200 |
| $x_3$ | 0 | $-\frac{1}{2}$ | 1 | $-\frac{1}{2}$ | 1 | 0 | 300 |
| $P$ | 0 | 1 | 0 | 2 | 3 | 1 | 4400 |

---

[6]Remember the original constraints were modified to assume the equality form by using the slack variables.

13, 14]. In this section, the KKT conditions for linear optimization problems are formalized:

| Primal problem | Dual problem |
|---|---|
| minimize $\mathbf{c}^T\mathbf{x}$ | maximize $\mathbf{w}^T\mathbf{b}$ |
| subject to $A\mathbf{x} \geq \mathbf{b}$ | subject to $\mathbf{w}^T A \leq \mathbf{c}$ |
| $\mathbf{x} \geq \mathbf{0}$ | $\mathbf{w} \geq \mathbf{0}$ |

In summary, the KKT conditions are employed to provide guarantees for all primal constraints, i.e. the *primal feasibility*:

$$A\mathbf{x} \geq \mathbf{b}, \mathbf{x} \geq \mathbf{0},$$

and for all dual constraints, i.e. the *dual feasibility*:

$$\mathbf{w}^T A \leq \mathbf{c}, \mathbf{w} \geq \mathbf{0}.$$

In addition, for linear optimization problems, KKT conditions prove the *strong duality*, i.e., both objective functions produce the same output, as follows:

$$\mathbf{c}^T\mathbf{x} = \mathbf{w}^T\mathbf{b}.$$

In order to prove the strong duality, consider the primal constraint:

$$A\mathbf{x} \geq \mathbf{b},$$

in terms of equality (what could be also obtained using some slack variable):

$$A\mathbf{x} = \mathbf{b},$$

thus, we substitute $\mathbf{b}$:

$$\mathbf{c}^T\mathbf{x} = \mathbf{w}^T\mathbf{b}$$
$$\mathbf{c}^T\mathbf{x} = \mathbf{w}^T A\mathbf{x}$$
$$\mathbf{c}^T\mathbf{x} - \mathbf{w}^T A\mathbf{x} = 0$$
$$(\mathbf{c}^T - \mathbf{w}^T A)\mathbf{x} = 0,$$

so if and only if $\mathbf{w}^T A = \mathbf{c}^T$, then this property is held, ensuring the objective functions output is the same on both forms.

Looking from the perspective of the dual constraint $\mathbf{w}^T A \leq \mathbf{c}$ and still having the same equality, we wish to hold:

$$\mathbf{c}^T \mathbf{x} = \mathbf{w}^T \mathbf{b}.$$

Considering the dual constraint holds the equality (what can be also ensured by adding slack variables), we have:

$$\mathbf{w}^T A = \mathbf{c},$$

and, then, $\mathbf{c}^T$ is substituted as follows:

$$A^T \mathbf{w} \mathbf{x} = \mathbf{w}^T \mathbf{b}$$

$$A^T \mathbf{w} \mathbf{x} - \mathbf{w}^T \mathbf{b} = 0$$

$$\mathbf{w}^T (A\mathbf{x} - \mathbf{b}) = 0.$$

As consequence, note $\mathbf{c}^T \mathbf{x} = \mathbf{w}^T \mathbf{b}$ holds, thus the following two equations must also hold:

$$\mathbf{c}^T = \mathbf{w}^T A$$

$$A\mathbf{x} = \mathbf{b},$$

confirming the solution is found by respecting the primal and the dual constraints.

However, our optimization problem has inequality constraints, so we need to reformulate it by adding slack variables and taking into account the possibility of a weak duality. This means we assume the outputs of the two objective functions may not be the same:

$$\mathbf{c}^T \mathbf{x} \geq \mathbf{w}^T \mathbf{b},$$

thus, the primal form to be minimized may produce greater values than the dual to be maximized. **This is what happens when the duality is weak**: the results provided by both objective functions may be different even for the best as possible solution found from each form.

Therefore, a slack variable $s \geq 0$ is added to ensure the equality:

$$\mathbf{c}^T \mathbf{x} \geq \mathbf{w}^T \mathbf{b}$$

$$\mathbf{c}^T \mathbf{x} - s = \mathbf{w}^T \mathbf{b}; \, s \geq 0,$$

and, then, proceed as follows:

$$\mathbf{c}^T \mathbf{x} - \mathbf{w}^T \mathbf{b} = s; \, s \geq 0.$$

Now the primal feasibility (primal constraint) can be written in equality form:

$$Ax = b,$$

once the slack variable accounts for the divergence between the primal and dual objective functions. Reformulating the equation:

$$\mathbf{c}^T \mathbf{x} - \mathbf{w}^T \mathbf{b} = s$$

$$\mathbf{c}^T \mathbf{x} - \mathbf{w}^T A \mathbf{x} = s$$

$$(\mathbf{c}^T - \mathbf{w}^T A)\mathbf{x} = s; s \geq 0,$$

as consequence:

$$\mathbf{c}^T - \mathbf{w}^T A = s\mathbf{x}; s \geq 0.$$

To ensure the strong duality, $s\mathbf{x} = 0$ for $s \geq 0$, and, finally, the KKT conditions are:

1. Primal feasibility: $Ax \geq \mathbf{b}, \mathbf{x} \geq \mathbf{0}$;
2. Dual feasibility: $\mathbf{w}^T A \leq \mathbf{c}, \mathbf{w} \geq \mathbf{0}$;
3. Complementary slackness: $\mathbf{w}^T (A\mathbf{x} - \mathbf{b}) = 0$; $s\mathbf{x} = 0$.

Finally, observe the first two KKT conditions are used to simply ensure all constraints (primal and dual), while the third condition connects both forms in terms of their objective functions.

### 5.4.4.1  Applying the Rules

The KKT conditions are applied to explain how to tackle the following practical scenario:

$$\text{minimize} \quad -x_1 - 3x_2$$
$$\text{subject to} \quad x_1 - 2x_2 \geq -4$$
$$-x_1 - x_2 \geq -4$$
$$x_1, x_2 \geq 0,$$

whose constraint functions, the gradient of the objective function, and the feasible set are illustrated in Fig. 5.22. Note the corners that the Simplex algorithm assesses to find the solution are given by the following pairs $(x_1, x_2)$: $(0, 0)$, $(4, 0)$, $(0, 2)$, and $\left(\frac{4}{3}, \frac{8}{3}\right)$. Now we attempt to apply the KKT conditions on those candidate solutions.

At first, consider corner $x_1 = 0$, $x_2 = 0$ as the candidate solution to be assessed. Beforehand, we need to define matrix $A$ as the quantities associated with the

**Fig. 5.22** Illustrating the linear optimization problem to be analyzed using the KKT conditions

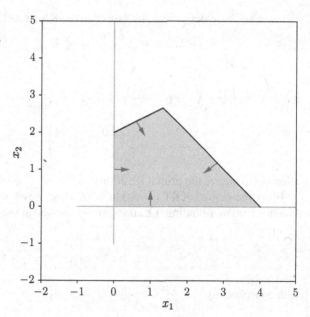

constraint functions:

$$A = \begin{bmatrix} 1 & -2 \\ -1 & -1 \end{bmatrix},$$

and vector **b** defined by the constraint bounds:

$$\mathbf{b} = \begin{bmatrix} -4 \\ -4 \end{bmatrix},$$

having vector **c** with the objective function quantities:

$$\mathbf{c} = \begin{bmatrix} -1 \\ -3 \end{bmatrix},$$

vector **x** defined by the pair $(x_1, x_2)$:

$$\mathbf{x} = \begin{bmatrix} x_1 \\ x_2 \end{bmatrix},$$

vector **w** is associated to the Lagrange multipliers (see Sect. 4.5.1) and it comes up when building the dual form. Finally, vector **s** corresponds to the slack variables to ensure equality constraints, as discussed in the previous section.

From all those definitions, by assessing the KKT conditions for:

$$\mathbf{x} = \begin{bmatrix} 0 \\ 0 \end{bmatrix},$$

the first condition $A\mathbf{x} \geq \mathbf{b}$, given $\mathbf{x} \geq \mathbf{0}$, is:

$$A\begin{bmatrix} 0 \\ 0 \end{bmatrix} \geq \begin{bmatrix} -4 \\ -4 \end{bmatrix},$$

allowing to ensure the primal feasibility.

From the second KKT condition $\mathbf{w}^T A \leq \mathbf{c}$, given $\mathbf{w} \geq \mathbf{0}$, the two constraint functions are non-binding, i.e., they are not applied at the point $(0, 0)$, therefore:

$$\mathbf{w} = \begin{bmatrix} 0 \\ 0 \end{bmatrix},$$

and, consequently:

$$\mathbf{w}A + \mathbf{s} = \mathbf{c}; \mathbf{w} \geq \mathbf{0}; \mathbf{s} \geq \mathbf{0}$$

$$\begin{bmatrix} 0 \\ 0 \end{bmatrix} A + \mathbf{s} = \mathbf{c}$$

$$\mathbf{s} = \mathbf{c}$$

$$\mathbf{s} = \begin{bmatrix} -1 \\ -3 \end{bmatrix},$$

what violates the non-negativity property $\mathbf{s} \geq \mathbf{0}$, therefore the candidate solution $(0, 0)$ is not optimal. That means there is no need for evaluating the remaining KKT conditions.

Now we go directly to the optimal point $\left( \frac{4}{3}, \frac{8}{3} \right)$, and suggest the reader to test the other corners later. Starting with the first KKT condition $A\mathbf{x} \geq \mathbf{b}$, $\mathbf{x} \geq \mathbf{0}$:

$$A\begin{bmatrix} \frac{4}{3} \\ \frac{8}{3} \end{bmatrix} \geq \begin{bmatrix} -4 \\ -4 \end{bmatrix}$$

$$\begin{bmatrix} -4 \\ -4 \end{bmatrix} = \begin{bmatrix} -4 \\ -4 \end{bmatrix},$$

which is true.

From the second condition $\mathbf{w}A + \mathbf{s} = \mathbf{c}$, $\mathbf{w} \geq \mathbf{0}$, given $\mathbf{s} \geq \mathbf{0}$, vector $\mathbf{w}$ must have values greater than zero, provided both constraint functions are binding. However, since $\mathbf{x} > \mathbf{0}$, in order to satisfy part of the third condition $\mathbf{s}\mathbf{x} = 0$, then $\mathbf{s}$ must be equal to the zero vector, i.e., $\mathbf{s} = \mathbf{0}$:

$$\mathbf{w}A + \mathbf{s} = \mathbf{c}; \ \mathbf{w} \geq \mathbf{0}, \mathbf{s} \geq \mathbf{0}$$

$$\mathbf{w}A + \begin{bmatrix} 0 \\ 0 \end{bmatrix} = \mathbf{c}$$

$$\begin{cases} w_1 - w_2 = -1 \\ -2w_1 - w_2 = -3 \end{cases},$$

resulting in:

$$\mathbf{w} = \begin{bmatrix} w_1 \\ w_2 \end{bmatrix} = \begin{bmatrix} \frac{2}{3} \\ \frac{5}{3} \end{bmatrix},$$

satisfying the second condition.

Proceeding with the third KKT condition, $\mathbf{w}^T(A\mathbf{x} - \mathbf{b}) = 0$, $s\mathbf{x} = 0$, we already know that $s\mathbf{x} = 0$, and then:

$$\mathbf{w}^T(A\mathbf{x} - \mathbf{b}) = 0$$

$$\begin{bmatrix} \frac{2}{3} & \frac{5}{3} \end{bmatrix} \left( \begin{bmatrix} 1 & -2 \\ -1 & -1 \end{bmatrix} \begin{bmatrix} \frac{4}{3} \\ \frac{8}{3} \end{bmatrix} - \begin{bmatrix} -4 \\ -4 \end{bmatrix} \right) = \begin{bmatrix} \frac{2}{3} & \frac{5}{3} \end{bmatrix} \begin{bmatrix} 0 \\ 0 \end{bmatrix} = 0 \right),$$

showing the last condition is satisfied. Therefore, the optimal point is:

$$\mathbf{x} = \begin{bmatrix} x_1 \\ x_2 \end{bmatrix} = \begin{bmatrix} \frac{4}{3} \\ \frac{8}{3} \end{bmatrix}.$$

This simple linear optimization problem illustrates how the KKT conditions can be used to find and prove the optimal solution.

## 5.4.4.2  Graphical Interpretation of the KKT Conditions

In this section, we get back to the previous optimization problem illustrated in Fig. 5.22:

$$\begin{aligned} \text{minimize} \quad & -x_1 - 3x_2 \\ \text{subject to} \quad & x_1 - 2x_2 \geq -4 \\ & -x_1 - x_2 \geq -4 \\ & x_1, x_2 \geq 0, \end{aligned}$$

**Fig. 5.23** Illustration of the linear optimization problem used to graphically analyze the KKT conditions: step 2

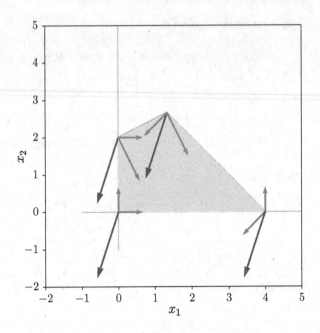

We start by computing the gradient vector for the objective function $-x_1 - 3x_2$:

$$\nabla f = \mathbf{c} = \begin{bmatrix} -1 \\ -3 \end{bmatrix},$$

and plotting it at every corner of the feasible set (see Fig. 5.23).

Afterwards, the gradient vectors of the constraint functions are computed (term $c^i$ is associated to the gradient vector for the $i$-th constraint):

$$\nabla c_1 = \begin{bmatrix} 1 \\ -2 \end{bmatrix}$$

$$\nabla c_2 = \begin{bmatrix} -1 \\ -1 \end{bmatrix},$$

as well as the gradient vectors bounding variables (i.e., $x_1, x_2 \geq 0$):

$$\nabla v_1 = \begin{bmatrix} 1 \\ 0 \end{bmatrix}$$

$$\nabla v_2 = \begin{bmatrix} 0 \\ 1 \end{bmatrix}.$$

**Fig. 5.24** Illustration of the linear optimization problem used to graphically analyze the KKT conditions: step 3

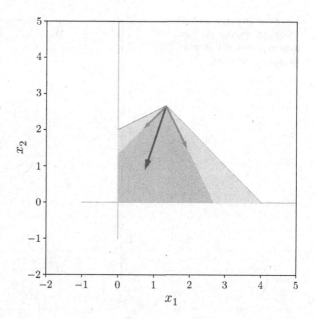

All those gradient vectors are plotted at the corners affected by such constraints (see Fig. 5.24). For example, at the point $\left(\frac{4}{3}, \frac{8}{3}\right)$, the gradients $\nabla c_1$ and $\nabla c_2$ are plotted; at the point $(4, 0)$, we plot the gradient vectors for the first constraint ($\nabla c_1$) and the one associated to variable $x_1$ ($\nabla v_1$).

Those gradient vectors allow a visual interpretation of the results evaluated at each candidate solution. Let us take point $(0, 2)$ and use its constraint gradients to define a cone, as depicted in Fig. 5.25. Observe the gradient of the objective function $\nabla f$ does not lay inside such a cone at that particular corner, meaning point $(0, 2)$ is not the optimal.

Now we inspect all other corners to look for the solution in which $\nabla f$ lies inside the convex cone. Observe $\left(\frac{4}{3}, \frac{8}{3}\right)$ is the only satisfying the criterion, and therefore represents the optimal (minimal) solution. If we had a maximization problem, we would instead check whether the negative of the gradient (i.e., $-\nabla f$) lays inside such a cone. This is the geometrical interpretation of the KKT conditions while ensuring the solution at a particular corner for linear problems. For more information about the KKT conditions, we suggest the reader to study the Farkas' lemma [1].

## 5.5 Convex Optimization Problems

Linear optimization problems can either be graphed, in terms of their gradient vectors, to be manually solved (when possible) or algorithmically approached using

**Fig. 5.25** Illustration of the
linear optimization problem
used to graphically analyze
the KKT conditions: step 4

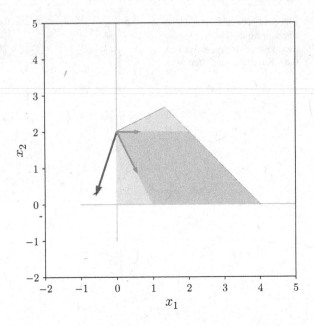

the Simplex method. However, nonlinear optimization problems require different
strategies. In this section, we are interested in convex nonlinear problems, in
particular defined by functions as follows:

1. Convex constraints and linear objective function;
2. Convex constraints and convex objective function.

Figure 5.26 illustrates the first scenario, in which the nonlinear optimization
problem has convex constraints and a linear objective function. Note the constraints
form a convex set, i.e., if a line segment is traced connecting two points inside the
feasible region, then all line points are within the same region. In this situation, we
can simply "project" the constraints onto the linear objective function to find either
the maximal or the minimal solution.

This problem was already discussed in Sect. 4.5.1, in the context of Lagrange
multipliers. Such method solves those optimization problems for equality constraints projected into the objective function. However, nonlinear optimization
problems with convex constraints and a convex objective function require more
advanced techniques. To illustrate, consider the primal problem for the Support
Vector Machines:

$$\underset{\mathbf{w}\in\mathbb{H},b\in\mathbb{R}}{\text{minimize}}\ \ \tau(\mathbf{w}) = \frac{1}{2}\left\|\mathbf{w}\right\|^2$$

subject to $\ y_i(<\mathbf{w}, \mathbf{x}_i> +b) \geq 1$, for all $i = 1, \ldots, m$,

**Fig. 5.26** Optimization
problem involving convex
constraints and a linear
objective function

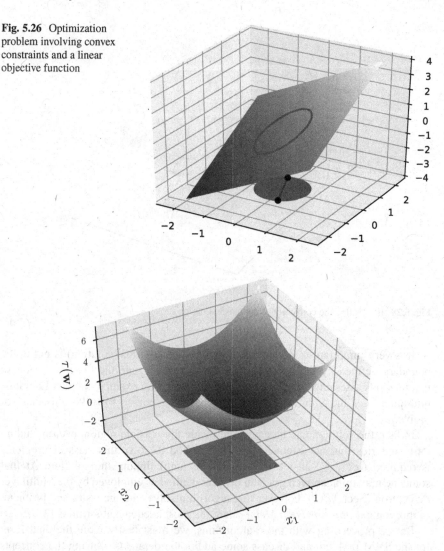

**Fig. 5.27** Optimization problem involving convex constraints and a convex objective function

in which the objective function is convex and the constraints are linear but form a convex feasible region, more details in [2]. Figure 5.27 illustrates this problem in a typical scenario in which constraints are defined in terms of two variables $\mathbf{x}_i = (x_{i,1}, x_{i,2})$, for every possible example $i$. By plugging $\mathbf{w}$ into the objective function, the convex surface is obtained whose values are represented by the third axis $\tau(\mathbf{w})$.

Notice linear constraints define a convex feasible set for $\mathbf{x}_i$. Figure 5.28 illustrates the convexity of this feasible region by using an affine function connecting any two points inside such set.

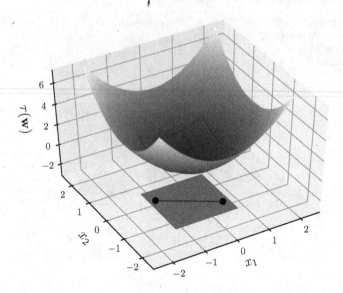

**Fig. 5.28** Illustrating the convexity of the feasible region

It is very important to notice that the minimum (optimal solution) is not at the boundary defined by the linear constraints, but it is inside the feasible region so that methods assessing boundaries are not sufficient. For example, neither Lagrange multipliers nor the Simplex algorithm would be adequate to tackle this class of problems.

In fact, the solution can be anywhere inside the feasible region, even including its boundaries (but not exclusively). The method to solve this class of problems is required to "walk" inside the feasible set until finding the solution. At first someone may think about a gradient descent method as employed by the Multilayer Perceptron (Sect. 1.5.2), however that algorithm solves an unconstrained problem. In this context, *Interior Point Methods* are the most adequate algorithms [2, 20].

Before proceeding with those algorithms, we must firstly write the dual form for the SVM problem and discuss some additional details (following the concepts provided in Sects. 4.6 and 4.7). Recall the SVM dual problem:

$$\underset{\alpha}{\text{maximize}} \ \ W(\boldsymbol{\alpha}) = -\frac{1}{2} \sum_{i=1}^{m} \sum_{j=1}^{m} \alpha_i \alpha_j y_i y_j < \mathbf{x}_i, \mathbf{x}_j > + \sum_{i=1}^{m} \alpha_i$$

subject to $0 \leq \alpha_i \leq C$, for all $i = 1, \ldots, m$,

$$\text{and} \ \ \sum_{i=1}^{m} \alpha_i y_i = 0,$$

after applying the Lagrange multipliers and the KKT conditions as problem adaptation tools. In addition, recall **x** is a vector defining the attributes for a given training example.

We now have sufficient background to discuss some properties:

1. About the number of constraints: $m$ defines both the number of examples and constraints. That means every additional training example imposes an additional linear constraint;
2. About the nonbinding constraints: assuming a binary classification problem without class overlapping, the equality for the primal constraints, $y_i(< \mathbf{w}, \mathbf{x}_i > +b) \geq 1$, must be ensured only for the support vectors. That is enough so the remaining points, far from the hyperplane, consequently satisfy constraints. Every constraint has a KKT multiplier associated with, whose value $\alpha_i > 0$ when binding. Observe that, by having $\alpha_i > 0$ for every $i$th support vector, we already define the SVM margin for the most restricted case, i.e. the points nearby the maximal margin hyperplane. As consequence, the remaining constraints are nonbinding, thus their $\alpha_j = 0$, for every example that is not a support vector;
3. About the objective functions: when the primal problem is represented by a convex function, its dual must form a concave function.

To illustrate the second property, suppose a two-class SVM problem (see Fig. 5.29), in which the best hyperplane is known, and $\mathbf{x}_i = (x_{i,1}, x_{i,2})$ for every training example $i$.

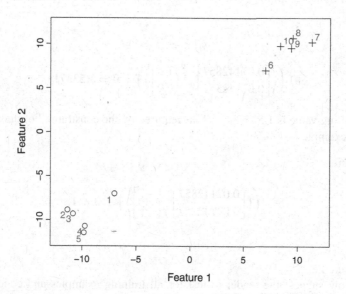

**Fig. 5.29** Illustration of a two-class SVM problem

Recall vector **w** for the SVM problem is given by:

$$\mathbf{w} = \sum_{i=1}^{m} \alpha_i y_i \mathbf{x}_i,$$

in which $\mathbf{x}_i$ is a vector in this bidimensional space, $y_i = -1$ for $i = 1, \ldots, 5$ and $y_i = +1$ for $i = 6, \ldots, 10$, and let $\alpha_1 = \alpha_6 = 0.005102041$ be the binding KKT multipliers, and $\alpha_i = 0$ the nonbinding ones, which are related to any other point in space. For this particular problem:

$$\mathbf{w} = \begin{bmatrix} 0.07142857 \\ 0.07142857 \end{bmatrix},$$

and $b = 0$. Then, let us observe what happens for the training example $(7, 7)$, which should be classified as $+1$ and respect the linear constraint function:

$$y_i(< \mathbf{w}, \mathbf{x}_i > +b) =$$

$$+1 \left( \left\langle \begin{bmatrix} 0.07142857 \\ 0.07142857 \end{bmatrix}, \begin{bmatrix} 7 \\ 7 \end{bmatrix} \right\rangle \right) + 0 = 1,$$

confirming the constraint function was held for this training example. Evaluating another training example, $(11.3, 10.2)$, we again conclude the linear constraint is respected:

$$y_i(< \mathbf{w}, \mathbf{x}_i > +b) =$$

$$+1 \left( \left\langle \begin{bmatrix} 0.07142857 \\ 0.07142857 \end{bmatrix}, \begin{bmatrix} 11.3 \\ 10.2 \end{bmatrix} \right\rangle \right) + 0 = 1.535714.$$

The resulting value is $1.535714 \geq 1$, as required by the constraint. For the negative training example $(-7, -7)$:

$$y_i(< \mathbf{w}, \mathbf{x}_i > +b) =$$

$$-1 \left( \left\langle \begin{bmatrix} 0.07142857 \\ 0.07142857 \end{bmatrix}, \begin{bmatrix} -7 \\ -7 \end{bmatrix} \right\rangle \right) + 0 = 1,$$

as required, i.e., $1 \geq 1$.

We now suggest the reader to test for all training examples and conclude all constraints are held for this linearly separable problem. Remember $\alpha_i = 0$ for 8 out of the 10 training examples. This is because only two constraints are necessary, related to the points closer to the ideal hyperplane, providing enough information

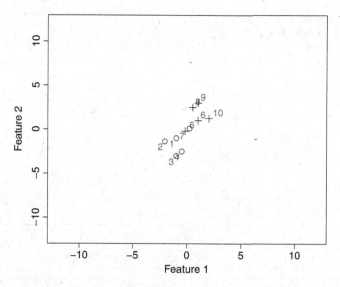

**Fig. 5.30** SVM problem without perfect linear separability

to ensure all 10 constraints, i.e., the two binding constraints defined by examples $(7, 7)$ and $(-7, -7)$. Moreover, in all linearly separable problems, many constraints are nonbinding and unnecessary.

Now, consider the problem illustrated in Fig. 5.30, which has no ideal linear hyperplane capable of separating all positive and negative examples.

Then, let $y_i = -1$, for $i = 1, \ldots, 5$, and $y_i = +1$, for $i = 6, \ldots, 10, \alpha_1 = \alpha_6 = 0.25$ and $\alpha_i = 0$ for any other example and, finally, $b = 0$. Thus, vector $\mathbf{w}$ is:

$$\mathbf{w} = \begin{bmatrix} 0.5 \\ 0.5 \end{bmatrix}.$$

Now consider the training example $(1, 1)$ and solve the constraint function as follows:

$$y_i(<\mathbf{w}, \mathbf{x}_i> +b) =$$

$$+1 \left( \left\langle \begin{bmatrix} 0.5 \\ 0.5 \end{bmatrix}, \begin{bmatrix} 1 \\ 1 \end{bmatrix} \right\rangle \right) + 0 = 1,$$

to confirm the constraint is held. We suggest the reader to solve for the training example $(-1, -1)$ and observe the result will be 1, which again confirms the linear constraint is satisfied. While attempting to solve for $(-0.2, -0.2)$, which is supposed to be classified as $+1$:

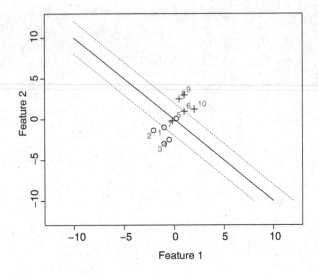

**Fig. 5.31** Illustrating the linear boundary defined by the support vectors

$$y_i(< \mathbf{w}, \mathbf{x}_i > +b) =$$

$$+1\left(\left\langle\begin{bmatrix}0.5\\0.5\end{bmatrix}, \begin{bmatrix}-0.2\\-0.2\end{bmatrix}\right\rangle\right) + 0 = -0.2,$$

from which the result does not respect the linear constraint $y_i(< \mathbf{w}, \mathbf{x}_i > +b) \geq 1$. However, this is acceptable when the soft-margin SVM is used, which means some examples may not respect constraints (see Sect. 4.7). In those situations, $\alpha_i = 0$ in order to disconsider those training examples and ensure $y_i(< \mathbf{w}, \mathbf{x}_i > +b) \geq 1$, as shown in Fig. 5.31.

From this, note linear constraints are only ensured for training examples laying above the positive support vector or below the negative one. There may have more than one positive or negative support vectors, but their tips must lie on the same support hyperplanes. Training examples in between those support hyperplanes provide uncertain results, that is why they are disconsidered while solving the optimization problem. Even in this more complex problem, which presents some mixture or uncertainty, the support vectors can be still used to define the necessary set of constraints:

$$y_i(< \mathbf{w}, \mathbf{x}_i > +b) > 1,$$

for more distant examples such as $(-2.06, -1.36)$, $(-1, -3)$, $(-0.5, -2.5)$, $(0.5, 2.5)$, $(1, 3)$, and $(2, 1.25)$.

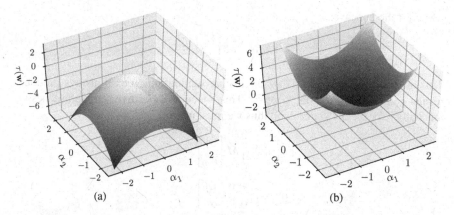

**Fig. 5.32** Illustrating concave (**a**) and convex (**b**) functions in terms of $\alpha$

To illustrate the third property, let the objective function in the primal form be:

$$\underset{\mathbf{w}\in\mathbb{H},b\in\mathbb{R}}{\text{minimize}} \;\; \tau(\mathbf{w}) = \frac{1}{2}\left\|\mathbf{w}\right\|^2,$$

and its dual form be:

$$\underset{\alpha}{\text{maximize}} \;\; W(\boldsymbol{\alpha}) = -\frac{1}{2}\sum_{i=1}^{m}\sum_{j=1}^{m}\alpha_i\alpha_j y_i y_j < \mathbf{x}_i,\mathbf{x}_j > + \sum_{i=1}^{m}\alpha_i.$$

In summary, the primal form attempts to minimize a convex function while the dual maximizes a concave one, being both required to ensure solution guarantee. For instance, consider the dual form and let term $y_i y_j < \mathbf{x}_i, \mathbf{x}_j >$ provide a convex function as required,[7] which is represented by:

$$M = \begin{bmatrix} 1 & 0 \\ 0 & 1, \end{bmatrix}$$

in which $\alpha_i, \alpha_j \in \mathbb{R}$. By adapting alphas and solving $\begin{bmatrix} \alpha_i & \alpha_j \end{bmatrix} M \begin{bmatrix} \alpha_i \\ \alpha_j \end{bmatrix}$, a convex function is obtained, as shown in Fig. 5.32a.

On the hand, the dual solves a maximization of $-\frac{1}{2}\sum_{i=1}^{m}\sum_{j=1}^{m}\alpha_i\alpha_j y_i y_j < \mathbf{x}_i, \mathbf{x}_j >$, in which a minus sign is used to produced the concave surface with a unique maximum, as shown in Fig. 5.32b.

For the convex function, we expect:

$$\begin{bmatrix} \alpha_i & \alpha_j \end{bmatrix} M \begin{bmatrix} \alpha_i \\ \alpha_j \end{bmatrix} \geq 0,$$

---

[7]This ends up as concave after applying the minus sign.

while for the concave:

$$- \begin{bmatrix} \alpha_i & \alpha_j \end{bmatrix} M \begin{bmatrix} \alpha_i \\ \alpha_j \end{bmatrix} \leq 0.$$

This inequality implies matrix $M$ is positive-semidefinite, i.e., it always produces results greater than or equal to zero.[8] This concept of positive-semidefiniteness is mandatory for the SVM problem, thus we must prove:

$$\begin{bmatrix} \alpha_i & \alpha_j \end{bmatrix} M \begin{bmatrix} \alpha_i \\ \alpha_j \end{bmatrix} =$$

$$\begin{bmatrix} \alpha_i & \alpha_j \end{bmatrix} \begin{matrix} 1 & 0 \\ 0 & 1, \end{matrix} \begin{bmatrix} \alpha_i \\ \alpha_j \end{bmatrix} =$$

$$\begin{bmatrix} 1\alpha_i + 0\alpha_j & 0\alpha_i + 1\alpha_j \end{bmatrix} \begin{bmatrix} \alpha_i \\ \alpha_j \end{bmatrix} = \alpha_i^2 + \alpha_j^2 \geq 0,$$

being equal to zero only when $\alpha_i = \alpha_j = 0$. This is essential to verify if matrix $M$ defines a convex primal objective function and a concave dual. In the next section, both SVM problem forms are algebraically reformulated, using the matrix form, in order to use well-known methods to study their convexity and concavity. To oppose this concept of positive-semidefiniteness, consider another matrix $M$:

$$M = \begin{bmatrix} -1 & 0 \\ 0 & 1, \end{bmatrix}$$

in such situation, we have:

$$\begin{bmatrix} \alpha_i & \alpha_j \end{bmatrix} M \begin{bmatrix} \alpha_i \\ \alpha_j \end{bmatrix} =$$

$$\begin{bmatrix} \alpha_i & \alpha_j \end{bmatrix} \begin{matrix} -1 & 0 \\ 0 & 1, \end{matrix} \begin{bmatrix} \alpha_i \\ \alpha_j \end{bmatrix} =$$

$$\begin{bmatrix} -1\alpha_i + 0\alpha_j & 0\alpha_i + 1\alpha_j \end{bmatrix} \begin{bmatrix} \alpha_i \\ \alpha_j \end{bmatrix} = -\alpha_i^2 + \alpha_j^2 \geq 0,$$

resulting in a saddle surface, as seen in Fig. 5.33. Such type of matrix cannot be solved for the SVM problem, given any attempt of maximization (dual form) would make the objective function go to positive infinity. There is neither a way to solve the primal, given it would go to negative infinity, confirming there is no point providing the optimal solution.

---

[8] A matrix $M$ is referred to as positive definite if $\begin{bmatrix} \alpha_i & \alpha_j \end{bmatrix} M \begin{bmatrix} \alpha_i \\ \alpha_j \end{bmatrix} > 0.$

**Fig. 5.33** Example of a
saddle surface

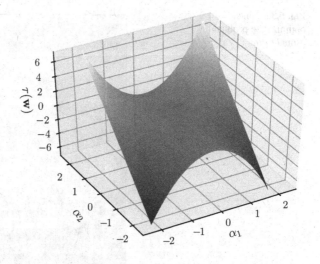

Going a step further, any kernel function must produce matrix $M = y_i y_j < \mathbf{x}_i, \mathbf{x}_j >$ as positive-semidefinite, so that the SVM problem has solution. There is also the scenario in which the kernel provides a locally convex/concave objective function what may be sufficient for some problems.

### 5.5.1 Interior Point Methods

At this point, we present a class of algorithms to solve convex optimization problems, referred to as Interior Point Methods. In the late 1940s, Dantzig proposed the Simplex method to approach linear optimization problems while other researchers proposed Interior Point Methods (IPM), such as Von Neumann [27], Hoffman et al. [10], and Frisch [3, 8, 25]. IPM is used to traverse across the interior of the feasible region in attempt to avoid the combinatorial complexities of vertex-following algorithms (such as Simplex). However, IPM required expensive computational steps and suffered from numerical instabilities, what discouraged the adoption of such methods in practical scenarios. In 1984, Karmakar [12] introduced a novel IPM approach to tackle practical problems. Gill et al. [9] then showed a formal relationship between that new IPM and the classical logarithmic barrier method.

There are three major types of IPM methods:

1. The *potential reduction algorithm* which most closely embodies the proposal by Karmakar;
2. The *affine scaling algorithm* which is probably the simplest to implement;
3. *Path-following algorithms* that, due to their arguably excellent behavior in theory and practice, are discussed in this book.

**Fig. 5.34** Linear
optimization problem in
primal form

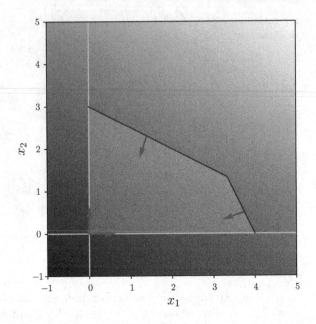

### 5.5.1.1   Primal-Dual IPM for Linear Problem

We start with a linear optimization problem to simplify the introduction of the path-following algorithm [25]. Let the following primal linear optimization problem:

$$\text{maximize} \quad f_0(\mathbf{x}) = 2x_1 + 3x_2$$

$$\text{subject to} \quad 2x_1 + x_2 \leq 8$$

$$x_1 + 2x_2 \leq 6$$

$$x_1 \geq 0, \ x_2 \geq 0,$$

illustrated in Fig. 5.34. It could be solved using the Simplex algorithm, by visiting constraint corners to find $x_1 = \frac{10}{3}$ and $x_2 = \frac{4}{3}$, and obtain $f_0(\mathbf{x}) = \frac{32}{3} \approx 10.66\ldots$..
   Consider also its dual form:

$$\text{minimize} \quad g_0(\boldsymbol{\pi}) = 8\pi_1 + 6\pi_2$$

$$\text{subject to} \quad 2\pi_1 + \pi_2 \geq 2$$

$$\pi_1 + 2\pi_2 \geq 3$$

$$\pi_1 \geq 0, \ \pi_2 \geq 0,$$

seen in Fig. 5.35. Again, the solution could be found using the Simplex algorithm, finding $\pi_1 = \frac{1}{3}$ and $\pi_2 = \frac{4}{3}$, producing $g_0(\boldsymbol{\pi}) = \frac{32}{3} \approx 10.66\ldots$..

**Fig. 5.35** Linear optimization problem in dual form

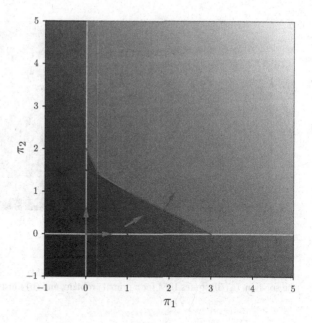

Simplex considers either the primal or the dual forms to solve the optimization problem, therefore there is no connection between them. On the other hand, the IPM path-following algorithm considers both forms [11]. Thus, we start by rewriting both forms using slack variables, similarly to the Simplex algorithm:

**Primal problem**

$$\text{maximize } f_0(\mathbf{x}) = 2x_1 + 3x_2$$

$$\text{subject to } 2x_1 + x_2 + x_3 = 8$$

$$x_1 + 2x_2 + x_4 = 6$$

$$x_1, x_2, x_3, x_4 \geq 0$$

**Dual problem**

$$\text{minimize } g_0(\boldsymbol{\pi}) = 8\pi_1 + 6\pi_2$$

$$\text{subject to } 2\pi_1 + \pi_2 - z_1 = 2$$

$$\pi_1 + 2\pi_2 - z_2 = 3$$

$$\pi_1 - z_3 = 0, \ \pi_2 - z_4 = 0$$

$$z_1, z_2, z_3, z_4 \geq 0,$$

having the slack variables $x_3$ and $x_4$ for the maximization, and $z_1, z_2, z_3, z_4$ for the minimization form.[9]

The plus sign in front of variables $x_3$ and $x_4$ make constraint functions reach the upper limit defined by $\leq$ (primal form), while the minus sign in front of $z_1, z_2, z_3$ and $z_4$ make constraints (dual form) equal to the right-side terms given the relational operator $\geq$.

---

[9]We could have added simply slack variables $\pi_3$ and $\pi_4$ to provide the same results in the minimization form, however we decided to use this formulation to follow the proposal [11].

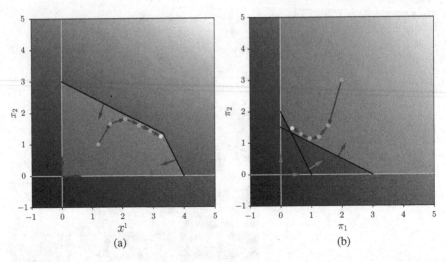

**Fig. 5.36** The Interior-Point Path-Following Method (IPM) traverses the feasible region to find the solution, (**a**) illustrates IPM for a primal problem, and (**b**) illustrates IPM for a dual problem

Instead of "walking" on the boundaries such as Simplex, this IPM path-following algorithm traverses inside the feasible region (see Fig. 5.36). Given this is a linear problem, the solution obviously lies on one constraint corner (boundary of the feasible region), as consequence IPM goes towards such corner without touching it, because it represents a barrier for the feasible set.

Before solving it, there is here the opportunity to provide a general formulation for this path-following algorithm. Consider the following general problem:

| **Primal problem** | **Dual problem** |
|---|---|
| maximize $\mathbf{c}^T \mathbf{x}$ | minimize $\boldsymbol{\pi}^T \mathbf{b}$ |
| subject to $A\mathbf{x} = \mathbf{b}$ | subject to $\boldsymbol{\pi} A - \mathbf{z} = \mathbf{c}$ |
| $\mathbf{x} \geq \mathbf{0}$ | $\mathbf{z} \geq \mathbf{0}$ |

having $\mathbf{z}$ as the slack variable vector. By building the barriers, IPM never reaches the boundary itself, otherwise the solution would be risked by assuming values outside the feasible set (they do not respect constraints).

Suppose a minimization problem for which variable $x \geq 0$, implying a boundary or barrier must be set to penalize solutions as they approach zero (Fig. 5.37a). Fiacco and McCormick [5] designed a logarithmic-based barrier to represent those variable boundaries. In this scenario, a logarithmic function applies a penalty to the objective function, pushing solutions back as they get to close to the barrier.

In another situation, given a maximization problem for which solutions must assume $x \geq 0$, as shown in Fig. 5.37b, a negative logarithmic function is employed.

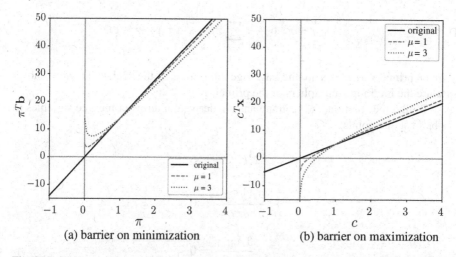

**Fig. 5.37** Illustration of the barrier method applied to an optimization problem. (**a**) Minimization: solutions avoid values close to the barrier while coming from the positive side, (**b**) Maximization: solution avoid values close to the barrier from the negative size

In our general formulation, we do not accept $\mathbf{x} < \mathbf{0}$ nor $\mathbf{z} \geq \mathbf{0}$, for the primal and dual forms, respectively. Therefore, we rewrite the original forms:

**Primal problem**

$$\text{maximize } \mathbf{c}^T \mathbf{x} + \mu \sum_{j=1}^{n} \log x_j$$

subject to $A\mathbf{x} = \mathbf{b}$

**Dual problem**

$$\text{minimize } \boldsymbol{\pi}^T \mathbf{b} - \mu \sum_{j=1}^{n} z_j$$

subject to $\boldsymbol{\pi} A - \mathbf{z} = \mathbf{c}$

allowing us to remove constraints $\mathbf{x} \geq \mathbf{0}$ and $\mathbf{z} \geq \mathbf{0}$, which were reformulated using the barrier method.

A new parameter $\mu$ appears in both primal and dual forms, which is responsible for controlling the relevance given to the barrier term along the algorithm iterations. If we give too much relevance for the barrier, the solution would never approach the boundary, avoiding the optimal solution when necessary. In practice, $\mu$ starts with some great value, which is reduced along iterations, allowing to get close enough boundaries.

Now we employ Lagrange multipliers to rewrite this equality-based general formulation (see Sect. 4.5.1):

$$\Lambda_{primal} = \mathbf{c}^T \mathbf{x} + \mu \sum_{j=1}^{n} \log x_j - \boldsymbol{\pi}^T (A\mathbf{x} - \mathbf{b})$$

$$\Lambda_{dual} = \pi^T \mathbf{b} - \mu \sum_{j=1}^{n} z_j - \mathbf{x}^T (\pi A - \mathbf{z} - \mathbf{c}),$$

given primal variables $\mathbf{x}$ are the Lagrange multipliers of the dual, and dual variables $\pi$ are the Lagrange multipliers of the primal.

To find the solution, the Lagrangians are derived in terms of the free variables to obtain stable points:

$$\frac{\partial \Lambda_{primal}}{\partial x_j} = 0$$

$$\frac{\partial \Lambda_{primal}}{\partial \pi_i} = 0$$

$$\frac{\partial \Lambda_{dual}}{\partial z_j} = 0$$

$$\frac{\partial \Lambda_{dual}}{\partial \pi_i} = 0$$

$$\frac{\partial \Lambda_{dual}}{\partial x_j} = 0.$$

Solving those derivatives, the following is found for the primal Lagrangian:

$$\frac{\partial \Lambda_{primal}}{\partial x_j} = c_j - \sum_{i=1}^{m} a_{ij} \pi_j + \frac{\mu}{x_j} = 0$$

$$\frac{\partial \Lambda_{primal}}{\partial \pi_i} = \sum_{j=1}^{n} a_{ij} x_j - b_i = 0,$$

and for the dual Lagrangian:

$$\frac{\partial \Lambda_{dual}}{\partial z_j} = -\frac{\mu}{z_j} + x_j = 0$$

$$\frac{\partial \Lambda_{dual}}{\partial \pi_i} = \sum_{j=1}^{n} a_{ij} x_j - b_i = 0$$

$$\frac{\partial \Lambda_{dual}}{\partial x_j} = \sum_{i=1}^{m} a_{ij} \pi_j - z_j - c_j = 0.$$

Then, this primal-dual path-following algorithm uses the derivatives obtained from both forms to find the solution. As first step, we compare the derivatives, starting with $\frac{\partial \Lambda_{primal}}{\partial x_j}$ and $\frac{\partial \Lambda_{dual}}{\partial z_j}$:

$$\frac{\partial \Lambda_{dual}}{\partial z_j} = -\frac{\mu}{z_j} + x_j = 0$$

$$z_j x_j = \mu, \text{ for } j = 1, \ldots, n,$$

thus, variable $\mu$ used in $\frac{\partial \Lambda_{primal}}{\partial x_j}$ must assume the same values as in the derivative of the dual, i.e.:

$$\frac{\partial \Lambda_{primal}}{\partial x_j} = c_j - \sum_{i=1}^{m} a_{ij} \pi_j + \frac{\mu}{x_j} = 0$$

$$\frac{\partial \Lambda_{primal}}{\partial x_j} = c_j - \sum_{i=1}^{m} a_{ij} \pi_j + \frac{z_j x_j}{x_j} = 0$$

$$\frac{\partial \Lambda_{primal}}{\partial x_j} = c_j - \sum_{i=1}^{m} a_{ij} \pi_j + z_j = 0.$$

The partial derivatives $\frac{\partial \Lambda_{primal}}{\partial x_j}$ and $\frac{\partial \Lambda_{dual}}{\partial z_j}$ produces what is referred to as the $\mu$-complementary slackness, whose equations must agree in both (primal and dual) forms.

In practice, we intend to make $\mu$ as close as possible to zero along iterations, allowing our algorithm to approach the boundary without crossing it. The objective functions for the primal and the dual forms will not produce the exact same results along iterations, because the optimal is at a corner in this linear optimization problem. Thus, while the primal form approaches a corner, the dual makes the same for its corresponding corner, in those steps, both objective functions provide different outputs. That difference can be computed as follows:

$$\text{Gap} = \sum_{j=1}^{n} z_j x_j,$$

given $z_j x_j = \mu$. This gap is typically used to control the iterations in terms of the stop criterion. Note the solution is at the corner, which would oblige $\mu = 0$. But since $\mu > 0$, this implies a gap between the objective functions of the primal and the dual forms.

One might ask why not set $\mu = 0$ since the beginning of our algorithm. That would most probably make it consider solutions outside the feasible region, because no relevance would be given for the barrier terms. Once outside, we may never get back to it.

Now compare $\frac{\partial \Lambda_{primal}}{\partial \pi_i}$ and $\frac{\partial \Lambda_{dual}}{\partial \pi_i}$. Both produce the same partial derivative $\sum_{j=1}^{n} a_{ij}x_j - b_i = 0$, which is another representation for the primal constraints. That is why those derivatives are referred to as the primal feasibility.

The dual form has an extra partial derivative:

$$\frac{\partial \Lambda_{dual}}{\partial x_j} = \sum_{i=1}^{m} a_{ij}\pi_j - z_j - c_j = 0$$

which corresponds to the dual constraints in a different representation, being referred to as the dual feasibility.

We now build a system including the three equations, that is used to solve the optimization problem. To simplify its formulation, we consider two diagonal matrices containing the elements of vectors $\mathbf{x}$ and $\mathbf{z}$, in form:

$$X = \text{diag}\{x_1, x_2, \ldots, x_n\} = \begin{bmatrix} x_1 & 0 & \ldots & 0 \\ 0 & x_2 & \ldots & 0 \\ \vdots & \ddots & \ldots & 0 \\ 0 & 0 & \ldots & x_n \end{bmatrix}$$

$$Z = \text{diag}\{z_1, z_2, \ldots, z_n\} = \begin{bmatrix} z_1 & 0 & \ldots & 0 \\ 0 & z_2 & \ldots & 0 \\ \vdots & \ddots & \ldots & 0 \\ 0 & 0 & \ldots & z_n \end{bmatrix},$$

and let $\mathbf{e} = \begin{bmatrix} 1 & 1 & \ldots & 1 \end{bmatrix}^T$ be a column vector with length $n$. Therefore, all derivatives are put together in the following system of equations:

$$\begin{cases} A\mathbf{x} - \mathbf{b} = \mathbf{0}, & \text{primal feasibility} \\ A^T\pi^T - \mathbf{z} - \mathbf{c}^T = \mathbf{0}, & \text{dual feasibility} \\ XZ\mathbf{e} - \mu\mathbf{e} = \mathbf{0}, & \mu - \text{complementary slackness.} \end{cases}$$

The solution of this system provides parallel gradient vectors for the objective functions of the primal and the dual forms. The third equation, i.e. $\mu$-complementary slackness, connects the solutions of both problems. In order to solve this system, the algorithm finds the zeros for those three equations using the Newton-Raphson method [24], which considers the following dynamical system:

$$x_{t+1} = x_t - \frac{f(x_t)}{f'(x_t)},$$

to modify variable $x$ along iterations indexed as $t$. This method estimates the zero for function $f(.)$ by taking its derivative $f'$ in the direction of $x$ (see Fig. 5.38).

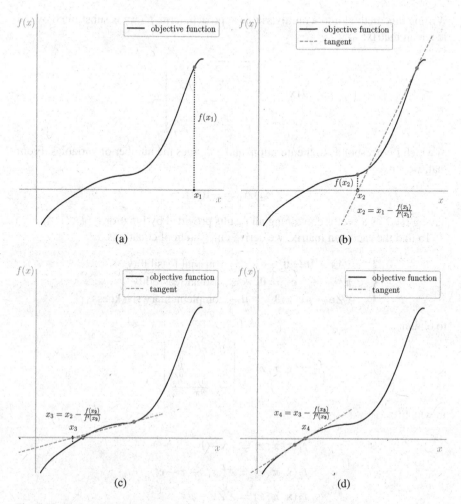

**Fig. 5.38**  Illustrating four iterations (**a–d**) of the Newton-Raphson method

We can reformulate the Newton-Raphson method, as follows:

$$x_{t+1} = x_t - \frac{f(x_t)}{f'(x_t)}$$

$$x_{t+1} - x_t = -\frac{f(x_t)}{f'(x_t)}$$

$$\Delta_x = -\frac{f(x_t)}{f'(x_t)},$$

to finally obtain:

$$f'(x_t)\Delta_x = -f(x_t).$$

Writing this method into a multivariable approach, term $f'(x_t)$ is substituted by the Jacobian matrix:

$$
J(\mathbf{x}_t) =
\begin{bmatrix}
\frac{\partial f_1}{\partial x_1} & \frac{\partial f_1}{\partial x_2} & \cdots & \frac{\partial f_1}{\partial x_n} \\
\frac{\partial f_2}{\partial x_1} & \frac{\partial f_2}{\partial x_2} & \cdots & \frac{\partial f_2}{\partial x_n} \\
\vdots & \vdots & \ddots & \vdots \\
\frac{\partial f_n}{\partial x_n} & \frac{\partial f_n}{\partial x_n} & \cdots & \frac{\partial f_n}{\partial x_n}
\end{bmatrix},
$$

in which $t$ corresponds to the iteration, and $n$ defines the number of variables. From that, we solve:

$$
J(\mathbf{x}_t)\Delta_x = -\mathbf{f}(\mathbf{x}_t),
$$

having $\mathbf{f}(\mathbf{x}_t)$ as a vector containing all results provided by functions $f_1, \ldots, f_n$.

To find the Jacobian matrix, we derive our system of equations:

$$
\begin{cases}
A\mathbf{x} - \mathbf{b} = \mathbf{0}, & \text{primal feasibility} \\
A^T \pi^T - \mathbf{z} - \mathbf{c}^T = \mathbf{0}, & \text{dual feasibility} \\
XZ\mathbf{e} - \mu\mathbf{e} = \mathbf{0}, & \mu - \text{complementary slackness}
\end{cases},
$$

to obtain:

$$
J(\mathbf{x}, \pi, \mathbf{z}) =
\begin{bmatrix}
\frac{\partial f_1}{\partial \mathbf{x}} & \frac{\partial f_1}{\partial \pi} & \frac{\partial f_1}{\partial \mathbf{z}} \\
\frac{\partial f_2}{\partial \mathbf{x}} & \frac{\partial f_2}{\partial \pi} & \frac{\partial f_2}{\partial \mathbf{z}} \\
\frac{\partial f_3}{\partial \mathbf{x}} & \frac{\partial f_3}{\partial \pi} & \frac{\partial f_3}{\partial \mathbf{z}}
\end{bmatrix},
$$

in which:

$$
f_1(\mathbf{x}, \pi, \mathbf{z}) = A\mathbf{x} - \mathbf{b}
$$

$$
f_2(\mathbf{x}, \pi, \mathbf{z}) = A^T \pi^T - \mathbf{z} - \mathbf{c}^T
$$

$$
f_3(\mathbf{x}, \pi, \mathbf{z}) = XZ\mathbf{e} - \mu\mathbf{e}.
$$

Consequently:

$$
\begin{array}{lll}
\frac{\partial f_1}{\partial \mathbf{x}} = A & \frac{\partial f_2}{\partial \mathbf{x}} = \mathbf{0} & \frac{\partial f_3}{\partial \mathbf{x}} = Z \\
\frac{\partial f_1}{\partial \pi} = \mathbf{0} & \frac{\partial f_2}{\partial \pi} = A^T & \frac{\partial f_3}{\partial \pi} = \mathbf{0} \\
\frac{\partial f_1}{\partial \mathbf{z}} = \mathbf{0} & \frac{\partial f_2}{\partial \mathbf{z}} = -I & \frac{\partial f_3}{\partial \mathbf{z}} = X.
\end{array}
$$

Therefore, the Jacobian matrix is:

$$
J(\mathbf{x}, \pi, \mathbf{z}) =
\begin{bmatrix}
A & \mathbf{0} & \mathbf{0} \\
\mathbf{0} & A^T & -I \\
Z & \mathbf{0} & X
\end{bmatrix}.
$$

Now we must consider some valid initial values, i.e., respecting the constraints, for variables in order to start the optimization process:

$$\mathbf{f}(\mathbf{x}_0) = A\mathbf{x}_0 - \mathbf{b}$$

$$\mathbf{f}(\boldsymbol{\pi}_0) = -\mathbf{c}^T + A^T(\boldsymbol{\pi}_0)^T - \mathbf{z}_0$$

$$\mathbf{f}(\mathbf{z}_0) = X_0 Z_0 \mathbf{e} - \mu\mathbf{e},$$

in which the subscript index refers to the iteration (first iteration is zero).

Using the Newton-Raphson's formulation:

$$J(\mathbf{x}_t)\Delta_x = -\mathbf{f}(\mathbf{x}_t),$$

to obtain:

$$J(\mathbf{x}, \boldsymbol{\pi}, \mathbf{z}) = \begin{bmatrix} A & 0 & 0 \\ 0 & A^T & -I \\ Z & 0 & X \end{bmatrix} \begin{bmatrix} \Delta_\mathbf{x} \\ \Delta_\pi \\ \Delta_\mathbf{z} \end{bmatrix} = - \begin{bmatrix} \mathbf{f}(\mathbf{x}_0) \\ \mathbf{f}(\boldsymbol{\pi}_0) \\ \mathbf{f}(\mathbf{z}_0) \end{bmatrix},$$

and in open-form:

$$J(\mathbf{x}, \boldsymbol{\pi}, \mathbf{z}) = \begin{bmatrix} A & 0 & 0 \\ 0 & A^T & -I \\ Z & 0 & X \end{bmatrix} \begin{bmatrix} \Delta_\mathbf{x} \\ \Delta_\pi \\ \Delta_\mathbf{z} \end{bmatrix} = - \begin{bmatrix} A\mathbf{x} - \mathbf{b} \\ -\mathbf{c}^T + A^T(\boldsymbol{\pi})^T - \mathbf{z} \\ XZ\mathbf{e} - \mu\mathbf{e} \end{bmatrix}.$$

Note this problem has analytical solution given the Jacobian matrix is sufficiently sparse.[10] Using the first row of the Jacobian, observe matrix $A$ helps us to find $\Delta_\mathbf{x}$, the second row helps us to find $\Delta_\pi$, and, at last, the third take both previous results to solve $\Delta_\mathbf{z}$. From a simple matrix multiplication, we find:

$$\Delta_\mathbf{x} = (AZ^{-1}XA^T)^{-1}(-\mathbf{b} + \mu AZ^{-1}\mathbf{e} + AZ^{-1}X\mathbf{f}(\boldsymbol{\pi}))$$

$$\Delta_\pi = -\mathbf{f}(\boldsymbol{\pi}) + A^T\Delta_\pi$$

$$\Delta_\mathbf{z} = Z^{-1}(\mu\mathbf{e} - XZ\mathbf{e} - X\Delta_\mathbf{z}),$$

which are the equations used to find the best values for $\mathbf{x}$, $\boldsymbol{\pi}$ and $\mathbf{z}$.

Listing 5.1 provides the implementation of Primal-Dual Path-Following optimization algorithm for this particular linear problem. Comments throughout the source code match every formulation step with the corresponding implementation.

---

[10]Other scenarios may require a solver to approximate the solution.

**Listing 5.1** Primal-Dual Path-Following optimization algorithm to solve the first linear problem

```
1   # This package is required to find matrix inverses (function
        ginv).
2   # The use of inverses is only for didactical purposes.
3   require(MASS)
4
5   ########## PRIMAL FORM ###########
6
7   # Defining vector c to multiply 2*x1+3*x2+0*x3+0*x4
8   c = matrix(c(2,3,0,0), nrow=1)
9
10  # Defining matrix A which defines the constraint functions:
11  #
12  #   2*x1 +   x2 + x3 = 8
13  #    x1 + 2*x2 + x4 = 6
14  #
15  A = matrix(c(2,1,1,0,
16                1,2,0,1), nrow=2, byrow=T)
17
18  # Defining the right-side terms for constraint functions
19  b = matrix(c(8,6), ncol=1)
20
21  # Defining some initialization for x1, x2, x3 and x4.
22  # This is made respecting the constraint functions
23  # for the primal form as follows:
24  #
25  # i) We simply decided to set x1=1, which respects the
        constraint x1 >=0
26  # ii) Then, we applied x1=1 into the first constraint
        function:
27  #
28  #         2*x1 +   x2 + x3 = 8
29  #         2*(1)+   x2 + x3 = 8
30  # iii) So, we also decided to set x2=1 (respecting the
        constraint x2>=0), thus:
31  #
32  #         2*(1)+   x2 + x3 = 8
33  #         2*(1)+   (1) + x3 = 8
34  #         3 + x3 = 8
35  #         x3 = 8 - 3 = 5
36  #
37  # iv) Allowing us to find x3=5, which also respects the
        constraint x3 >= 0
38  #
39  # v) Next, we had to find an acceptable value for x4 which
        must be x4 >= 0,
40  #       as defined by the primal form. So we got the second
        constraint function:
41  #
42  #     x1 + 2*x2 + x4 = 6
43  #
44  #     Given x1=1, x2=1 and x3=5 we found:
45  #
```

```
46  #     x1  + 2*x2  + x4 = 6
47  #     (1) + 2*(1) + x4 = 6
48  #     3 + x4 = 6
49  #     x4 = 6 - 3 = 3
50  #
51  # vi) Next, we organized x1, x2, x3 and x4 in a diagonal
        matrix as follows.
52  X = diag(c(1,1,5,3))
53
54  ########## DUAL FORM ###########
55
56  # From the dual side, we had to set pi1, pi2. For that, we
        simply selected
57  # two values as follows:
58  Pi = matrix(c(2,2), ncol=1)
59
60  # Then, we ensured the constraint functions for the dual
        form were respected.
61  #
62  # i) First constraint function:
63  #
64  #     2*pi1 + pi2 - z1 = 2
65  #     2*(2) + (2) - z1 = 2
66  #     6 - z1 = 2
67  #     - z1 = 2 - 6
68  #     z1 = 4
69  #
70  #         respecting the constraint z1 >= 0, which is defined
        for this dual form.
71  #
72  # ii) Second constraint function:
73  #
74  #     pi1 + 2*pi2 - z2 = 3
75  #
76  #         substituting pi1=2 and pi2=2, we have:
77  #
78  #     (2) + 2*(2) - z2 = 3
79  #     6 - z2 = 3
80  #     - z2 = 3 - 6
81  #     z2 = 3
82  #
83  #         respecting the constraint z2 >= 0, which is defined
        for this dual form.
84  #
85  # iii) Third constraint function:
86  #
87  #     pi1 - z3 = 0
88  #     (2) - z3 = 0
89  #     z3 = 2
90  #
91  #         respecting the constraint z3 >= 0, which is defined
        for this dual form.
92  #
93  # iv) Fourth constraint function:
```

```
94  #
95  #          pi2 - z4 = 0
96  #          (2) - z4 = 0
97  #          z4 = 2
98  #
99  #          respecting the constraint z4 >= 0, which is defined
        for this dual form.
100 #
101 # v) Next, we simply set the diagonal matrix Z with those
        values we found (remember
102 #      they respect all constraint functions, otherwise we
        would not obtain the solution).
103 Z = diag(c(4,3,2,2))
104
105 # Starting mu with some positive value.
106 mu = 1.4375
107
108 # Defining a column vector filled with 1s.
109 e = matrix(rep(1,4), ncol=1)
110
111 # Variable eta defines the rate of change for the primal and
        dual variables along iterations.
112 eta = 0.995
113
114 # Defining vectors Delta_x, Delta_pi and Delta_z. Combined
        they define the column vector
115 # on which the Jacobian matrix will be applied to.
116 dX = rep(0, 4)
117 dPi = rep(0, 2)
118 dZ = rep(0, 4)
119
120 # Setting a counter to know the current iteration of this
        algorithm.
121 counter = 1
122
123 # Defining a stop criterion. While the gap term is greater
        than such threshold,
124 # this algorithm keeps running, otherwise it will stop and
        print the solution out.
125 threshold = 1e-5
126
127 # Computing the current gap term for the solution we defined
        , i.e., for the current
128 # values of x1, x2, x3 and x4 in the primal form and z1, z2,
        z3 and z4 in the dual.
129 gap = t(e) %*% X %*% Z %*% e
130
131 # While the gap is greater than acceptable, run:
132 while (gap > threshold) {
133
134         # Printing out the current iteration and the gap
135         cat("Iteration:_", counter, "_with_Gap_=_", gap, "\n
                ")
136
```

```
137              # Solving the linear system of equations
138              deltaD = t(A)%*%dPi − dZ
139              dPi = ginv(A%*%ginv(Z)%*%X%*%t(A))%*%(−b+mu*A%*%ginv
                     (Z)%*%e+A%*%ginv(Z)%*%X%*%deltaD)
140              dZ = −deltaD + t(A)%*%dPi
141              dX = ginv(Z)%*%(e%*%mu−X%*%Z%*%e−X%*%dZ)
142
143              # Changing variables for the next iteration (only
                     the diagonal in here).
144              # The algorithm walks according to the gradient
                     vector
145              X = X + eta * diag(as.vector(dX))
146              Pi = Pi + eta * dPi
147              Z = Z + eta * diag(as.vector(dZ))
148
149              # Computing the gap again to verify if we will carry
                     on running
150              gap = t(e) %*% X %*% Z %*% e
151
152              # Reducing the influence of the barrier term, so we
                     can get closer to a vertex
153              # if the solution is eventually there (in this case,
                     it is!)
154              mu = as.numeric(gap / counter^2)
155
156              # Counting the number of iterations
157              counter = counter + 1
158    }
159
160    cat("Constraint_functions_must_be_equal_to_zero:\n")
161
162    cat("Primal_feasibility:\n")
163    print(A%*%diag(X)−b)
164
165    cat("Dual_feasibility:\n")
166    print(t(A)%*%Pi−diag(Z)−t(c))
167
168    cat("u−complementary_slackness:\n")
169    print(diag(X)%*%Z−mu)
170
171    cat("Values_found_for_X:\n")
172    print(diag(X))
173
174    cat("Values_found_for_Z:\n")
175    print(diag(Z))
```

After running the function of Listing 5.1, the following output is obtained:

**Listing 5.2** Text output produced by Listing 5.1

```
Loading required package: MASS
Iteration:  1  with Gap =  23
Iteration:  2  with Gap =  5.83625
Iteration:  3  with Gap =  23.25746
```

```
Iteration:    4   with  Gap  =    23.25746
Iteration:    5   with  Gap  =    10.40125
Iteration:    6   with  Gap  =    2.639318
Iteration:    7   with  Gap  =    0.4333759
Iteration:    8   with  Gap  =    0.050079
Iteration:    9   with  Gap  =    0.004318036
Iteration:    10  with  Gap  =    0.000290118
Iteration:    11  with  Gap  =    1.570577e−05
Constraint  functions  must  be  equal  to  zero:
Primal  feasibility:
                    [,1]
[1,]   −2.664535e−15
[2,]   −1.776357e−15
Dual  feasibility:
          [,1]
[1,]      0
[2,]      0
[3,]      0
[4,]      0
u−complementary  slackness:
               [,1]            [,2]            [,3]            [,4]
[1,]  1.701224e−07 1.700303e−07 1.700874e−07 1.701183e−07
Values  found  for  X:
[1]  3.333333e+00 1.333333e+00 5.277074e−07 1.319500e−07
Values  found  for  Z:
[1]  5.278124e−08 1.318840e−07 3.333333e−01 1.333333e+00
```

Observe the gap reduces along iterations, confirming the algorithm is getting closer to the solution from the primal and the dual sides simultaneously, as illustrated in Fig. 5.36. At the end, the primal constraints (primal feasibility), the dual constraints (dual feasibility), and the $\mu$-complementary slackness are close to zero as expected.

The solution is found so that the variables $X$ and $Z$ are: $x_1 = 3.33\ldots$, $x_2 = 1.33\ldots$, $x_3 = 5.27 \times 10^{-7}$, $x_4 = 1.31 \times 10^{-7}$, $z_1 = 5.27 \times 10^{-8}$, $z_2 = 1.31 \times 10^{-7}$, $z_3 = 0.33\ldots$, $z_4 = 1.33\ldots$. Observe all of them respect the constraints, once they are equal or greater than zero.

Next, notice $x_3$ and $x_4$ approach zero as desired, meaning the slack variables for the primal form have a minimal effect. Also notice $z_1$ and $z_2$ approach zero for the same reasons, but for the dual problem. Assessing the linear optimization problem using the solution:

$$x_1 = 3.33\ldots, x_2 = 1.33\ldots$$

$$z_3 = 0.33\ldots, z_4 = 1.33\ldots,$$

which (approximately) corresponds to the solution graphically found in Fig. 5.36, i.e., $x_1 = \frac{10}{3}$, $x_2 = \frac{4}{3}$, $\pi_1 = \frac{1}{3}$, and $\pi_2 = \frac{4}{3}$. The only difference is that now we have $z_3$ and $z_4$ as slack variables to make $\pi_1$ and $\pi_2$ equal to zero, respectively, thus from the dual variable constraints:

$$\pi_1 - z_3 = 0$$

$$\pi_1 - 0.33\ldots = 0$$

$$\pi_1 = 0.33\ldots$$

$$\pi_2 - z_4 = 0$$

$$\pi_2 - 1.333333 = 0$$

$$\pi_2 = 1.333333.$$

### 5.5.2 IPM to Solve the SVM Optimization Problem

In this section, we consider the Interior Point Method proposed in [6] as basis to address the SVM optimization problem.[11] Let the soft-margin SVM optimization problem:

**Primal problem**

$$\underset{\mathbf{w}\in\mathbb{H}, b\in\mathbb{R}, \boldsymbol{\xi}\in\mathbb{R}^m}{\text{minimize}} \quad \tau(\mathbf{w}) = \frac{1}{2}\|\mathbf{w}\|^2 + C\sum_{i=1}^m \xi_i$$

subject to $y_i(<\mathbf{w}, \mathbf{x}_i> +b) \geq 1 - \xi$, for all $i = 1, \ldots, m$,

$\xi_i \geq 0$, for all $i = 1, \ldots, m$

**Dual problem**

$$\underset{\alpha}{\text{maximize}} \quad W(\boldsymbol{\alpha}) = -\frac{1}{2}\sum_{i=1}^m\sum_{j=1}^m \alpha_i\alpha_j y_i y_j <\mathbf{x}_i, \mathbf{x}_j> + \sum_{i=1}^m \alpha_i$$

subject to $0 \leq \alpha_i \leq C$, for all $i = 1, \ldots, m$,

$$\text{and} \quad \sum_{i=1}^m \alpha_i y_i = 0.$$

We start by rewriting the primal and dual forms for the soft-margin SVM optimization problem in matrix-form, as detailed next. Such step requires the terms found after the primal Lagrangian (as seen in Sect. 4.3):

$$\mathbf{w} = \sum_{i=1}^m \alpha_i y_i \mathbf{x}_i,$$

---

[11] We detail and implement most of such paper, but we do not consider its rank reduction.

which is substituted in the primal:

$$\underset{\alpha \in \mathbb{R}^m}{\text{minimize}} \ \tau(\mathbf{w}) = \frac{1}{2} < \sum_{i=1}^{m} \alpha_i y_i \mathbf{x}_i, \sum_{j=1}^{m} \alpha_j y_j \mathbf{x}_j > + C \sum_{i=1}^{m} \xi_i$$

$$\text{subject to} \ \ y_i (< \sum_{j=1}^{m} \alpha_j y_j \mathbf{x}_j, \mathbf{x}_i > + b) \geq 1 - \xi, \ \text{for all } i = 1, \ldots, m,$$

$$\xi_i \geq 0, \ \text{for all } i = 1, \ldots, m,$$

and then:

$$\underset{\alpha \in \mathbb{R}^m}{\text{minimize}} \ \tau(\mathbf{w}) = \frac{1}{2} \sum_{i=1}^{m} \sum_{j=1}^{m} \alpha_i \alpha_j y_i y_j < \mathbf{x}_i, \mathbf{x}_j > + C \sum_{i=1}^{m} \xi_i$$

$$\text{subject to} \ \sum_{j=1}^{m} \alpha_j y_j y_i < \mathbf{x}_j, \mathbf{x}_i > + y_i b + \xi_i \geq 1, \ \text{for all } i = 1, \ldots, m,$$

$$\xi_i \geq 0, \ \text{for all } i = 1, \ldots, m.$$

Next, we define $\mathbf{e} = \begin{bmatrix} 1 & 1 & \ldots & 1 \end{bmatrix}^T$ to be a column vector with length $m$, in order to produce:

$$\underset{\alpha \in \mathbb{R}^m}{\text{minimize}} \ \tau(\mathbf{w}) = \frac{1}{2} \sum_{i=1}^{m} \sum_{j=1}^{m} \alpha_i \alpha_j y_i y_j < \mathbf{x}_i, \mathbf{x}_j > + C \sum_{i=1}^{m} \xi_i$$

$$\text{subject to} \ \sum_{j=1}^{m} \alpha_j y_j y_i < \mathbf{x}_j, \mathbf{x}_i > + y_i b + \xi_i \geq \mathbf{e}_i, \ \text{for all } i = 1, \ldots, m,$$

$$\xi_i \geq 0, \ \text{for all } i = 1, \ldots, m.$$

Consider matrix $X$ contains every vector $\mathbf{x}_i$ along its rows, and the column vector $\mathbf{y}$ with the corresponding classes $y_i$. From that, matrix $Q = (\mathbf{y}\mathbf{y}^T) \times (\mathbf{X}\mathbf{X}^T)$ allows us to simplify the primal form:

$$\underset{\alpha \in \mathbb{R}^m}{\text{minimize}} \ \tau(\mathbf{w}) = \frac{1}{2} \alpha^T Q \alpha + C \sum_{i=1}^{m} \xi_i$$

$$\text{subject to} \ Q\alpha + \mathbf{y}b + \boldsymbol{\xi} \geq \mathbf{e}$$

$$\xi_i \geq 0, \ \text{for all } i = 1, \ldots, m,$$

in which:

$$\alpha = \begin{bmatrix} 1 \\ 1 \\ \vdots \\ 1 \end{bmatrix},$$

and the constraint function is provided in matrix form.

We now remove variable $\mathbf{w}$ from the formulation and, simultaneously, add a vector $\mathbf{s}$ with slack variables, permitting us to reformulate the constraint function in terms of equalities:

$$\underset{\alpha \in \mathbb{R}^m}{\text{minimize}} \quad \frac{1}{2}\alpha^T Q\alpha + C \sum_{i=1}^{m} \xi_i$$

$$\text{subject to} \quad Q\alpha + yb + \xi + \mathbf{s} = \mathbf{e}$$

$$\xi \geq \mathbf{0}, \mathbf{s} \geq \mathbf{0},$$

having the minimization in terms of the slack variables.[12] Also notice all variables $\xi_i$ and $s_i$ are represented as vectors. This is the final matrix form for this primal problem.

Thus, we proceed with the same representation for the dual form:

$$\underset{\alpha}{\text{maximize}} \quad W(\alpha) = -\frac{1}{2}\sum_{i=1}^{m}\sum_{j=1}^{m}\alpha_i\alpha_j y_i y_j < \mathbf{x}_i, \mathbf{x}_j > + \sum_{i=1}^{m}\alpha_i$$

$$\text{subject to} \quad 0 \leq \alpha_i \leq C, \text{ for all } i = 1,\ldots,m,$$

$$\text{and} \quad \sum_{i=1}^{m}\alpha_i y_i = 0,$$

to obtain:

$$\underset{\alpha}{\text{maximize}} \quad W(\alpha) = -\frac{1}{2}\alpha^T Q\alpha + \mathbf{e}^T\alpha$$

$$\text{subject to} \quad \mathbf{0} \leq \alpha \leq \mathbf{C},$$

$$\mathbf{y}^T\alpha = 0,$$

---

[12]Meaning we wish them to have the least relevance as possible for our problem, once they are associated to relaxation terms.

having all constraints in vector form, including vector $\mathbf{C}$, which is given by:

$$\mathbf{C} = \begin{bmatrix} C \\ C \\ \vdots \\ C \end{bmatrix}.$$

We then present the primal and the dual forms in the matrix form:

**Primal problem**

$$\underset{\alpha \in \mathbb{R}^m}{\text{minimize}} \quad \frac{1}{2}\alpha^T Q \alpha + C \sum_{i=1}^{m} \xi_i$$

subject to $\quad Q\alpha + \mathbf{y}b + \boldsymbol{\xi} + \mathbf{s} = \mathbf{e}$

$$\boldsymbol{\xi} \geq \mathbf{0}, \mathbf{s} \geq \mathbf{0}$$

**Dual problem**

$$\underset{\alpha}{\text{maximize}} \quad -\frac{1}{2}\alpha^T Q \alpha + \mathbf{e}^T \alpha$$

subject to $\quad \mathbf{0} \leq \alpha \leq \mathbf{C},$

$$\mathbf{y}^T \alpha = 0$$

and proceed with the formulation of Lagrangians to later build the Primal-Dual Path Following algorithm:

$$\Lambda_{\text{primal}} = \frac{1}{2}\alpha^T Q \alpha + C \sum_{i=1}^{m} \xi_i - \lambda(Q\alpha + \mathbf{y}b + \boldsymbol{\xi} + \mathbf{s} - \mathbf{e}) - \mu \sum_{i=1}^{m} \log \xi_i$$

$$- \mu \sum_{i=1}^{m} \log s_i,$$

which contains two barrier terms to bound vectors $\boldsymbol{\xi}$ and $\mathbf{s}$. Remember the constraint function must be equal to zero and considered as part of the Lagrangian (see more in Sects. 4.5.1 and 4.5.2), having $\lambda$ as the KKT multiplier.

Then, we derive the Lagrangian $\Lambda_{\text{primal}}$ in order to find the stationary point:

$$\frac{\partial \Lambda_{\text{primal}}}{s_i} = -\mu \frac{1}{s_i} + \lambda = 0$$

$$\frac{\partial \Lambda_{\text{primal}}}{b} = -\mathbf{y}^T \lambda = 0$$

$$\frac{\partial \Lambda_{\text{primal}}}{\xi_i} = C - \mu \frac{1}{\xi_i} - \lambda = 0$$

$$\frac{\partial \Lambda_{\text{primal}}}{\lambda} = Q\alpha + \mathbf{y}b + \boldsymbol{\xi} - \mathbf{s} - \mathbf{e} = 0,$$

and, then, we represent vectors $\mathbf{s}$ and $\boldsymbol{\xi}$ as diagonal matrices $S$ and $\boldsymbol{\Xi}$:

$$S = \begin{bmatrix} s_1 & 0 & \ldots & 0 \\ 0 & s_2 & \ldots & 0 \\ \vdots & \vdots & \ddots & 0 \\ 0 & 0 & \ldots & s_m \end{bmatrix},$$

and:

$$\boldsymbol{\Xi} = \begin{bmatrix} \xi_1 & 0 & \ldots & 0 \\ 0 & \xi_2 & \ldots & 0 \\ \vdots & \vdots & \ddots & 0 \\ 0 & 0 & \ldots & \xi_m \end{bmatrix}.$$

from which derivatives are rewritten as:

$$\frac{\partial \Lambda_{\text{primal}}}{S} = -\mu S^{-1} + \lambda = 0$$

$$\frac{\partial \Lambda_{\text{primal}}}{b} = -\mathbf{y}^T \lambda = 0$$

$$\frac{\partial \Lambda_{\text{primal}}}{\boldsymbol{\Xi}} = C - \mu \boldsymbol{\Xi}^{-1} - \lambda = 0$$

$$\frac{\partial \Lambda_{\text{primal}}}{\lambda} = Q\boldsymbol{\alpha} + \mathbf{y}b + \boldsymbol{\xi} - \mathbf{s} - \mathbf{e} = 0.$$

We now build the Lagrangian for the dual form:

$$\Lambda_{\text{dual}} = -\frac{1}{2}\boldsymbol{\alpha}^T Q\boldsymbol{\alpha} + \mathbf{e}^T \boldsymbol{\alpha} - \beta(\mathbf{y}^T \boldsymbol{\alpha}) + \mu \sum_{i=1}^{m} \log \alpha_i + \mu \sum_{i=1}^{m} m \log(-\alpha_i + C),$$

in which $\beta$ is the KKT multiplier, and two barrier terms were added to ensure $\alpha_i \geq 0$ for $i = 1, \ldots, m$ and $\alpha_i \leq C$. This second constraint was obtained as follows:

$$\alpha_i \leq C$$

$$\alpha_i - C \leq 0$$

Multiplying both sides by $-1$ we find:

$$-\alpha_i + C \geq 0,$$

which provides the barrier term.

We then derive $\Lambda_{\text{dual}}$:

$$\frac{\partial \Lambda_{\text{dual}}}{\partial \alpha_i} = -\frac{1}{2}Q\alpha + \mathbf{e} - \beta\mathbf{y} + \mu\frac{1}{\alpha_i} - \mu\frac{1}{\alpha_i - C} = 0$$

$$\frac{\partial \Lambda_{\text{dual}}}{\partial \beta} = -\mathbf{y}^T\alpha = 0,$$

and consider $\alpha$ to be a diagonal matrix:

$$\alpha = \begin{bmatrix} \alpha_1 & 0 & \dots & 0 \\ 0 & \alpha_2 & \dots & 0 \\ \vdots & \vdots & \ddots & 0 \\ 0 & 0 & \dots & \alpha_m \end{bmatrix},$$

allowing to rewrite derivatives:

$$\frac{\partial \Lambda_{\text{dual}}}{\partial \alpha} = -\frac{1}{2}Q\alpha + \mathbf{e} - \beta\mathbf{y} + \mu\alpha^{-1} - \mu(\alpha - C)^{-1} = 0$$

$$\frac{\partial \Lambda_{\text{dual}}}{\partial \beta} = -\mathbf{y}^T\alpha = 0.$$

As next step, the Primal-Dual Path Following algorithm firstly assesses which of the derivatives must be considered in the system of equations. Initially, we select only the derivatives that are different from both primal and dual forms to compose the system. Given that condition, we have:

$$\frac{\partial \Lambda_{\text{primal}}}{b} = -\mathbf{y}^T\lambda = 0,$$

and:

$$\frac{\partial \Lambda_{\text{dual}}}{\partial \beta} = -\mathbf{y}^T\alpha = 0,$$

from the primal and dual forms, respectively. Thus observe $\lambda$ must be equal to $\alpha$, so both derivatives are equivalent. This also reflects in the following derivatives:

$$\frac{\partial \Lambda_{\text{primal}}}{S} = -\mu S^{-1} + \lambda = 0$$

$$\mu = S\lambda,$$

and:

$$\frac{\partial \Lambda_{\text{dual}}}{\partial \alpha} = -\frac{1}{2} Q\alpha + \mathbf{e} - \beta \mathbf{y} + \mu \alpha^{-1} - \mu(\alpha - C)^{-1} = 0,$$

what is necessary to ensure both forms (primal and dual) provide the same solution. Also, observe the derivatives:

$$\frac{\partial \Lambda_{\text{primal}}}{\lambda} = Q\alpha + \mathbf{y}b + \boldsymbol{\xi} - \mathbf{s} - \mathbf{e} = 0,$$

and:

$$\frac{\partial \Lambda_{\text{dual}}}{\partial \alpha} = -\frac{1}{2} Q\alpha + \mathbf{e} - \beta \mathbf{y} + \mu \alpha^{-1} - \mu(\alpha - C)^{-1} = 0,$$

allow us to conclude that:

1. vector $\mathbf{s}$, organized as a diagonal matrix $S$, must be equal to $\mu \alpha^{-1}$;
2. vector $\boldsymbol{\xi}$, organized as a diagonal matrix $\Xi$, must be equal to $\mu(C-\alpha)^{-1}$, so that both forms provide the same solution.

From those remarks, the smallest system of equations to be solved is:

$$\begin{cases} Q\alpha + \mathbf{y}b + \boldsymbol{\xi} - \mathbf{s} - \mathbf{e} = 0 \\ -\mathbf{y}^T \alpha = 0 \\ S\lambda - \mu = 0 \\ (C - \alpha)\Xi - \mu = 0, \end{cases}$$

in which the last equation ensures $\Xi = \mu(C - \alpha)^{-1}$, a result from:

$$\Xi = -\mu(C - \alpha)^{-1}$$

$$(C - \alpha)\Xi = \mu$$

$$(C - \alpha)\Xi - \mu = 0.$$

Next, we solve this system using Newton-Raphson's method, which requires the Jacobian matrix:

$$J(\alpha, b, \mathbf{s}, \boldsymbol{\xi}) = \begin{bmatrix} \frac{\partial f_1}{\partial \alpha} & \frac{\partial f_1}{\partial b} & \frac{\partial f_1}{\partial s} & \frac{\partial f_1}{\partial \xi} \\ \frac{\partial f_2}{\partial \alpha} & \frac{\partial f_2}{\partial b} & \frac{\partial f_2}{\partial s} & \frac{\partial f_2}{\partial \xi} \\ \frac{\partial f_3}{\partial \alpha} & \frac{\partial f_3}{\partial b} & \frac{\partial f_3}{\partial s} & \frac{\partial f_3}{\partial \xi} \\ \frac{\partial f_4}{\partial \alpha} & \frac{\partial f_4}{\partial b} & \frac{\partial f_4}{\partial s} & \frac{\partial f_4}{\partial \xi} \end{bmatrix},$$

which is given by:

$$J(\alpha, b, \mathbf{s}, \boldsymbol{\xi}) = \begin{bmatrix} Q & \mathbf{y} & -I & I \\ -\mathbf{y}^T & 0 & 0 & 0 \\ S & 0 & \alpha & 0 \\ -\varXi & 0 & 0 & (C-\alpha) \end{bmatrix}.$$

So, applying the Newton-Raphson's method:

$$J(\mathbf{x}_t)\Delta_x = -\mathbf{f}(\mathbf{x}_t),$$

we solve:

$$J(\alpha, b, \mathbf{s}, \boldsymbol{\xi})\Delta_{\alpha,b,s,\xi} = - \begin{bmatrix} Q\alpha + \mathbf{y}b + \boldsymbol{\xi} - \mathbf{s} - \mathbf{e} \\ -\mathbf{y}^T\alpha \\ S\lambda - \mu \\ (C-\alpha)\varXi - \mu \end{bmatrix}.$$

Let the training set contain 200 examples organized as row vectors $\mathbf{x}_i$ in matrix $X$, a column vector $\mathbf{y}$ containing all respective classes in $\{-1, +1\}$, and, finally, recall matrix $Q = (\mathbf{y}\mathbf{y}^T) \times (XX^T)$. The Jacobian matrix must be squared, contain the same number of rows and columns as variables of the problem,[13] in which blocks represent each term: $I$ is the identity matrix, and $-I$ its negative; and terms $S$, $\alpha$, $\varXi$, and $(C - \alpha)$ are diagonal matrices. The free problem variables are $\alpha$ (vector with $m$ elements), $b$ (scalar value), $\mathbf{s}$ (vector with $m$ slack variables), and $\boldsymbol{\xi}$ (vector with $m$ elements). In total, we must adapt $3m + 1$ free variables, which indeed define the number of rows and columns of this Jacobian matrix. Parameter $m$ refers to the number of examples in the training set, which is $m = 200$ for this particular instance. After applying this Jacobian matrix on a column vector $\Delta_{\alpha,b,s,\xi}$, the necessary modifications ($\Delta_\alpha$, $\Delta_b$, $\Delta_s$ and $\Delta_\xi$) are computed for this problem.

Applying the linear transformation provided by the $(3m+1) \times (3m+1)$-Jacobian matrix on the $(3m + 1) \times 1$ vector $\Delta_{\alpha,b,s,\xi}$, a resultant vector with $3m + 1$ rows and one column is obtained.

The Primal-Dual Path Following algorithm still requires the following:

1. initial values for $\alpha$, respecting the feasible region, i.e., $\alpha \geq \mathbf{0}$;
2. initial values for $\boldsymbol{\xi}$;
3. an initial value for $b$;
4. and, finally, set matrix $S$ in terms of $\alpha$, $\varXi$ and $b$.

---

[13] Jacobian matrices must be squared so that the input and the output spaces have the same dimensionality, allowing inverse transformations.

Listing 5.3 details the SVM Primal-Dual Path Following algorithm, with additional comments. After loading this code in the R Statistical Software, the user should run function *ipm.svm()* to test the algorithm.

**Listing 5.3**  Primal-Dual Path Following algorithm to address the SVM optimization problem

```
1   # This function builds up a very simple linearly separable
        dataset
2   simpleDataset <- function() {
3
4           # Producing a two-dimensional dataset using the
                Normal distribution.
5           # Negative examples are defined to have mean (0,0)
                with standard deviation (1,1).
6           # Positive examples are defined to have mean (10,10)
                with standard deviation (1,1).
7
8           # These are the training examples
9           train <- cbind(rnorm(mean=0, sd=1, n=100), rnorm(
                mean=0, sd=1, n=100))
10          train <- rbind(train, cbind(rnorm(mean=10, sd=1, n
                =100), rnorm(mean=10, sd=1, n=100)))
11          train <- cbind(train, c(rep(-1, 100), rep(1, 100)))
12
13          # These are the test examples
14          test <- cbind(rnorm(mean=0, sd=1, n=10), rnorm(mean
                =0, sd=1, n=10))
15          test <- rbind(test, cbind(rnorm(mean=10, sd=1, n=10)
                , rnorm(mean=10, sd=1, n=10)))
16          test <- cbind(test, c(rep(-1, 10), rep(1, 10)))
17
18          # Returning the training and test sets using a list
19          return (list(train=train, test=test))
20  }
21
22  # This function outputs the classification (labels) for a
        given set.
23  # In our case, we use it to print out the labels predicted
        for the test set.
24  discrete.classification <- function(X, Y, alpha, b, X.test,
        Y.test, threshold = 1e-5) {
25          all.labels = NULL
26          alphas = diag(alpha)
27          alphas[alphas < threshold] = 0
28
29          for (i in 1:nrow(X.test)) {
30                  label = sum(alphas * as.vector(Y) * (X.test[
                        i,] %*% t(X))) + b
31                  if (label >= 0)
32                          label = 1
33                  else
34                          label = -1
35                  expected_label = Y.test[i,]
```

```
36                        all.labels = rbind(all.labels, cbind(
                              expected_label, label))
37                }
38
39             colnames(all.labels) = c("Expected_class", "
                   Predicted_class")
40
41             return (all.labels)
42     }
43
44     # This function implements the Primal−Dual Path Following
           algorithm
45     # for the simple dataset (linearly separable).
46     ipm.svm <- function() {
47
48             # Building up the linearly separable dataset
49             dataset = simpleDataset()
50
51             # Creating matrix X to represent all vector x_i (
                   training examples)
52             X = dataset$train[,1:2]
53
54             # This vector y with labels −1 and +1
55             # (we in fact created a matrix with a single column)
56             Y = matrix(dataset$train[,3], ncol=1)
57
58             # Number of training examples
59             npoints = nrow(X)
60
61             # Computing matrix Q as we formulated
62             Q = (Y %*% t(Y)) * (X %*% t(X))
63
64             # Defining values to start the execution
65             C = 1                              # Upper limit for
                   alpha_i
66             eta = 0.1                          # This will be used
                   to adapt the variables of our problem
67             b = runif(min=−1, max=1, n=1)     # Initial b
68             iteration = 1                      # Counter of
                   iteration
69             threshold = 1e−5                   # This parameter
                   defines the stop criterion
70
71             # Vector filled out with ones (one's vector)
72             e = rep(1, nrow(Q))
73
74             # Identity matrix
75             I = diag(rep(1, nrow(Q)))
76
77             # Setting all alphas as half of C into a diagonal
                   matrix
78             Alpha = diag(rep(C/2, nrow(Q)))
79
```

```
80          # Defining the diagonal matrix Xi using the same
                values of alphas,
.81         # what makes constraints respected according to the
                values of the
82          # slack variables
83          Xi = Alpha
84

85          # Computing the diagonal matrix S using the first
                equation, as follows:
86          # Q alpha + y b + Xi − s − e = 0
87          # Q alpha + y b + Xi − e = s
88          S = diag(as.vector(Q%*%diag(Alpha) + Y*b + diag(Xi)
                − e))

89
90          # This value found for S helps us to compute
                equation:
91          #
92          # S alpha − mu = 0
93          # S alpha = mu
94          #
95          # allowing to find the current Gap for our solution
96          # (please refer to the first Primal−Dual Path
                Following algorithm
97          #  used to tackle the linear optimization problem)
98          gap = e%*%S%*%Alpha%*%e
99
100         # This is the initial mu
101         mu = as.numeric(gap)
102
103         # Factor to reduce mu along iterations and
                eventually get closer to barriers.
104         # This is necessary if the solution is close to one
                of the barrier terms.
105         reducing.factor = 0.9
106
107         # Identity matrix
108         I = diag(nrow(Q))
109
110         # Building up the Jacobian matrix.
111         # First and second rows will not change anymore,
112         # therefore they are initialized before the
                iterative process.

113
114         # Jacobian matrix: first row
115         A = matrix(0, nrow=(3*npoints+1), ncol=(3*npoints+1)
                )
116         A[1:nrow(Q), 1:ncol(Q)] = Q
117         A[1:nrow(Q), ncol(Q)+1] = Y
118         A[1:nrow(Q), (ncol(Q)+2):(2*npoints+1)] = −I
119         A[1:nrow(Q), (2*npoints+2):(3*npoints+1)] = I
120
121         # Jacobian matrix: second row
122         A[nrow(Q)+1, 1:length(Y)] = −t(Y)
123
```

```
124        while (gap > threshold) {
125
126               # Jacobian matrix: third row
127               A[(npoints+2):(2*npoints+1), 1:npoints] = S
128               A[(npoints+2):(2*npoints+1), (npoints+2):(2*
                     npoints+1)] = Alpha
129
130               # Jacobian matrix: fourth row
131               A[(2*npoints+2):(3*npoints+1), 1:npoints] =
                     -Xi
132               A[(2*npoints+2):(3*npoints+1), (2*npoints+2)
                     :(3*npoints+1)] = diag(rep(C, npoints))-
                     Alpha
133
134               # Building up vector b
135               B = matrix(0, nrow=2*npoints+1, ncol=1)
136
137               # First function
138               f1 = - Q%*%diag(Alpha) - Y*b - diag(Xi) +
                     diag(S) + e
139
140               # Second function
141               f2 = diag(Alpha)%*%Y
142
143               # Third function
144               f3 = -diag(S%*%Alpha) + mu
145
146               # Fourth function
147               f4 = -(diag(rep(C, npoints))-Alpha)%*%diag(Xi
                     ) + mu
148
149               B[1:npoints] = f1
150               B[npoints+1] = f2
151               B[(npoints+2):(2*npoints+1)] = f3
152               B[(2*npoints+2):(3*npoints+1)] = f4
153
154               # Solving the system (this solver comes with
                     the package base)
155               d = solve(A, B)
156
157               # Cutting out the corresponding Deltas for
                     Alpha, b, S and Xi
158               # to be later used as updating factors
159               d_alpha = d[1:npoints]
160               d_b = d[npoints+1]
161               d_S = d[(npoints+2):(2*npoints+1)]
162               d_Xi = d[(2*npoints+2):(3*npoints+1)]
163
164               # Updating the variables for our problem.
165               # Parameter eta defines the update step
166               diag(Alpha) = diag(Alpha) + eta * d_alpha
167               b = b + eta * d_b
168               diag(S) = diag(S) + eta * d_S
169               diag(Xi) = diag(Xi) + eta * d_Xi
```

```
170
171                         # Counting iterations
172                         iteration = iteration + 1
173
174                         # Recalculating the Gap for the next
                                iteration
175                         gap = e%*%S%*%Alpha%*%e
176
177                         # We decrease the value of mu to allow our
                                algorithm to get
178                         # closer to barriers whenever necessary
179                         mu = mu * reducing.factor
180
181                         cat("Current_Gap_is_", gap, "\n")
182                 }
183
184         # Plotting the dataset and the support vectors.
185         # Support vectors correspond to every x_i that was
                found to help
186         # our algorithm define the maximal-margin hyperplane
187         colors = rep(1, nrow(Q))
188         ids = which(diag(Alpha) > 1e-5)
189         colors[ids] = 2
190         plot(X, col=colors, main="Dataset_and_support_
                vectors")
191         locator(1)
192
193         # Plotting the classification results.
194         # Creating matrix X to represent all vectors x_i (
                test examples)
195         X.test = dataset$test[,1:2]
196
197         # This vector y contains labels -1 and +1
198         # (we created a matrix with a single column)
199         Y.test = matrix(dataset$test[,3], ncol=1)
200
201         print(discrete.classification(X, Y, Alpha, b, X.test
                , Y.test, threshold = 1e-5))
202 }
```

The output provided by Listing 5.3 is similar to the output below, which contains information about the current gap, and the expected versus the predicted classes for test examples. Figure 5.39 illustrates the solution produced by this algorithm, in which every gray dot corresponds to a training example, while black dots to estimated support vectors.

**Listing 5.4** Text output produced by Listing 5.3

```
Current Gap is    1.701278e-05
Current Gap is    1.53115e-05
Current Gap is    1.378035e-05
Current Gap is    1.240231e-05
Current Gap is    1.116208e-05
```

**Fig. 5.39** Solution produced
by the Primal-Dual Path
Following algorithm for the
SVM problem

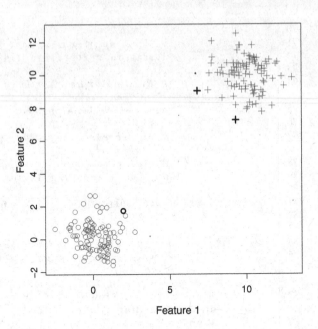

| Current | Gap | is | 1.004587e−05 | |
|---|---|---|---|---|
| Current | Gap | is | 9.041287e−06 | |
| | Expected | class | Predicted | class |
| [1,] | | −1 | | −1 |
| [2,] | | −1 | | −1 |
| [3,] | | −1 | | −1 |
| [4,] | | −1 | | −1 |
| [5,] | | −1 | | −1 |
| [6,] | | −1 | | −1 |
| [7,] | | −1 | | −1 |
| [8,] | | −1 | | −1 |
| [9,] | | −1 | | −1 |
| [10,] | | −1 | | −1 |
| [11,] | | 1 | | 1 |
| [12,] | | 1 | | 1 |
| [13,] | | 1 | | 1 |
| [14,] | | 1 | | 1 |
| [15,] | | 1 | | 1 |
| [16,] | | 1 | | 1 |
| [17,] | | 1 | | 1 |
| [18,] | | 1 | | 1 |
| [19,] | | 1 | | 1 |
| [20,] | | 1 | | 1 |

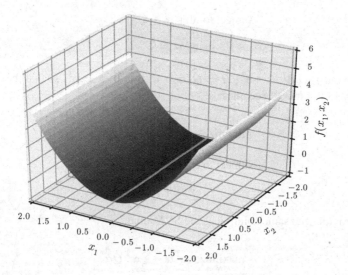

**Fig. 5.40**  Half-cylinder: all points on the light gray line represent a possible solution

### 5.5.3  Solving the SVM Optimization Problem Using Package LowRankQP

Now we introduce an optimized version of the SVM Primal-Dual Path Following algorithm available at package LowRankQP with the R Statistical Software [15]. LowRankQP implements the features described in [4, 6, 16, 20] to solve the following convex optimization problem:

$$\text{minimize } \mathbf{d}^T \alpha + \frac{1}{2}\alpha^T H \alpha$$

$$\text{subject to } A\alpha = \mathbf{b}$$

$$0 \leq \alpha \leq \mathbf{u},$$

in which $H$ is either a positive definite matrix, i.e., $\alpha^T H \alpha > 0$, or a positive semi-definite matrix, i.e., $\alpha^T H \alpha \geq 0$. If positive definite, term $\frac{1}{2}\alpha^T H \alpha$ forms a convex function, which is known to have a single minimum. If positive semi-definite, it also forms a convex function, but eventually having an infinite set of equivalent solutions such as in a half-cylinder (see Fig. 5.40).

The minimization problem solved by LowRankQP is equivalent to the SVM dual optimization problem, given by:

$$\underset{\alpha}{\text{maximize }} -\frac{1}{2}\alpha^T Q\alpha + \mathbf{e}^T \alpha$$

$$\text{subject to } \mathbf{0} \le \boldsymbol{\alpha} \le \mathbf{C},$$

$$\mathbf{y}^T \boldsymbol{\alpha} = 0.$$

Thus, we must set $\mathbf{d}^T = -\mathbf{e}^T$, $H = Q$, $\mathbf{u} = \mathbf{C}$, $\mathbf{b} = \mathbf{0}$, $A = \mathbf{y}^T$, so that the problem solved by LowRankQP becomes:

$$\underset{\alpha}{\text{minimize}} \quad -\mathbf{e}^T \boldsymbol{\alpha} + \frac{1}{2} \boldsymbol{\alpha}^T Q \boldsymbol{\alpha}$$

$$\text{subject to } \mathbf{y}^T \boldsymbol{\alpha} = \mathbf{0}$$

$$\mathbf{0} \le \boldsymbol{\alpha} \le \mathbf{C},$$

which provides the same solution as the SVM dual:

$$\underset{\alpha}{\text{maximize}} \quad \mathbf{e}^T \boldsymbol{\alpha} - \frac{1}{2} \boldsymbol{\alpha}^T Q \boldsymbol{\alpha}$$

$$\text{subject to } \mathbf{0} \le \boldsymbol{\alpha} \le \mathbf{C},$$

$$\mathbf{y}^T \boldsymbol{\alpha} = 0,$$

due to the inverse signs to translate the maximization to a minimization, as required by such a package.

Listing 5.5 details the SVM optimization problem solved using LowRankQP. Function *testSimpleDataset()* runs all steps, outputting plots and results.

**Listing 5.5** Approaching the SVM optimization problem by using the package LowRankQP

```
1    # Loading package LowRankQP
2    library(LowRankQP)
3
4    # This is the main function which is responsible for solving
            the optimization
5    # problem using linear or a kth−order polynomial kernel. It
            receives the training
6    # set X and its corresponding classes Y in {−1, +1}. We also
            set the upper limit C
7    # for every value contained in vector alpha.
8    svm.polynomial <- function(X, Y, C = Inf, polynomial.order =
            2, threshold = 1e−8) {
9
10            # Building up matrix Q. Observe the kernel function
                    is defined
11            # in here. If polynomial.order=1, then we are
                    considering the
12            # original input space of examples in X, otherwise
                    we are applying
13            # some nonlinear space transformation.
```

```
14        Qmat <- (Y %*% t(Y)) * (1+(X %*% t(X)))^polynomial.
          order
15
16        # Defining d as a vector containing values equal to
              -1
17        # to ensure the problem solved by LowRankQP is the
              same as ours
18        dvec <- rep(-1, nrow(X))
19
20        # Defining matrix A as the transpose of vector y
21        Amat <- t(Y)
22
23        # Defining b as a zero vector
24        bvec <- 0
25
26        # Setting the upper limit vector with values defined
              by C
27        uvec <- rep(C, nrow(X))
28
29        # Running the LowRankQP function to find vector
              alpha for which
30        # constraints are satisfied. Thus, we minimize the
              functional
31        # defined by LowRankQP
32        res <- LowRankQP(Qmat, dvec, Amat, bvec, uvec,
          method="CHOL")
33
34        # This is vector alpha found after the optimization
              process
35        alphas <- res$alpha
36
37        # Here we define which are the support vectors using
              the values
38        # in vector alpha. Values above some threshold are
              taken as more
39        # relevant (remember these are the KKT multipliers)
              to define
40        # constraints
41        support.vectors <- which(alphas > threshold)
42
43        # Finally, we define the identifiers of support
              vectors so we
44        # know who they are
45        support.alphas <- alphas[support.vectors]
46
47        # Now we define the margin using the support vectors
48        margin <- support.vectors
49
50        # and then compute the value for b
51        b <- Y[margin] - t(support.alphas*Y[support.vectors
          ]) %*% (1+(X[support.vectors,] %*% t(X[margin,])
          ))^polynomial.order
52
```

```
53           # Returning the whole model found during the
                 optimization process
54           return (list(X=X, Y=Y, polynomial.order=polynomial.
                 order, support.vectors=support.vectors, support.
                 alphas=support.alphas, b=mean(b), all.alphas=as.
                 vector(alphas)))
55      }
56
57      # This is a simple function to provide the discrete
             classification for unseen examples
58      discrete.classification <- function(model, testSet) {
59
60           # Creating a vector to store labels
61           all.labels = c()
62
63           # For every unseen example in this test set
64           for (i in 1:nrow(testSet)) {
65
66                   # Use the model found through function svm.
                         polynomial to
67                   # obtain the classification output
68                   label = sum(model$all.alphas * model$Y *
                         (1+(testSet[i,] %*% t(model$X)))^model$
                         polynomial.order) + model$b
69
70                   # If label >= 0, so the test example lies on
                         the positive side
71                   # of the hyperplane, otherwise it lies on
                         the negative one
72                   if (label >= 0)
73                           label = 1
74                   else
75                           label = -1
76
77                   # Storing labels
78                   all.labels = c(all.labels, label)
79           }
80
81           # Returning the labels found
82           return (all.labels)
83      }
84
85      # This is a simple function to provide the continuous
             classification for unseen examples
86      continuous.classification <- function(model, testSet) {
87
88           # Creating a vector to store labels
89           all.labels = c()
90
91           # For every unseen example in this test set
92           for (i in 1:nrow(testSet)) {
93
94                   # Use the model found through function svm.
                         polynomial to
```

```
95                          # obtain the classification output
96                          label = sum(model$all.alphas * model$Y *
                                 (1+(testSet[i,] %*% t(model$X)))^model$
                                 polynomial.order) + model$b
97
98                          # Storing labels
99                          # The signal associated with this value
                                 indicates the label, i.e., - corresponds
100                         # to class -1 and + to class +1. In addition
                                 , the magnitude of this variable
101                         # 'label' informs us how close or far the
                                 unseen example is from the hyperplane
102                         all.labels = c(all.labels, label)
103                    }
104
105               # Returning the labels found
106               return (all.labels)
107       }
108
109       # This is a simple function to plot the hyperplane found,
               but only for bidimensional training
110       # and test sets
111       plotHyperplane <- function(model, x.axis=c(-1,1), y.axis=c
               (-1,1), resolution=100, continuous=TRUE) {
112
113               # Producing a set of values for the two dimensions
                       of the training/test sets
114               x = seq(x.axis[1], x.axis[2], len=resolution)
115               y = seq(y.axis[1], y.axis[2], len=resolution)
116
117               # This is a matrix to store what we refer to plot
                       set.
118               # It is bidimensional as the training and test sets
119               plotSet = NULL
120               for (i in 1:length(x)) {
121                       for (j in 1:length(y)) {
122                               plotSet = rbind(plotSet, c(x[i], y[j
                                       ]))
123                       }
124               }
125
126               # This is a matrix to save labels for plotting
127               labels = NULL
128               if (continuous) {
129                       # Running the continuous classification
130                       labels = matrix(continuous.classification(
                               model, plotSet), nrow=length(x), ncol=
                               length(y), byrow=T)
131               } else {
132                       # or the discrete classification
133                       labels = matrix(discrete.classification(
                               model, plotSet), nrow=length(x), ncol=
                               length(y), byrow=T)
134               }
```

```
135
136                  # Plotting the hyperplane found
137                  filled.contour(x,y,labels)
138     }
139
140     # This function produces very simple linearly separable
               training/test sets
141     simpleDataset <- function() {
142
143                  # Building up the training set with 100 examples
144                  train <- cbind(rnorm(mean=0, sd=1, n=100), rnorm(
                        mean=0, sd=1, n=100))
145                  train <- rbind(train, cbind(rnorm(mean=10, sd=1, n
                        =100), rnorm(mean=10, sd=1, n=100)))
146                  train <- cbind(train, c(rep(-1, 100), rep(1, 100)))
147
148                  # Building up the test set with 10 examples
149                  test <- cbind(rnorm(mean=0, sd=1, n=10), rnorm(mean
                        =0, sd=1, n=10))
150                  test <- rbind(test, cbind(rnorm(mean=10, sd=1, n=10)
                        , rnorm(mean=10, sd=1, n=10)))
151                  test <- cbind(test, c(rep(-1, 10), rep(1, 10)))
152
153                  # Returning both sets
154                  return (list(train=train, test=test))
155     }
156
157     # This is a very simple function to test function svm.
               polynomial
158     testSimpleDataset <- function() {
159
160                  # Building up a very simple linearly separable
                        training set
161                  dataset = simpleDataset()
162
163                  # Optimizing values for alpha, given this simple
                        dataset
164                  model = svm.polynomial(dataset$train[,1:2], dataset$
                        train[,3], C=10000, polynomial.order=1)
165
166                  # Plotting all values in vector alpha to check them
                        out
167                  # and observe which are the most relevant ones
168                  plot(model$all.alphas)
169                  locator(1)
170
171                  # Plotting data space in black and support vectors
                        in red
172                  plot(dataset$train[,1:2])
173                  points(dataset$train[model$support.vectors,1:2], col
                        =2)
174                  locator(1)
175
176                  # Printing labels -1 and +1 for verification
```

```
177        labels = discrete.classification(model, dataset$test
                [,1:2])
178        result = cbind(dataset$test[,3], labels)
179        colnames(result) = c("Expected_class", "Obtained_
                class")
180        cat("Discrete_classification:\n")
181        print(result)
182
183        # Printing the continuous classification out
184        labels = continuous.classification(model, dataset$
                test[,1:2])
185        result = cbind(dataset$test[,3], labels)
186        colnames(result) = c("Expected_class", "Obtained_
                class")
187        cat("Continuous_classification:\n")
188        print(result)
189
190        # Plotting the hyperplane found
191        plotHyperplane(model, x.axis=c(-1,11), y.axis=c
                (-1,11), resolution=100, continuous=FALSE)
192    }
```

The output below shows an example of results from function *testSimpleDataset()*. Notice that package LowRankQP shows information about the KKT conditions, which ideally must be equal to zero (in fact they simply tend to zero, otherwise the algorithm may cross barriers and make constraints unfeasible). The term "Duality Gap" informs how close the results of the objective functions (primal and the dual) are. The closer the duality gap is from zero, the better the solution is.

**Listing 5.6** Text output produced by function testSimpleDataset() from Listing 5.5

```
LowRankQP CONVERGED IN 22 ITERATIONS

        Primal Feasibility      =    1.3990730e-13
        Dual Feasibility        =    2.3485572e-17
        Complementarity Value   =    1.4534699e-12
        Duality Gap             =    1.4534309e-12
        Termination Condition   =    1.4167561e-12
Discrete classification:
        Expected class Obtained class
 [1,]          -1              -1
 [2,]          -1              -1
 [3,]          -1              -1
 [4,]          -1              -1
 [5,]          -1              -1
 [6,]          -1              -1
 [7,]          -1              -1
 [8,]          -1              -1
 [9,]          -1              -1
[10,]          -1              -1
[11,]           1               1
[12,]           1               1
[13,]           1               1
```

| [14,] | 1 | 1 |
| [15,] | 1 | 1 |
| [16,] | 1 | 1 |
| [17,] | 1 | 1 |
| [18,] | 1 | 1 |
| [19,] | 1 | 1 |
| [20,] | 1 | 1 |

Continuous classification:

| | Expected class | Obtained class |
| --- | --- | --- |
| [1,] | $-1$ | $-1.483436$ |
| [2,] | $-1$ | $-1.450068$ |
| [3,] | $-1$ | $-1.908535$ |
| [4,] | $-1$ | $-1.450721$ |
| [5,] | $-1$ | $-1.435895$ |
| [6,] | $-1$ | $-1.474640$ |
| [7,] | $-1$ | $-1.217873$ |
| [8,] | $-1$ | $-1.473798$ |
| [9,] | $-1$ | $-1.456945$ |
| [10,] | $-1$ | $-1.599042$ |
| [11,] | 1 | 1.933268 |
| [12,] | 1 | 1.674008 |
| [13,] | 1 | 1.551290 |
| [14,] | 1 | 1.529009 |
| [15,] | 1 | 2.222672 |
| [16,] | 1 | 1.092659 |
| [17,] | 1 | 1.745056 |
| [18,] | 1 | 1.749871 |
| [19,] | 1 | 1.622522 |
| [20,] | 1 | 1.687724 |

Also notice the discrete classification results provided the labels as expected (first column is the expected class, while the second is the obtained by using the classifier). We also suggest the reader to pay special attention to the continuous classification, whose plus and minus signs are associated with the discrete label, while the magnitude informs us about how far an unseen example is from the maximal-margin hyperplane. For example, the first continuous classification produced $-1.483436$, meaning such unseen example is located at the negative side (class $-1$) but more distant from the hyperplane than the support vector used to define such negative class. Figure 5.41a illustrates the training examples, with support vectors in black, while Fig. 5.41b illustrates the resulting hyperplane.

Listing 5.7 includes two other functions to test this SVM optimization program, but now using a second-order polynomial kernel. The original data space is illustrated in Fig. 5.42a, while Fig. 5.42b shows the features space projected into the original data space, i.e. after applying the second-order polynomial kernel. Our code runs on top of the features space, consequently a single hyperplane is enough to separate classes, which is equivalent to find a more complex decision boundary in the input space.

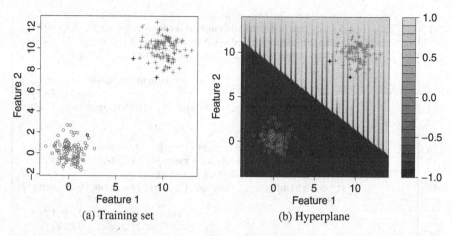

**Fig. 5.41** SVM optimization problem: (**a**) illustrates examples of the training set, with support vectors highlighted in black; (**b**) shows the hyperplane found by using the optimization algorithm

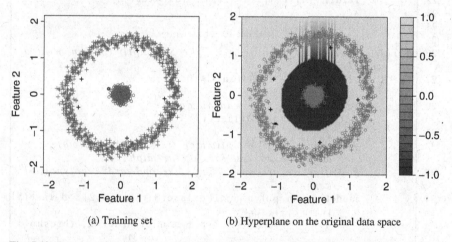

**Fig. 5.42** Data space requiring a second-order polynomial kernel. (**a**) Training set with the support vectors highlighted in black, (**b**) decision boundary mapped into the original data space, after applying the second-order polynomial kernel

**Listing 5.7** SVM optimization program using a second-order polynomial kernel

```
1  # This package is used to build up the radial dataset
2  require(tseriesChaos)
3
4  # Loading the source code to start the optimization process
5  source("lowrankqp-svm-simple.r")
6
7  # This function builds up a radial dataset
8  radialDataset <- function() {
9
```

```
10          # Building up the training set with 1000 examples .
11          train <- rbind(cbind(rnorm(mean=0, sd=0.1, n=1000),
                 rnorm(mean=0, sd=0.1, n=1000)))
12          train <- rbind(train, embedd(2*sin(2*pi*seq(0,9,
                 length=1027))+
13                                            rnorm(mean=0, sd=0.1, n
                                                  =1027), m=2, d=27))
14          train <- cbind(train, c(rep(-1, 1000), rep(+1, 1000)
                 ))
15
16          # Building up the test set with 10 examples
17          test <- rbind(cbind(rnorm(mean=0, sd=0.1, n=10),
                 rnorm(mean=0, sd=0.1, n=10)))
18          test <- rbind(test, embedd(2*sin(2*pi*seq(0,9,length
                 =37))+
19                                            rnorm(mean=0, sd=0.1, n
                                                  =37), m=2, d=27))
20          test <- cbind(test, c(rep(-1, 10), rep(+1, 10)))
21
22          return (list(train=train, test=test))
23
24 }
25
26 # This function is used to test the SVM optimization with a
        radial dataset
27 testRadialDataset <- function(C=10) {
28
29          # Building up the radial dataset
30          dataset = radialDataset()
31
32          # Running the SVM optimizer, so we can estimate
                 adequate values for vector alpha.
33          # Notice we are now using a second-order polynomial
                 kernel.
34          model = svm.polynomial(dataset$train[,1:2], dataset$
                 train[,3], C=C,
35                                            polynomial.order=2, threshold
                                                  = 1e-3)
36
37          # Plotting all values contained in vector alpha in
                 order to check them out
38          # and conclude on which are the most relevant ones
39          plot(model$all.alphas)
40          locator(1)
41
42          # Plotting the data space in black and support
                 vectors in red
43          plot(dataset$train[,1:2])
44          points(dataset$train[model$support.vectors,1:2], col
                 =2)
45          locator(1)
46
47          # Printing labels -1 and +1 for verification
```

```
48        labels = discrete.classification(model, dataset$test
             [,1:2])
49        result = cbind(dataset$test[,3], labels)
50        colnames(result) = c("Expected_class", "Obtained_
             class")
51        cat("Discrete_classification:\n")
52        print(result)
53
54        # Printing the continuous classification out
55        labels = continuous.classification(model, dataset$
             test[,1:2])
56        result = cbind(dataset$test[,3], labels)
57        colnames(result) = c("Expected_class", "Obtained_
             class")
58        cat("Continuous_classification:\n")
59        print(result)
60
61        # Plotting the hyperplane found
62        plotHyperplane(model, x.axis=c(-5,5), y.axis=c(-5,5)
             , resolution=100, continuous=FALSE)
63   }
```

Next, an example of output of Listing 5.7 is shown. Notice the "Duality Gap" is close to zero, and that all KKT conditions (primal, dual and complementary slackness) are similarly close to zero, and therefore to be fully satisfied. The expected and obtained classes indeed confirm good classification results for unseen examples. As before, the continuous classification provides additional information about the relative position of test examples to the hyperplane. Such position can be seen as a distance, from which an Entropy measure may be computed to analyze the complexity of classifying some dataset [7, 21, 22].

**Listing 5.8** Text output produced by Listing 5.7

```
LowRankQP CONVERGED IN 22 ITERATIONS

      Primal Feasibility     =   4.7949031e-12
      Dual Feasibility       =   1.3756696e-15
      Complementarity Value  =   3.6998986e-11
      Duality Gap            =   3.6999463e-11
      Termination Condition  =   2.4646533e-11
Discrete classification:
      Expected class Obtained class
 [1,]          -1              -1
 [2,]          -1              -1
 [3,]          -1              -1
 [4,]          -1              -1
 [5,]          -1              -1
 [6,]          -1              -1
 [7,]          -1              -1
 [8,]          -1              -1
 [9,]          -1              -1
[10,]          -1              -1
```

| [11,] | 1 | 1 |
| [12,] | 1 | 1 |
| [13,] | 1 | 1 |
| [14,] | 1 | 1 |
| [15,] | 1 | 1 |
| [16,] | 1 | 1 |
| [17,] | 1 | 1 |
| [18,] | 1 | 1 |
| [19,] | 1 | 1 |
| [20,] | 1 | 1 |

Continuous classification:

| | Expected class | Obtained class |
| --- | --- | --- |
| [1,] | −1 | −1.084941 |
| [2,] | −1 | −1.091868 |
| [3,] | −1 | −1.045681 |
| [4,] | −1 | −1.049770 |
| [5,] | −1 . | −1.079689 |
| [6,] | −1 | −1.086685 |
| [7,] | −1 | −1.086326 |
| [8,] | −1 | −1.062396 |
| [9,] | −1 | −1.085283 |
| [10,] | −1 | −1.087076 |
| [11,] | 1 | 1.760680 |
| [12,] | 1 | 2.257556 |
| [13,] | 1 | 1.712013 |
| [14,] | 1 | 1.685414 |
| [15,] | 1 | 1.536826 |
| [16,] | 1 | 1.421677 |
| [17,] | 1 | 1.744846 |
| [18,] | 1 | 1.766647 |
| [19,] | 1 | 1.459362 |
| [20,] | 1 | 1.422031 |

## 5.6   Concluding Remarks

This chapter introduced the necessary optimization concepts and tools for the design, from scratch, of an Interior Point Method to approach the SVM optimization problem. In the next chapter, we address some aspects of kernels and Linear Algebra. Up until this point, we just employed such kernels to exemplify practical scenarios, without theoretical discussion. By understanding how the input spaces can be transformed, one can better study such spaces in order to improve classification results, while keeping learning guarantees.

## 5.7 List of Exercises

1. Implement the Simplex method described in Sect. 5.4.3, and apply it to solve the linear problems discussed throughout this chapter.
2. Implement the Interior Point Method for the SVM optimization problem, and apply it on different datasets, including: linearly separable data, linearly separable data with random noise (try uniform and Gaussian noises, for example), and class-overlapped datasets.

# References

1. M.S. Bazaraa, J.J. Jarvis, H.D. Sherali, *Linear Programming and Network Flows* (Wiley, Hoboken, 2010)
2. S. Boyd, L. Vandenberghe, *Convex Optimization* (Cambridge University Press, New York, 2004)
3. G.B. Dantzig, M.N. Thapa, *Linear Programming 2: Theory and Extensions* (Springer, Berlin, 2006)
4. M.C. Ferris, T.S. Munson, Interior-point methods for massive support vector machines. SIAM J. Optim. **13**(3), 783–804 (2002)
5. A.V. Fiacco, G.P. McCormick, *Nonlinear Programming: Sequential Unconstrained Minimization Techniques*. Classics in Applied Mathematics (Society for Industrial and Applied Mathematics, Philadelphia, 1990)
6. S. Fine, K. Scheinberg, Efficient svm training using low-rank kernel representations. J. Mach. Learn. Res. **2**, 243–264 (2002)
7. J.A. Freeman, D.M. Skapura, *Neural Networks: Algorithms, Applications, and Programming Techniques*. Addison-Wesley Computation and Neural Systems Series (Addison-Wesley, Boston, 1991)
8. R. Frisch, The logarithmic potential method of convex programming with particular application to the dynamics of planning for national development: synopsis of a communication to be presented at the international colloquium of econometrics in Paris 23–28 May 1955, Technical report, University (Oslo), Institute of Economics, 1955
9. P.E. Gill, W. Murray, M.A. Saunders, J.A. Tomlin, M.H. Wright, On projected Newton barrier methods for linear programming and an equivalence to Karmarkar's projective method. Math. Program. **36**(2), 183–209 (1986)
10. A.J. Hoffmann, M. Mannos, D. Sokolowsky, N. Wiegmann, Computational experience in solving linear programs. J. Soc. Ind. Appl. Math. **1**, 17–33 (1953)
11. P.A. Jensen, J.F. Bard, *Operations Research Models and Methods*. Operations Research: Models and Methods (Wiley, Hoboken, 2003)
12. N. Karmarkar, A new polynomial-time algorithm for linear programming. Combinatorica **4**(4), 373–396 (1984)
13. W. Karush, Minima of functions of several variables with inequalities as side conditions, Master's thesis, Department of Mathematics, University of Chicago, Chicago, IL, 1939
14. H.W. Kuhn, A.W. Tucker, Nonlinear programming, in *Proceedings of the Second Berkeley Symposium on Mathematical Statistics and Probability, Berkeley, CA* (University of California Press, Berkeley, 1951), pp. 481–492
15. J.T. Ormerod, M.P. Wand, *Low Rank Quadratic Programming* (R Foundation for Statistical Computing, Vienna, 2015)
16. J.T. Ormerod, M.P. Wand, I. Koch, Penalised spline support vector classifiers: computational issues. Comput. Stat. **23**(4), 623–641 (2008)

17. PatrickJMT, Linear programming (2008). https://youtu.be/M4K6HYLHREQ
18. PatrickJMT, Linear programming word problem - example 1 (2010). https://youtu.be/2ACJ9ewUC6U
19. PatrickJMT, The simplex method - finding a maximum/word problem example (part 1 to 5) (2010). https://youtu.be/gRgsT9BB5-8
20. B. Scholkopf, A.J. Smola, *Learning with Kernels: Support Vector Machines, Regularization, Optimization, and Beyond* (MIT Press, Cambridge, 2001)
21. C.E. Shannon, Prediction and entropy of printed English. Bell Syst. Tech. J. **30**, 50–64 (1951)
22. C.E. Shannon, W. Weaver, *A Mathematical Theory of Communication* (University of Illinois Press, Champaign, 1963)
23. G. Strang, *Introduction to Linear Algebra* (Wellesley-Cambridge Press, Wellesley, 2009)
24. E. Süli, D.F. Mayers, *An Introduction to Numerical Analysis* (Cambridge University Press, 2003)
25. T. Terlaky, *Interior Point Methods of Mathematical Programming*, 1st edn. (Springer, Berlin, 1996)
26. M. videos de Matemáticas, Dualidad
27. Mini-videos de Matemáticas, Dualidad (2013). https://youtu.be/KMmgF3ZaBRE

# Chapter 6
# A Brief Introduction on Kernels

In the previous chapters, we described the Support Vector Machines as a method that creates an optimal hyperplane separating two classes by minimizing the loss via margin maximization. This maximization led to a dual optimization problem resulting in a Lagrangian function which is quadratic and requires simple inequality constraints. The support vectors are responsible for defining the hyperplane, resulting in the support vector classifier $f$ which not only provides a unique solution to the problem, but also ensures learning with tighter guarantees. However this is only possible if the input space is sufficiently linearly separable. On the other hand, many input spaces are, in fact, not linearly separable. In order to overcome this restriction, nonlinear transformations can be used to implicitly obtain a more adequate space.

In this chapter, we address the design of **kernels** to transform, or map, data into a space that is as linearly separable as possible. The optimization problem is formulated in terms of dot products, so that we can replace such operations with a transformation function (kernel) that maps the input space into a higher dimensional space. The main property of kernel functions is that they can be seen as providing dot products in some Hilbert space in which the original input space is virtually embedded.

The use of such strategy is often referred to as the "*kernel trick*" or "*kernel substitution*", as no modification is necessary in the primal and dual forms for the SVM problem when the kernel application produces a Gram matrix. Designing a kernel often requires prior knowledge about the problem, but allows space $\mathscr{F}$ to be more restricted, leading to a faster convergence, and stronger learning guarantees.

In this context, we present the most typical kernel functions, practical aspects, and the interpretation of such transformation in the light of Linear Algebra concepts. Then, the kernel trick is discussed in terms of the SVM optimization problem.

© Springer International Publishing AG, part of Springer Nature 2018
R. Fernandes de Mello, M. Antonelli Ponti, *Machine Learning*,
https://doi.org/10.1007/978-3-319-94989-5_6

## 6.1   Definitions, Typical Kernels and Examples

A kernel $k(x, y)$ is a function that takes two elements, typically vectors from some space $\mathbb{R}^n$, and outputs a measure of similarity between them. Considering the metric space, we usually have something in form $k : \mathbb{R}^n \times \mathbb{R}^n \rightarrow \mathbb{R}$. The reader can think of it as computing dot products that quantify the similarity between examples—this function has to obey certain mathematical properties that will be discussed later.

The simplest kernel function is the identity map, that is often called linear kernel, defined as: $k(x, y) = < \mathbf{x}, \mathbf{y} > = \mathbf{x}^T \mathbf{y}$, given $x$ and $y$ are represented by vectors $\mathbf{x}$ and $\mathbf{y}$, respectively. The dot product between a pair of data points is a linear combination between $\mathbf{x}$ and $\mathbf{y}$, as depicted in Fig. 6.1. However, notice that the linear kernel has no additional effect because it does not map input data into a higher dimensional space.

Among the kernels that map data into higher dimensional spaces, the most known and widely used are: the polynomial, the radial basis function (RBF), and the sigmoidal. In the following sections, we briefly describe those typical kernels throughout examples. Later in the chapter, a more detailed discussion in the context of SVM optimization is provided.

**Fig. 6.1** Space formed by the dot product between a pair of vectors

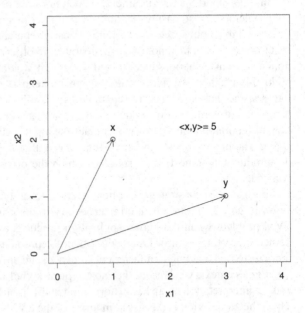

### 6.1.1 The Polynomial Kernel

The polynomial kernel is expressed by:

$$k(x_i, x_j) = < \mathbf{x}_i, \mathbf{x}_j + c >^d,$$

in which the arguments of function $k(.,.)$ are non-bold, because those are not necessarily vectors in a Hilbert space; however the corresponding vectors $\mathbf{x}_i, \mathbf{x}_j$ must be in a space in which dot products are defined, i.e. the Hilbert space; $d$ is the kernel order; and, finally, $c \geq 0$ is a parameter that trades off the influence of higher-order versus lower-order terms in the polynomial. When $c = 0$ the kernel is homogeneous polynomial, in which all terms have equal influence:

$$k(x_i, x_j) = < \mathbf{x}_i, \mathbf{x}_j >^d .$$

Considering a second-order homogeneous polynomial kernel ($d = 2$), and an input space $\mathbb{R}^2$ containing the following vectors:

$$\mathbf{x} = [x_1, x_2]^T$$

$$\mathbf{y} = [y_1, y_2]^T,$$

then, the polynomial kernel is:

$$k(x, y) = < \mathbf{x}, \mathbf{y} >^2 = (x_1 y_1 + x_2 y_2)^2$$

$$= x_1^2 y_1^2 + 2 x_1 y_1 x_2 y_2 + x_2^2 y_2^2$$

Note the same result is obtained by representing the function as a dot product in another space. The three last terms in the previous equation can be rewritten as:

$$[x_1^2, x_1 x_2, x_1 x_2, x_2^2]^T \cdot [y_1^2, y_1 y_2, y_1 y_2, y_2^2]^T. \tag{6.1}$$

Equation (6.1) describes a space with **four** dimensions, in other words, it is a map from a 2-dimensional **data space** to a 4-dimensional **features space**. But, in fact, two of the features space dimensions are the same, what allow us to simplify it to three dimensions, as follows:

$$k(x, y) = x_1^2 y_1^2 + 2 x_1 y_1 x_2 y_2 + x_2^2 y_2^2$$

$$= [x_1^2, \sqrt{2} x_1 x_2, x_2^2]^T . [y_1^2, \sqrt{2} y_1 y_2, y_2^2]^T, \tag{6.2}$$

but yet product the same dot-product result. By using this definition each 2-dimensional vector $\mathbf{x}$ in the input space is mapped into a 3-dimensional vector in the features space, as depicted in Fig. 6.2.

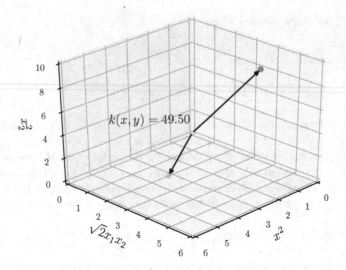

**Fig. 6.2** Space formed by second-order polynomial kernel applied to a nonlinear problem

### 6.1.2   The Radial Basis Function Kernel

Another typical and widely used kernel is the Radial Basis Function (RBF) kernel, whose most common example is the Gaussian kernel, defined as:

$$k(x, y) = \exp(-\gamma ||\mathbf{x}, \mathbf{y}||^2), \tag{6.3}$$

in which $\gamma > 0$ is commonly parametrized as $\gamma = 1/(2\sigma^2)$, with $\sigma$ as the standard deviation of the Gaussian distribution. Considering the data space $\mathbb{R}^2$:

$$\mathbf{x} = [x_1, x_2]^T,$$

$$\mathbf{y} = [y_1, y_2]^T,$$

the RBF kernel is:

$$k(x, y) = \exp(-\gamma ||\mathbf{x}, \mathbf{y}||^2)$$

$$= \exp\left(-\gamma ||[x_1, x_2]^T - [y_1, y_2]^T||^2\right)$$

$$= \exp\left(-\gamma ||[x_1, x_2]^T - [y_1, y_2]^T||^2\right). \tag{6.4}$$

Notice this operation first computes the norm of the difference vector. For each possible pair of vectors $\mathbf{x}$ and $\mathbf{y}$ in the data space, a difference vector is created in the **features space**. For a data space in $\mathbb{R}^2$, the features space is also in $\mathbb{R}^2$, as depicted in Fig. 6.3.

Fig. 6.3 Space formed by the difference vectors. Notice that the length of the difference vector carries the important information: the shorter the vector is, the more similar vectors **x** and **y** are. $k(x, y)$ is computed using *gamma* $= 1/2$

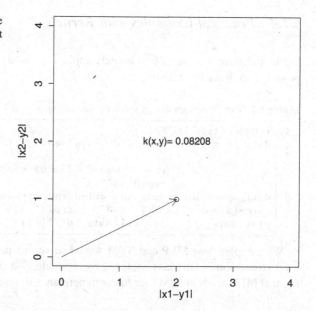

The Gaussian function is then applied over the length of the difference vector in order to weigh the distances, so that greater relevance is given to nearby instances. Due to the shape of the Gaussian distribution, it produces different weights for vectors **y** lying on different radii with respect to a reference vector **x**. Remember the parameter $\gamma$ is related to the standard deviation of the Gaussian distribution, and it defines how fast the weights drop as the length of the difference vector increases.

### 6.1.3 The Sigmoidal Kernel

As last example of a typical kernel, the sigmoidal produces a space that is similar to the identity operation, but bounding magnitudes, i.e.:

$$k(x, y) = \tanh(-\kappa \mathbf{x} \cdot \mathbf{y} + c), \tag{6.5}$$

for some $\kappa > 0$ and $c < 0$. Considering again the data space $\mathbb{R}^2$, the sigmoidal kernel produces a space that is the same as the linear kernel, but with $\kappa$ as a modifier for the dot product magnitudes, and $c$ as a shifting parameter for the hyperbolic tangent curve. The most relevant aspect of this kernel is that it produces a continuous output in a bounding range $[-1, 1]$.

### 6.1.4  Practical Examples with Kernels

Let us illustrate the use of two kernels applied on synthetic data examples in $\mathbb{R}^2$, produced as describe in Listing 6.1:

**Listing 6.1**  Two kernels applied on synthetic data examples in $\mathbb{R}^2$

```
1   require(tseriesChaos)
2   data_sp = cbind(embedd(0.6*sin(seq(0,9,len=1000)),m=2,d=175)
        +
3                   rnorm(mean=0,sd=0.01,n=825),
4                   rep(0, 825))
5   data_sp = rbind(data_sp, cbind(rnorm(mean=0, sd=0.1, n=825),
6   rnorm(mean=0, sd=0.1, n=825), rep(1, 825)))
7   plot(data_sp[,1:2], col=data_sp[,3]+1)
```

We compare how MLP and SVM would solve this problem, in which we have a Gaussian-distributed class enclosed by a circularly-distributed class. In Fig. 6.4, see that MLP needs at least four linear hyperplanes to separate the two distributions

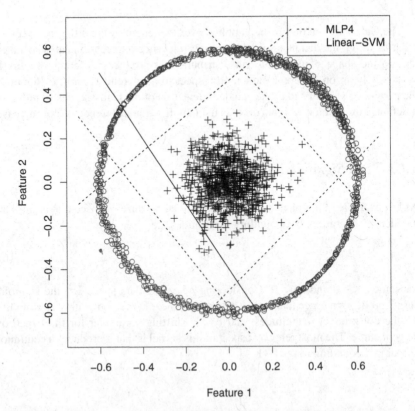

**Fig. 6.4**  Comparison between the results provided by MLP with 4 hyperplanes and SVM

(meaning the MLP model must have at least four neurons in the hidden layer). Those hyperplanes do not guarantee maximal margin, which may hinder generalization guarantees. On the other hand, the SVM optimizes the maximal margin for a single hyperplane, but the results are poor because the original data space is not ideal.

However, by having prior knowledge about the class distributions, one can design a second-order polynomial kernel to produce a features space in which a single hyperplane is enough to separate classes. Such features space is computed and visualized using Listing 6.2:

**Listing 6.2**  Designing a second-order polynomial kernel

```
1  require(rgl)
2  feature_sp = cbind(data_sp[,1]^2, sqrt(2)*data_sp[,1]*data_
     sp[,2],
3                    data_sp[,2]^2)
4  plot3d(feature_sp, col=data_sp[,3]+1)
```

Figure 6.5 depicts the resulting features space, in which classes are now linearly separable, emphasizing the importance of kernels. Now both MLP (still with 4 hyperplanes) and SVM are able to solve the problem, but providing different VC dimensions and, consequently, generalization guarantees. When comparing the output of those two inferred classifiers, one may only assess the numerical results, e.g. in terms of accuracies, disregarding the theoretical aspects. As covered by previous chapters, SVM has lower VC dimension due to the large-margin bounds, and therefore stronger learning guarantees.

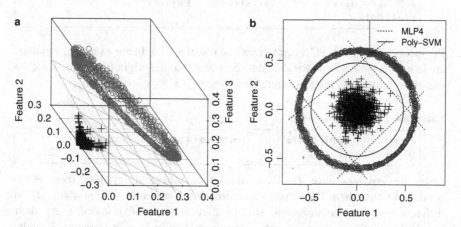

**Fig. 6.5**  Second-order polynomial kernel: (**a**) resulting 3D features space, (**b**) hyperplane after projecting the classification into the original 2D input space

This example illustrates in practice how important is to consider kernels when addressing classification problems that are originally non-linearly separable. When comparing SVM with another classification algorithm such as MLP, if the latter produces higher accuracies then most probably the data space is not adequate.

## 6.2   Principal Component Analysis

The Principal Component Analysis (PCA) is a method to linearly transform a domain into a co-domain whose basis vectors are orthogonal. This transformation is often considered for dimensionality reduction via feature selection, for studying the data variance along reference axes, as well as for analyzing data subspaces. Given this method is based on a linear transformation, it considers the linear combination of reference axes (a.k.a. space dimensions) to form vectors in a target space, in which linearly correlated dimensions are simplified in terms of an orthogonal and centralized representation.

As an example, consider the synthetic dataset illustrated in Fig. 6.6 produced using Listing 6.3, in which there is a linear correlation between students' grades in the subjects of Physics and Statistics.

**Listing 6.3**  Producing a synthetic dataset

```
1  require(splus2R)
2  data = as.data.frame( rmvnorm(100, rho=0.9 ) * 5+70 )
3  colnames(data) = c("Statistics", "Physics")
4  plot(data)
```

**Pre-processing**  First, PCA centralizes and rescales data. In our example, variables are in interval $[50, 85]$, so they must be centralized to the origin $(0, 0)$, i.e., let $\mathbf{x}_i$ be the $i$th variable in $X$, then its centralized version is $\mathbf{x}'_i = \mathbf{x}_i - \text{mean}(\mathbf{x}_i)$:

**Listing 6.4**  Centralizing input data

```
1  central = apply(data, 2, function(x_i) { x_i-mean(x_i) })
2  plot(central)
```

Once the variables have the same unit of measurement (grades), there is no need to rescale them. However, there are situations in which variables are in different units of measurement, so that they need to be rescaled using their standard deviation, i.e., $\mathbf{x}''_i = (\mathbf{x}_i - \text{mean}(\mathbf{x}_i))/\text{sd}(\mathbf{x}_i)$ or in terms of R code: `(x_i-mean(x_i))/sd(x_i)`.

**Covariance Matrix**  After preprocessing data, the data covariance matrix must be computed. The definition of the covariance of $X$ is:

$$(\mathbf{X} - \text{column means}(\mathbf{X}))^T (\mathbf{X} - \text{column means}(\mathbf{X}))$$

**Fig. 6.6** Data with linear correlation between dimensions

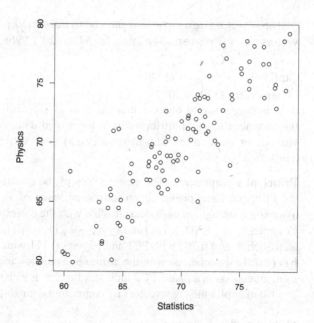

Considering the centralized version of data $X' = \mathbf{X} - \text{column means}(\mathbf{X})$:

$$\text{cov}(X') = X'^T X',$$

as the covariance matrix along its attributes. To compute it using the R Statistical Software:

**Listing 6.5** Computing the covariance matrix

```
1   central.cov = cov(central)
```

**Statistics Physics**

resulting in: **Statistics** 29.4139 27.7378
**Physics** 27.7378 30.0590

**Eigenvalues and Eigenvectors** As a third step, PCA computes the eigenvalues and eigenvectors for the correlation matrix:

**Listing 6.6** Computing the eigenvalues and eigenvectors

```
1   eigens = eigen(central.cov)
2   rownames(eigens$vectors) = c("Statistics","Physics")
3   colnames(eigens$vectors) = c("PC1","PC2")
```

The object `eigens` contains the eigenvalues $\lambda_1$, $\lambda_2$, and the respective eigenvectors $\mathbf{v}_1$, $\mathbf{v}_2$: `eigens$values = 57.47620 1.99683 eigens$vectors =`

|            | **PC1**   | **PC2**   |
|------------|-----------|-----------|
| **Statistics** | 0.70298 | −0.71120 |
| **Physics**    | 0.71120 | 0.70298  |

It is interesting to observe that the sum of eigenvalues is actually the sum of the variances for both attributes from the scaled data, i.e., $\lambda_1 + \lambda_2 = \text{var}(\mathbf{x}_1'') + \text{var}(\mathbf{x}_2'')$, or in R: `sum(eigens$values) == var(scaled[,1]) + var(scaled[,2])`

**Principal Components** The eigenvectors of the covariance matrix are known as the Principal Components. Each individual value of an eigenvector indicates the association strength of each data attribute with the corresponding component. In our example, $\mathbf{v}_1 = [0.702, 0.711]$ and $\mathbf{v}_2 = [-0.711, 0.702]$, meaning that Statistics has an association of 0.702 with PC1 and Physics 0.711 with the same component. PC1 has positive associations with the attributes Statistics and Physics, i.e., they pull to same directions in a plane. PC2 indicates inverse associations with each attribute.

One may plot the eigenvectors by computing their slopes:

**Listing 6.7** Plotting the eigenvectors

```
1   pc1.slope = eigens$vectors[1,1] / eigens$vectors[2,1]
2   pc2.slope = eigens$vectors[1,2] / eigens$vectors[2,2]
3   abline(0,pc1.slope, col="red")
4   abline(0,pc2.slope, col="blue")
```

As expected, eigenvectors are orthogonal (Fig. 6.7). The first principal component always corresponds to the axis with the greater data variance. PCA considers

**Fig. 6.7** Eigenvectors related to the principal components of data. Note the data is centralized, i.e. with zero mean

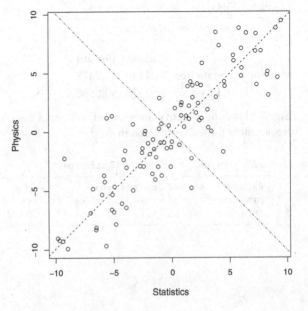

**Fig. 6.8** Data obtained after the rotation employed by the eigenvectors

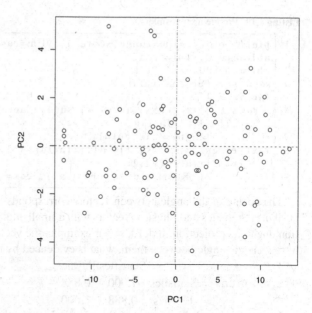

that components with the greatest variance are the most relevant. The percentage of variation in each axis is given by the corresponding eigenvalues. Given $\lambda_1 = 57.47620$ and $\lambda_2 = 1.99683$, then computing:

**Listing 6.8** Percentage of data variances along the principal components

```
1  eigens$values / (sum(eigens$values)*100)
```

we find the relative percentual relevance of each axis: 94.45926 5.54073 Therefore, the first principal component PC1 carries more than 94% of the data variance, but remember it is not Physics nor Statistics, it is indeed a linear combination of both. As matter of fact, PCA computes the orthogonal basis for some input data, so that the main directions can be used to describe the data space.

**Space Transformation** In order to remove the linear correlation, PCA rotates data so that eigenvectors (principal components) become parallel to cartesian axes (Fig. 6.8). This allows us to illustrate the *score plot*, computed after changing the basis of the data matrix:

**Listing 6.9** Changing basis to obtain the score plot

```
1  scores = central %*% eigens$vectors
2  plot(scores)
3  abline(0,0, col="red")
4  abline(0,90, col="blue")
```

The Biplot illustrates the relationship among the score plot and data attributes, which is produced as follows:

**Listing 6.10** Producing the Biplot

```
1  plot(scores, xlim=range(scores), ylim=range(scores))
2  abline(0,0, col="red")
3  abline(0,90, col="blue")
4  sd = sqrt(eigens$values)
5  factor = 1
6  arrows(0,0,eigens$vectors[,1]*sd[1]/factor, eigens$vectors
       [,2]*sd[2]/factor,
7          length=0.1, col=1)
8  text(eigens$vectors[,1]*sd[1]/factor*1.2, eigens$vectors[,2]
       *sd[2]/factor*1.2,
9          c("Statistics","Physics"), cex=1.6, col=2)
```

The cosine of the angle between vectors corresponds to the correlation between attributes Statistics and Physics. Vectors with a small angle present a greater correlation due to its projection. Indeed, in our example, the vectors representing attributes have an acute angle between them, what is evidenced by the correlation of 0.888:

$$
\texttt{cor(data)} = \begin{array}{l} \\ \text{Statistics} \\ \text{Physics} \end{array} \begin{array}{cc} \text{Statistics} & \text{Physics} \\ 1.000 & 0.888 \\ 0.888 & 1.000 \end{array}
$$

In the Biplot shown in Fig. 6.9, points close to the tip of vector Statistics correspond to greater grades in that subject, many of the coincide with the points close to Physics. Points in the opposite direction correspond to lower grades. Given there is correlation, attribute vectors pull to nearby directions, that is why there is an acute angle between them. An anti-correlation is also possible, with vectors indicating opposite directions. Note that eigenvectors are orthogonal but attribute values can assume other organizations.

For problems requiring dimensionality reduction, i.e., in which one wishes to consider less variables, one may simply use the first principal components to retain the desired relative percentual relevance. In our example, the use of PC1 is equivalent to projecting any vector in the original space (after centralization) into the direction of the first principal component: $\mathbf{v}_i^T \mathbf{x}$, with $i = 1$. Provided eigenvectors form an orthogonal matrix, its inverse is trivial and given by its transpose. In that manner, one may map vectors back to the original space. A practical aspect to be considered is to compute the covariance matrix on training data, so that the eigenvectors may be used to transform test examples to this new space.

## 6.3  Kernel Principal Component Analysis

PCA assumes input data, represented by position vectors, are represented by linear combinations. Alternatively, the Kernel Principal Component Analysis (KPCA) assumes some kernel function must be applied on input data to make such position vectors be represented through linear combinations.

For example, consider the bidimensional input data which has two distributions, in a similar way as in Fig. 6.5, one drawn from a Gaussian distribution, another surrounding this first in a circular way. Suppose some specialist wishes to find

**Fig. 6.9** Biplot illustrating the correlation between variables

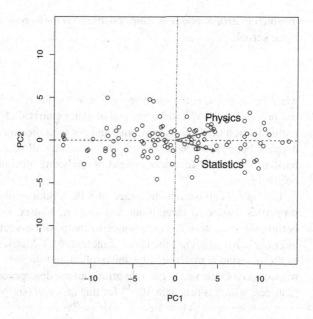

an alternative representation to vectors. After applying a second-order polynomial kernel:

$$k(\mathbf{x}_i, \mathbf{x}_j) = < \Phi(\mathbf{x}_i), \Phi(\mathbf{x}_j) >,$$ (6.6)

which operates in $\mathbb{R}^2$ to produce:

$$\Phi\left(\begin{bmatrix} x_{i,1} \\ x_{i,2} \end{bmatrix}\right) = \begin{bmatrix} x_{i,1}^2 \\ \sqrt{2}x_{i,1}x_{i,2} \\ x_{i,2}^2 \end{bmatrix}$$

resulting in a function to map every input vector $\mathbf{x}_i$, composed of two scalars $x_{i,1}$ and $x_{i,2}$, into a features space, as illustrated in Fig. 6.5b. On this new space, PCA may be used to provide the best as possible linear representation for data.

There is no need to explicitly apply some kernel function as defined in Eq. (6.6). Instead, one can simply ensure the similarity matrix among pairs $\mathbf{x}_i, \mathbf{x}_j \in X$ is Gramian. A Gram matrix $M$ is positive semidefinite, i.e.:

$$\mathbf{x}^T M \mathbf{x} \geq 0,$$

ensuring a convex function. As consequence, the kernel transformation may be directly computed on top of the covariance matrix (which is already Gramian):

$$C = XX^T,$$

in which matrix $X$ contains every position vector as row vectors, and then we apply some kernel:

$$M = (XX^T)^2,$$

which in this case is the same second-order polynomial kernel as before. Then, we can proceed with PCA on $M$ instead of using matrix $C$. Note that here we employ a multiplication of $X$ by its transpose and not the other way around as for PCA. This happens because we need the dot product among all vectors in the original space to build up the kernel function, instead of analyzing attributes such as for PCA in its original form.[1]

Listing 6.11 illustrates the usage of KPCA on a synthetic dataset with the same properties discussed throughout this section. Matrix $X$ is the same as discussed before, while matrix $K$ corresponds to the position vectors in the features space, after explicitly applying the kernel function $\Phi(.)$. Matrices $M_1$ and $M_2$ correspond to the covariance matrices after the explicit and implicit application of the kernel, respectively. Observe the last line prints out the divergence between both covariance matrices, which is less than $10^{-14}$ for this instance (simply a numerical error).

**Listing 6.11** Example of the Kernel principal component analysis

```
1   # Required packages
2   require(rgl)
3   require(tseriesChaos)
4
5   # Building up the dataset
6   X = embedd(sin(2*pi*seq(0,9,len=500)), m=2, d=14)
7   X = rbind(X, cbind(rnorm(mean=0, sd=0.1, n=500),
8                      rnorm(mean=0, sd=0.1, n=500)))
9   plot(X)
10
11  # Manually creating the kernel space.
12  # This is the impact of the 2-order polynomial kernel
13  K = cbind(X[,1]^2, sqrt(2)*X[,1]*X[,2], X[,2]^2)
14  plot3d(K)
15
16  # Computing the covariance matrix using the explicit
17  # transformation of vectors using the kernel function
18  M_1 = K%*%t(K)
19
20  # Computing the covariance matrix for the 2nd polynomial
21  # kernel, using implicit application of the kernel function
22  M_2 = (X%*%t(X))^2
23
24  cat("Difference:_", sqrt(sum((M_1 - M_2)^2)), "\n")
```

---

[1] There are variations for PCA to study data subspaces which employ $XX^T$ in order to analyze the correlations among position vectors.

We suggest the reader to employ PCA on $M_1$ and $M_2$ to observe eigenvectors and eigenvalues are the same, apart some numerical error divergence. As a result, PCA supports the separation of both data groups using the eigenvector with the maximal variance. Section 6.5 discusses more about the need for such Gram matrix in the context of SVM.

**Other Interesting Remarks** The following methods are also very useful to tackle data decompositions:

- **Singular Value Decomposition (SVD)**: if we decompose the centralized data matrix $X'$ using Singular Value Decomposition, then we have $X' = U \Sigma V^T$, in which $U = XX^T$ is a $d \times d$ matrix ($d$ is the number of variables), $V = X^T X$ is a matrix $n \times n$ ($n$ is the number of data points), and $\Sigma$ is a diagonal matrix containing the eigenvalues of $U$. This means one could perform SVD in the centralized data matrix and then apply the eigen decomposition in the resulting matrix $U$;
- **Linear Discriminant Analysis (LDA)**: this is a generalization of Fisher's linear discriminant, a method used in Machine Learning, Statistics, Engineering, etc. to find a linear combination of features that allows the characterization or the separation of examples labeled under two or more classes;
- **Kernel Discriminant Analysis (KDA)**: this is for LDA, as KPCA is to PCA, meaning some kernel maps examples from the input space to another features space in which LDA is employed;
- **Low-Rank Representation (LRR)**: is a technique that performs a similar analysis to PCA, but computing the covariance matrix for the data rows instead of columns.

## 6.4 Exploratory Data Analysis

As previously discussed the typical kernels are: linear, polynomial, radial basis and sigmoidal. The linear kernel simply consider the dot product of input vectors, in form:

$$k_{\text{linear}}(x_i, x_j) = <\mathbf{x}_i, \mathbf{x}_j>,$$

having $x_i$ and $x_j$ as data elements, while $\mathbf{x}_i$ and $\mathbf{x}_j$ as their vectorial representations in some Hilbert space $\mathscr{H}$. The polynomial kernel computes the power of $d$ (kernel order) for such dot product and it may add a constant $c$, as follows:

$$k(x_i, x_j) = <\mathbf{x}_i, \mathbf{x}_j + c>^d. \tag{6.7}$$

The radial basis function (RBF) kernel considers some real value $\gamma > 0$ to compute:

$$k(x_i, x_j) = \exp(-\gamma ||\mathbf{x}_i, \mathbf{x}_j||^2),$$

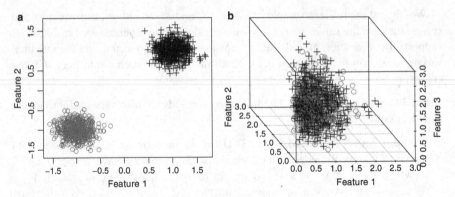

**Fig. 6.10** Space provided by a two-class linearly separable dataset. (**a**) Original space before kernel function is applied, (**b**) space after transformation via kernel function

and, finally, the sigmoidal kernel takes two real values $\kappa > 0$ and $c < 0$, in form:

$$k(x_i, x_j) = \tanh(-\kappa \mathbf{x}_i \cdot \mathbf{x}_j + c).$$

Each kernel may be more adequate to tackle a given scenario, so, in order to illustrate their usefulness, next we discuss about some classification tasks in terms of different data spaces.

### 6.4.1   How Does the Data Space Affect the Kernel Selection?

In this section, we use some datasets to illustrate the selection of SVM kernels. We start with the two-class linearly separable dataset shown in Fig. 6.10. Listing 6.12 details the use of SVM to assess this input space. Observe the space is already linearly separable, therefore the best choice is to use the linear kernel, instead of the polynomial one, as confirmed by the output provided in Listing 6.13.

This code was implemented using the R package e1071, which contains function *svm*. In Listing 6.12, we set parameter $x = X$ and $y = as.factor(Y)$ that correspond to the examples (attributes along columns) and labels, respectively. Function *as.factor* is necessary due to labels must be translated to such R data type. We set no scale modification (*scale = FALSE*) for attributes, otherwise they would be modified to zero mean and unit variance (recall the PCA preprocessing stage). Parameter *degree* informs us that a second-order polynomial kernel is used, term *d* in Eq. (6.7), while *coef0* is associated with term *c* in the same equation. Argument *cross=10* means the SVM will proceed with a tenfold cross validation strategy to train and test the classifiers found. Variable *cost=1000* defines the upper bound for the cost constant $C$ (more information in Sect. 4.7).

**Listing 6.12** Space provided by a two-class linearly separable dataset: assessing a linear and a second-order polynomial kernel

```
1   require(e1071)
2   require(tseriesChaos)
3
4   # Building the dataset
5   X = cbind(rnorm(mean=-1, sd=0.1, n=1000),
6             rnorm(mean=-1, sd=0.1, n=1000))
7   X = rbind(X, cbind(rnorm(mean=1, sd=0.1, n=1000),
8                      rnorm(mean=1, sd=0.1, n=1000)))
9
10  # Defining the class labels
11  Y = c(rep(-1, 1000), rep(+1, 1000))
12
13  # Plotting the input space
14  plot(X, xlim=c(min(X), max(X)), ylim=c(min(X), max(X)), col=
        Y+2)
15
16  # Using a linear kernel
17  model1 = svm(x = X, y = as.factor(Y), scale=FALSE,
18               kernel="linear", cost=1000, cross=10)
19
20  cat("Accuracies for each one of the ten classifiers found:\n
        ")
21  print(model1$accuracies)
22
23  # Using a second-order polynomial kernel
24  model2 = svm(x = X, y = as.factor(Y), scale=FALSE,
25               kernel="polynomial", degree=2,
26               coef0=0, cost=1000, cross=10)
27
28  cat("Accuracies for each one of the ten classifiers found:\n
        ")
29  print(model2$accuracies)
```

The results provided by Listing 6.12 is shown next, which presents the ten accuracies obtained along each one of the ten classifiers induced along the tenfold cross validation strategy. Notice the linear hyperplane was already enough to separate this first dataset as expected. The bias provided by the second-order polynomial kernel was not good for this situation.

**Listing 6.13** Text output produced by Listing 6.12

```
Accuracies for each one of the ten classifiers found:
[1] 100 100 100 100 100 100 100 100 100 100
Accuracies for each one of the ten classifiers found:
[1] 49.0 53.0 39.0 51.0 49.0 50.5 47.5 46.5 48.0 50.5
```

As a next instance, consider the input data illustrated in Fig. 6.11a, whose examples are in a bidimensional space and centered at (0, 0). Such data organization requires some second-order polynomial kernel to make it linearly separable through an additional dimension. Listing 6.14 details a program to assess the same two

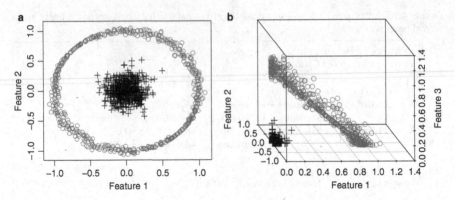

**Fig. 6.11** Space requiring a second-order polynomial kernel. (**a**) Original space before kernel function is applied, (**b**) space after transformation via kernel function

kernels used in the previous example, but now the usefulness of the second-order polynomial kernel is clear. Figure 6.11b illustrates the third-dimensional space obtained after applying the second-order polynomial kernel, in which this features space is then linearly separable.

**Listing 6.14** Space provided by a two-class second-order polynomially separable dataset: assessing a linear and a second-order polynomial kernel

```
1   require(rgl)
2   require(e1071)
3   require(tseriesChaos)
4
5   # Building the dataset
6   X = cbind(rnorm(mean=0, sd=0.1, n=1000),
7               rnorm(mean=0, sd=0.1, n=1000))
8   X = rbind(X, embedd(sin(2*pi*seq(0,9,len=1027)), m=2, d=27))
9
10  # Defining the class labels
11  Y = c(rep(-1, 1000), rep(+1, 1000))
12
13  # Plotting the input space
14  plot(X, xlim=c(min(X), max(X)), ylim=c(min(X), max(X)), col=
        Y+2)
15
16  # Using a linear kernel
17  model1 = svm(x = X, y = as.factor(Y), scale=FALSE,
18               kernel="linear", cost=1000, cross=10)
19
20  cat("Accuracies_for_each_one_of_the_ten_classifiers_found:\n
        ")
21  print(model1$accuracies)
22
23  # Using a second-order polynomial kernel
24  model2 = svm(x = X, y = as.factor(Y), scale=FALSE,
```

```
25                   kernel="polynomial", degree=2, coef0=0,
26                   cost=1000, cross=10)
27
28   cat("Accuracies for each one of the ten classifiers found:\n
     ")
29   print(model2$accuracies)
30
31   # Effect of the second-order polynomial kernel
32   after.kernel = cbind(X[,1]^2, sqrt(2)*X[,1]*X[,2], X[,2]^2)
33   plot3d(after.kernel, col=Y+2)
```

The outputs below confirm the second-order polynomial kernel provides good results when the input data has the same characteristics as in Fig. 6.11. Users may run a linear SVM on this type of data space and question the abilities of SVM as a classification algorithm. In all those situations, it is worth to remember that SVM provides the strongest learning guarantees as possible, so when one observes low classification performance that is due to the lack of linear separability of the provided data space. In that context, SVM is still the best supervised algorithm, however it requires an adequate kernel to transform input examples into linearly separable data. As a practical example, suppose Multilayer Perceptron performs better on some original input space while SVM provides poor results. The reason is simple, the input space is not adequate for SVM, while MLP may apply several hyperplanes to produce some better result. In this instance we have just studied, a second-order polynomial kernel is necessary to produce space transformations and allow SVM to work properly.

**Listing 6.15** Text output produced by Listing 6.14

```
Accuracies for each one of the ten classifiers found:
 [1] 45.0 65.5 66.0 65.5 69.5 63.0 73.5 44.5 64.5 69.5
Accuracies for each one of the ten classifiers found:
 [1] 100 100 100 100 100 100 100 100 100 100
```

At a next instance, consider the dataset illustrated in Fig. 6.12a, which motivates us to employ different kernels and analyze results. Listing 6.16 details the program used to design a dataset as well as all SVM settings. We employed a linear kernel (model1), an homogeneous second-order polynomial kernel (here referred to as homogeneous given $c = 0$ in Eq. (6.7)—model2), an homogeneous third-order polynomial kernel (model3), another homogeneous but fourth-order polynomial kernel (model4) and, finally, a nonhomogeneous third-order polynomial kernel (because $c \neq 0$ in Eq. (6.7)—model5). In Fig. 6.12b we show the result of applying the second order polynomial kernel to illustrate its effect. The remaining ones are not plotted due to the their higher dimensionalities.

For the sake of comparison, we analyze the homogeneous versus the nonhomogeneous third-order polynomial kernel and their effects in the features space. Let vectors:

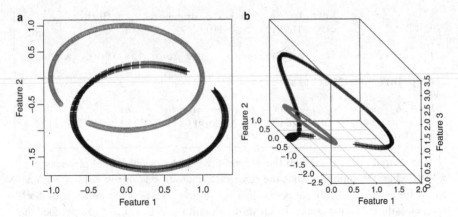

**Fig. 6.12** Space provided by a two-class complex dataset. (**a**) Original space before kernel function is applied, (**b**) space after transformation via kernel function

$$\mathbf{x} = \begin{bmatrix} x_1 \\ x_2 \end{bmatrix} \quad \text{and} \quad \mathbf{y} = \begin{bmatrix} y_1 \\ y_2 \end{bmatrix}.$$

The homogeneous third-order polynomial kernel produces the following dot product:

$$(< \mathbf{x}, \mathbf{y} >)^3 = (x_1 y_1 + x_2 y_2)^3$$

$$x_1^3 y_1^3 + 3x_1^2 y_1^2 x_2 y_2 + 3x_1 y_1 x_2^2 y_2^2 + x_2^3 y_2^3,$$

which can be represented in terms of two vectors so one can understand the features space:

$$\dot{\mathbf{x}} = \begin{bmatrix} x_1^3 \\ \sqrt{3}x_1^2 x_2 \\ \sqrt{3}x_1 x_2^2 \\ x_2^3 \end{bmatrix} \quad \text{and} \quad \dot{\mathbf{y}} = \begin{bmatrix} y_1^3 \\ \sqrt{3}y_1^2 y_2 \\ \sqrt{3}y_1 y_2^2 \\ y_2^3 \end{bmatrix},$$

therefore, $< \dot{\mathbf{x}}, \dot{\mathbf{y}} >= (< \mathbf{x}, \mathbf{y} >)^3$. However, $< \dot{\mathbf{x}}, \dot{\mathbf{y}} >$ allows us to analyze the dimensions of the features space, which contains four axes as defined by those vectors. If the reader is interested in visualizing the impacts of this features space, we suggest him/her to plot every three possible axes and empirically check up the effects of this kernel on the input space.

By using the nonhomogeneous third-order kernel:

$$(< \mathbf{x}, \mathbf{y} > +1)^3 = (x_1 y_1 + x_2 y_2 + 1)^3$$

$$x_1^3 y_1^3 + 3x_1^2 y_1^2 x_2 y_2 + 3x_1^2 y_1^2 + 3x_1 y_1 x_2^2 y_2^2$$

$$+ 6x_1y_1x_2y_2 + 3x_1y_1 + x_2^3y_2^3 + 3x_2^2y_2^2 \qquad +3x_2y_2 + 1,$$

the dot product in the features space becomes an operation on vectors:

$$\dot{\mathbf{x}} = \begin{bmatrix} x_1^3 \\ \sqrt{3}x_1^2x_2 \\ \sqrt{3}x_1^2 \\ \sqrt{3}x_1x_2^2 \\ \sqrt{6}x_1x_2 \\ \sqrt{3}x_1 \\ x_2^3 \\ \sqrt{3}x_2^2 \\ \sqrt{3}x_2 \\ 1 \end{bmatrix} \qquad \dot{\mathbf{y}} = \begin{bmatrix} y_1^3 \\ \sqrt{3}y_1^2y_2 \\ \sqrt{3}y_1^2 \\ \sqrt{3}y_1y_2^2 \\ \sqrt{6}y_1y_2 \\ \sqrt{3}y_1 \\ y_2^3 \\ \sqrt{3}y_2^2 \\ \sqrt{3}y_2 \\ 1 \end{bmatrix}.$$

In this situation, the features space has ten dimensions instead of four. In addition, notice the four dimensions obtained through the homogeneous kernel are included in this nonhomogeneous space, as well as $\sqrt{3}y_1^2$, $\sqrt{6}y_1y_2$, $\sqrt{3}y_1$, $\sqrt{3}y_2^2$, $\sqrt{3}y_2$, and 1. Notice dimensions $\sqrt{3}y_1^2$, $\sqrt{6}y_1y_2$, and $\sqrt{3}y_2^2$ are associated with a second-order polynomial kernel, while dimensions $\sqrt{3}y_1$ and $\sqrt{3}y_2$ correspond to the linear kernel (just multiplied by a scalar). The last dimension, i.e., 1, comes from the nonhomogeneous constant $c$ but has no effect, since all vectors will have the same scalar value for it.

The effect of the nonhomogeneous third-order polynomial kernel is to bring information from all previous orders in terms of axes in the target features space. In that sense, one has access to the third order, but also to the second and the linear ones as result of such nonhomogeneous kernel. This happens whenever someone decides to employ a $k$th nonhomogeneous kernel, allowing the features space to bring information from all previous orders.

**Listing 6.16** Space provided by a two-class complex dataset: assessing different kernels

```
1   require(e1071)
2   require(tseriesChaos)
3
4   # Building the dataset
5   X1 = embedd(sin(2*pi*seq(0,9,len=1000)), m=2, d=27)
6   ids = which(X1[,1] < -0.5 & X1[,2] < -0.5)
7   X1 = X1[-ids,]
8
9   X2 = embedd(sin(2*pi*seq(0,9,len=1000)), m=2, d=27)
10  ids = which(X2[,1] > 0.5 & X2[,2] > 0.5)
11  X2 = X2[-ids,]
12  X2[,1]=X2[,1]+0.3
13  X2[,2]=X2[,2]-0.75
14
15  # Defining the class labels
16  X = rbind(X1, X2)
```

```
17  Y = c(rep(-1, nrow(X1)), rep(+1, nrow(X2)))
18
19  # Plotting the input space
20  plot(X, col=Y+2)
21
22  # Using a linear kernel
23  model1 = svm(x = X, y = as.factor(Y), scale=FALSE,
24              kernel="linear", cost=10, cross=10)
25  cat("Accuracy_with_a_linear_kernel:_", model1$tot.accuracy,
        "\n")
26
27  # Using a second-order polynomial kernel
28  model2 = svm(x = X, y = as.factor(Y), scale=FALSE,
29              kernel="polynomial", degree=2, coef0=0,
30              cost=10, cross=10)
31  cat("Accuracy_with_a_second-order_polynomial_kernel:_",
32              model2$tot.accuracy, "\n")
33
34  # Using a third-order polynomial kernel
35  model3 = svm(x = X, y = as.factor(Y), scale=FALSE,
36              kernel="polynomial", degree=3, coef0=0,
37              cost=10, cross=10)
38  cat("Accuracy_with_a_third-order_polynomial_kernel:_",
39              model3$tot.accuracy, "\n")
40
41  # Using a fourth-order polynomial kernel
42  model4 = svm(x = X, y = as.factor(Y), scale=FALSE,
43              kernel="polynomial", degree=4, coef0=0,
44              cost=10, cross=10)
45  cat("Accuracy_with_a_fourth-order_polynomial_kernel:_",
46              model4$tot.accuracy, "\n")
47
48  # Using a third-order polynomial kernel with coef0=1
49  model5 = svm(x = X, y = as.factor(Y), scale=FALSE,
50              kernel="polynomial", degree=3, coef0=1,
51              cost=10, cross=10)
52  cat("Accuracy_with_a_3rd-order_polynomial_kernel_coef0=1:_",
53              model5$tot.accuracy, "\n")
```

The output below confirms the third-order polynomial kernel has already provided fair accuracy results ($\approx 82$), however only after assessing its nonhomogeneous version ($\approx 99$) we notice the effects of adding the previous orders in the features space. This is a very useful strategy to assess some unknown input space, i.e., by increasing the polynomial order until classifiers reduce their accuracy, what happened for the fourth-order kernel. Then, we may attempt to use the previous orders, less than four, in nonhomogeneous kernel versions and assess the results. This brings us information about the inherent data complexity, or if one prefers, about the necessary axes/dimensions to make the original space linearly separable.

**Listing 6.17** Text output produced by Listing 6.16

```
Accuracy with a linear kernel:   69.082
```

```
Accuracy with a second-order polynomial kernel:  77.633
Accuracy with a third-order polynomial kernel:  82.049
Accuracy with a fourth-order polynomial kernel:  79.841
Accuracy with a third-order polynomial kernel coef0=1:
  99.660
```

As next scenario, we attempt to classify the input space illustrated in Fig. 6.13a, in which examples are displaced from the center, i.e., their average is not centered at the coordinate (0, 0). This brings up a simple but interesting analysis (Listing 6.18), from which we notice the second-order polynomial kernel is not adequate anymore. Note for example Fig. 6.13b.

**Listing 6.18** Space provided by a two-class and uncentered dataset: assessing different kernels

```
1  require(e1071)
```

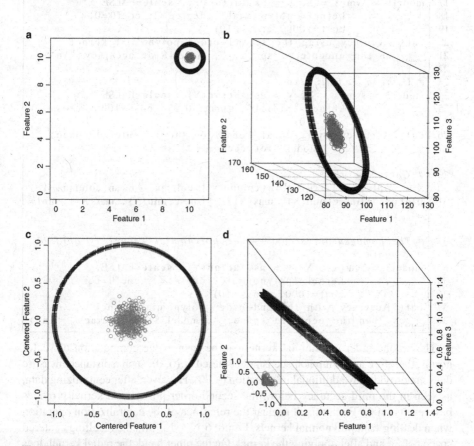

**Fig. 6.13** Space provided by a two-class and uncentered dataset. (**a**) Uncentered input space, (**b**) illustrates how the use of a polynomial kernel for uncentered data does not provide an adequate transformation, (**c**) input space after centering, (**d**) the features space after kernel transformation is now linearly separable

```
 2  require(tseriesChaos)
 3
 4  # Building the dataset
 5  X = cbind(rnorm(mean=10, sd=0.1, n=1000),
 6              rnorm(mean=10, sd=0.1, n=1000))
 7  X = rbind(X, embedd(sin(2*pi*seq(0,9,len=1027)), m=2, d=27)
            + 10)
 8
 9  # Defining the class labels
10  Y = c(rep(-1, 1000), rep(+1, 1000))
11
12  # Plotting the input space
13  par(mfrow=c(1,2))
14  plot(X, xlim=c(0, max(X)), ylim=c(0, max(X)), col=Y+2)
15
16  # Using an homogeneous second-order polynomial kernel
17  model1 = svm(x = X, y = as.factor(Y), scale=FALSE,
18              kernel="polynomial", degree=2, coef0=0,
19              cost=1000, cross=10)
20  cat("Accuracy using the second-order polynomial kernel
21      on the uncentered space: ", model1$tot.accuracy,"\n")
22
23  # Using a radial kernel
24  model2 = svm(x = X, y = as.factor(Y), scale=FALSE,
25              kernel="radial", gamma=0.25, cost=1000, cross
                =10)
26  cat("Accuracy with radial kernel on the uncentered space: ",
27                  model2$tot.accuracy, "\n")
28
29  # Centering the dataset
30  X = apply(X, 2, function(column) { column - mean(column) } )
31  plot(X, xlim=c(min(X), max(X)), ylim=c(min(X), max(X)), col=
        Y+2)
32
33  # The homogeneous second-order polynomial kernel is applied
        again
34  model3 = svm(x = X, y = as.factor(Y), scale=FALSE,
35              kernel="polynomial", degree=2, coef0=0,
36              cost=1000, cross=10)
37  cat("Accuracy with the 2nd-order polynomial kernel
38      on the centered space: ", model3$tot.accuracy, "\n")
```

The second-order polynomial kernel can be seen as the computation of vector norms, therefore as the input space is not centered at $(0, 0)$, such a distance measure does not provide any additional information ($\approx 90$). However, after centralizing data, as seen in the third accuracy below, the second-order polynomial kernel provides best results (100). From this, it is clear the relevance of the centralization operation when dealing with polynomial kernels. Figure 6.13c, d shows the resulting centered space before and after applying the kernel. On the other hand, the radial kernel does not require centralization, since its parameters allow it to be centered anywhere in the space, so that vector norms are locally computed. Besides the simplicity of this problem instance, it is very significative to compare polynomial kernels against the radial basis one: (1) second-order polynomial kernels and radials have the same usefulness when data is centered; and (2) polynomial kernels always measure vector norms in terms of the original input space, but assuming different factors according to the selected kernel.

**Listing 6.19** Text output produced by Listing 6.18

```
Accuracy  with  second−order  polynomial  kernel ,  uncentered
    space:  90.95
Accuracy  with  radial  kernel ,  uncentered  space:  100
Accuracy  with  second−order  polynomial  kernel ,  centered  space
    :  100
```

## 6.4.2  Kernels on a 3-Class Problem

The SVM classifier and kernel functions can be employed also in multi-class problems. In this section, we study the space with 3 classes illustrated in Fig. 6.14, in which two classes, similar to those presented in Fig. 6.12, are surrounded by a circularly shaped third class. This is interesting because SVM tackles multi-class problems often using a one-versus-all approach. Listing 6.20 presents the source code to generate the data and carry out experiments with this problem, whose results are shown in Listing 6.21. In this case, a homogeneous second-order polynomial kernel is also not adequate, but non-homogeneous polynomial kernels seem to improve results (see Fig. 6.15 for some examples of 3d spaces using the features space), in particular the third-order non-homogeneous polynomial kernel is the best fit. The radial basis kernel is also adequate in this scenario.

**Listing 6.20** Space with 3 classes: assessing different kernels

```
1   require(e1071)
2   require(tseriesChaos)
3   x = seq(−1.0,1.0,len=1000)
4
5   # class 1
6   X1 = 1*cos(x*2*pi)
7   X1 = cbind(X1, 1.1*sin(x*2*pi))
8   X1 = X1[ −which((X1[,1] < −0.5) & (X1[,2] < 0.5)), ]
9   X1 = X1 + cbind(rep(0.1,nrow(X1)), rep(0.4,nrow(X1)))
10  X1 = X1 + rnorm(mean=0,sd=0.1,n=nrow(X1))
11
12  # class 2
13  X2 = 1.1*cos(x*2*pi)
14  X2 = cbind(X2, 1*sin(x*2*pi))
15  X2 = X2[ −which((X2[,1] > 0.6) & (X2[,2] < 0.4)), ]
16  X2 = X2 − 0.4
17  X2 = X2 + rnorm(mean=0,sd=0.1,n=nrow(X2))
18
```

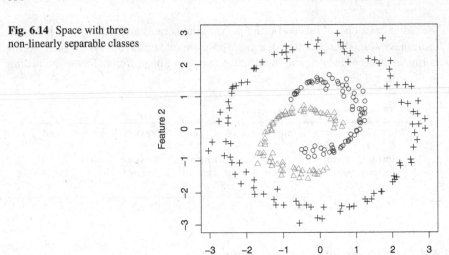

**Fig. 6.14** Space with three non-linearly separable classes

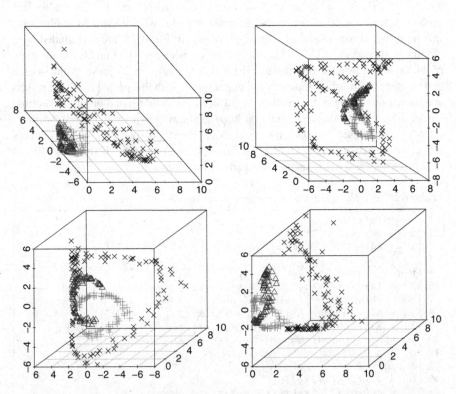

**Fig. 6.15** 3d projections combining every distinct triple of features produced after the second-order nonhomogeneous polynomial kernel with coef0=2. The top-left figure shows the same resultant space as a second-order homogeneous kernel. The additional figures, illustrate the other subspaces contained in the nonhomogeneous function

```
19  # class 3
20  X3 = 2.6*cos(x*2*pi)
21  X3 = cbind(X3, 2.6*sin(x*2*pi))
22  X3 = X3 + rnorm(mean=0,sd=0.2,n=nrow(X3))
23
24  # Defining the class labels
25  X = rbind(X1, X2, X3)
26  Y = c(rep(0, nrow(X1)), rep(1, nrow(X2)), rep(2, nrow(X3)))
27
28  # Using a linear kernel
29  model1 = svm(x = X, y = as.factor(Y), scale=FALSE,
30              kernel="linear", cost=10, cross=10)
31  cat("Accuracy_with_linear_kernel:_", model1$tot.accuracy, "\
        n")
32
33  # Using a second-order polynomial kernel
34  model2 = svm(x = X, y = as.factor(Y), scale=FALSE,
35              kernel="polynomial", degree=2, coef0=0,
36              cost=10, cross=10)
37  cat("Accuracy_with_2nd-order_polynomial_kernel:_",
38              model2$tot.accuracy, "\n")
39
40  # Using a third-order polynomial kernel
41  model3 = svm(x = X, y = as.factor(Y), scale=FALSE,
42              kernel="polynomial", degree=3, coef0=0,
43              cost=10, cross=10)
44  cat("Accuracy_with_3rd-order_polynomial_kernel:_",
45              model3$tot.accuracy, "\n")
46
47  # Using a fourth-order polynomial kernel
48  model4 = svm(x = X, y = as.factor(Y), scale=FALSE,
49              kernel="polynomial", degree=4, coef0=0,
50              cost=10, cross=10)
51  cat("Accuracy_with_4th-order_polynomial_kernel:_",
52              model4$tot.accuracy, "\n")
53
54  # Using a radial kernel
55  model5 = svm(x = X, y = as.factor(Y), scale=FALSE,
56              kernel="radial", gamma=0.5, cost=10, cross=10)
57  cat("Accuracy_with_radial_kernel,_gamma=0.5:_",
58              model5$tot.accuracy, "\n")
59
60  # Using a 2nd-order polynomial kernel with coef0=2
61  model6 = svm(x = X, y = as.factor(Y), scale=FALSE,
62              kernel="polynomial", degree=2, coef0=2,
63              cost=10, cross=10)
64  cat("Accuracy_with_2nd-order_polynomial_kernel,_coef0=2:_",
65              model6$tot.accuracy, "\n")
66
67  # Using a 3rd-order polynomial kernel with coef0=1
68  model7 = svm(x = X, y = as.factor(Y), scale=FALSE,
69              kernel="polynomial", degree=3, coef0=1,
70              cost=10, cross=10)
71  cat("Accuracy_with_3rd-order_polynomial_kernel,_coef0=1:_",
```

```
72  |                     model7$tot.accuracy , "\n")
```

**Listing 6.21** Text output produced by Listing 6.20

```
Accuracy with linear kernel              :  40.5842
Accuracy with 2nd-order polynomial kernel:  74.9703
Accuracy with 3rd-order polynomial kernel:  39.4788
Accuracy with 4th-order polynomial kernel:  79.6683
Accuracy with radial kernel , gamma=0.5  :  99.8026
Accuracy with 2nd-order polynomial kernel , coef0=2:  92.8148
Accuracy with 3rd-order polynomial kernel , coef0=1:  99.8420
```

Observe the usefulness of assessing multiple kernels and their parameters to understand the input space. A grid search approach supports such assessment, however it is much more important to interpret the most adequate settings found because they provide relevant information on the features embedding.

### 6.4.3  Studying the Data Spaces in an Empirical Fashion

The same input space illustrated in Fig. 6.12 is considered in this section, so we can explore an empirical approach to assess different kernels and analyze data complexity. Listing 6.22 is used to analyze polynomial kernels under different orders, which are parametrized using five values for term $c$ in Eq. (6.7), from an homogeneous to four other nonhomogeneous kernels.

**Listing 6.22** Assessing polynomial kernels

```
 1  require(e1071)
 2  require(tseriesChaos)
 3
 4  # Building the dataset
 5  X1 = embedd(sin(2*pi*seq(0,9,len=1000)), m=2, d=27)
 6  ids = which(X1[,1] < -0.5 & X1[,2] < -0.5)
 7  X1 = X1[-ids ,]
 8
 9  X2 = embedd(sin(2*pi*seq(0,9,len=1000)), m=2, d=27)
10  ids = which(X2[,1] > 0.5 & X2[,2] > 0.5)
11  X2 = X2[-ids ,]
12  X2[,1]=X2[,1]+0.3
13  X2[,2]=X2[,2]-0.75
14
15  X = rbind(X1, X2)
16
17  # Defining the class labels
18  Y = c(rep(-1, nrow(X1)), rep(+1, nrow(X2)))
19
20  # Plotting the input space
21  plot(X, col=Y+2)
```

```
22
23   # Assessing several polynomial kernels
24   results = matrix(0, nrow=7, ncol=5)
25   coeffs = seq(0, 1, length=5)
26   for (d in 1:7) {
27           cat("Running for degree ", d, "\n")
28           col = 1
29           for (c in coeffs) {
30               results[d, col] = svm(x=X, y=as.factor(Y),
31                    scale=FALSE, kernel="polynomial",
32                    degree=d, coef0=c,
33                    cost=10, cross=10)$tot.accuracy
34                    col = col + 1
35           }
36   }
37
38   column.names = c()
39   for (c in coeffs) {
40       column.names = c(column.names, paste("coef=", c, sep="")
             )
41   }
42
43   order.names = c()
44   for (d in 1:7) {
45       order.names = c(order.names, paste("order=", d, " :",
             sep=""))
46   }
47
48   results = as.data.frame(results)
49   colnames(results) = column.names
50   rownames(results) = order.names
51
52   # Printing out the accuracies
53   print(results)
```

The outputs below confirm the usefulness of this basic empirical approach. Observe the tenfold cross validation strategy informs which kernels are better to address this particular classification task. We suggest the reader to expand this source code and make it evaluate other kernels and general enough to tackle any input space. The greater the polynomial order needed, the more complex the original data space is. In addition, term $c$ (indicated as coef in the results below) confirms the need for considering previous polynomial orders along the features space, as previously discussed. Observe the three-order nonhomogeneous polynomial kernel is already a good option ($\approx$99), but of course the fourth-order provides the best result (100) without adding unnecessary dimensions to the features space.

**Listing 6.23** Text output produced by Listing 6.22

|         | coef=0   | coef=0.25 | coef=0.5 | coef=0.75 | coef=1   |
|---------|----------|-----------|----------|-----------|----------|
| order=1 : | 69.25255 | 69.25255  | 69.13930 | 69.25255  | 69.25255 |
| order=2 : | 78.02945 | 80.35108  | 80.01133 | 79.89807  | 79.84145 |
| order=3 : | 81.59683 | 98.98075  | 99.49037 | 99.60362  | 99.83012 |

```
order=4 :  80.06795   98.75425  100.00000  100.00000  100.00000
order=5 :  83.57871   99.49037  100.00000  100.00000  100.00000
order=6 :  81.59683  100.00000  100.00000  100.00000  100.00000
order=7 :  83.12571  100.00000  100.00000  100.00000  100.00000
```

To finish this section, consider Listing 6.24 which makes the same as Listing 6.22, but using function **tune.svm** from the R package e1071. Such a function receives a list of parameters and produces the performance after a tenfold cross validation for all possible combinations, reflecting a grid-search strategy.

**Listing 6.24** Assessing polynomial kernels through function tune

```
1   require(e1071)
2   require(tseriesChaos)
3
4   # Building the dataset
5   X1 = embedd(sin(2*pi*seq(0,9,len=1000)), m=2, d=27)
6   ids = which(X1[,1] < -0.5 & X1[,2] < -0.5)
7   X1 = X1[-ids,]
8
9   X2 = embedd(sin(2*pi*seq(0,9,len=1000)), m=2, d=27)
10  ids = which(X2[,1] > 0.5 & X2[,2] > 0.5)
11  X2 = X2[-ids,]
12  X2[,1]=X2[,1]+0.3
13  X2[,2]=X2[,2]-0.75
14
15  X = rbind(X1, X2)
16
17  # Defining the class labels
18  Y = c(rep(-1, nrow(X1)), rep(+1, nrow(X2)))
19
20  # Plotting the input space
21  plot(X, col=Y+2)
22
23  # Using tune.svm to study kernels
24  model = tune.svm(x = X, y = as.factor(Y),
25                   kernel="polynomial", degree=1:7,
26                   coef0=seq(0, 1, length=5), cost=10)
27
28  # Printing out the results in terms of errors
29  print(model$performances)
```

The output below shows the summary of results provided with **tune.svm** for different polynomial kernel orders degree, its nonhomogeneous constant value coef and the SVM cost. Note the output is in terms of average error and its variance, instead of accuracy, e.g. the first scenario provides an error equal to $0.3125257$, therefore the accuracy is $1 - 0.3125257 = 0.6874743$. We conclude the third-order nonhomogeneous polynomial kernel provides very good results while being less complex than the ones at greater orders.

**Listing 6.25** Text output produced by Listing 6.24

| | degree | coef0 | cost | error | dispersion |
|---|---|---|---|---|---|
| 1 | 1 | 0.00 | 10 | 0.3125257 | 0.039928485 |
| 2 | 2 | 0.00 | 10 | 0.5481510 | 0.033711649 |
| 3 | 3 | 0.00 | 10 | 0.3063014 | 0.036855644 |
| 4 | 4 | 0.00 | 10 | 0.5022470 | 0.031115611 |
| 5 | 5 | 0.00 | 10 | 0.2564715 | 0.052094297 |
| 6 | 6 | 0.00 | 10 | 0.5022310 | 0.031628824 |
| 7 | 7 | 0.00 | 10 | 0.1981574 | 0.064456709 |
| 8 | 1 | 0.25 | 10 | 0.3125257 | 0.039928485 |
| 9 | 2 | 0.25 | 10 | 0.2015826 | 0.031714991 |
| 10 | 3 | 0.25 | 10 | 0.0000000 | 0.000000000 |
| 11 | 4 | 0.25 | 10 | 0.0118933 | 0.004188703 |
| 12 | 5 | 0.25 | 10 | 0.0000000 | 0.000000000 |
| 13 | 6 | 0.25 | 10 | 0.0000000 | 0.000000000 |
| 14 | 7 | 0.25 | 10 | 0.0000000 | 0.000000000 |
| 15 | 1 | 0.50 | 10 | 0.3125257 | 0.039928485 |
| 16 | 2 | 0.50 | 10 | 0.2015826 | 0.031714991 |
| 17 | 3 | 0.50 | 10 | 0.0000000 | 0.000000000 |
| 18 | 4 | 0.50 | 10 | 0.0000000 | 0.000000000 |
| 19 | 5 | 0.50 | 10 | 0.0000000 | 0.000000000 |
| 20 | 6 | 0.50 | 10 | 0.0000000 | 0.000000000 |
| 21 | 7 | 0.50 | 10 | 0.0000000 | 0.000000000 |
| 22 | 1 | 0.75 | 10 | 0.3125257 | 0.039928485 |
| 23 | 2 | 0.75 | 10 | 0.2015826 | 0.031714991 |
| 24 | 3 | 0.75 | 10 | 0.0000000 | 0.000000000 |
| 25 | 4 | 0.75 | 10 | 0.0000000 | 0.000000000 |
| 26 | 5 | 0.75 | 10 | 0.0000000 | 0.000000000 |
| 27 | 6 | 0.75 | 10 | 0.0000000 | 0.000000000 |
| 28 | 7 | 0.75 | 10 | 0.0000000 | 0.000000000 |
| 29 | 1 | 1.00 | 10 | 0.3125257 | 0.039928485 |
| 30 | 2 | 1.00 | 10 | 0.2015826 | 0.031714991 |
| 31 | 3 | 1.00 | 10 | 0.0000000 | 0.000000000 |
| 32 | 4 | 1.00 | 10 | 0.0000000 | 0.000000000 |
| 33 | 5 | 1.00 | 10 | 0.0000000 | 0.000000000 |
| 34 | 6 | 1.00 | 10 | 0.0000000 | 0.000000000 |
| 35 | 7 | 1.00 | 10 | 0.0000000 | 0.000000000 |

### 6.4.4 Additional Notes on Kernels

We here list some notes on SVM kernels. The sigmoidal kernel (Sect. 6.1.3) was not exemplified in the exploratory analysis since it produces the same classifier as the linear kernel. In practice, it multiplies the inner product by some value $-\kappa$, has a nonhomogeneity term $c$ and also applies a hyperbolic tangent after computing the main operation (Eq. (6.5)), whose outputs are similar as for the linear kernel but rescaled in range $[-1, 1]$ in a similar manner as the sigmoid function for the Perceptron and Multilayer Perceptron, both discussed in Sect. 1.5.

In addition, we also make an important note to motivate the reader to design his/her kernels using some basic principles. Given some input space, at first the reader should simply apply the linear kernel and evaluate results, as provided in Sect. 6.4.3. If no good result is obtained, we suggest to try handcrafting the features space. For example, consider the reader is interested in some sort of text classification task, in which word frequencies or some other measurement is available. Consider the interest in representing the word "machine" coming right before "learning". In that situation, one could use the frequency of both words together as another space axis.

In addition, one could also attempt the multiplication of some attribute in the input space by another, in attempt to represent some sort of correlation between them. One attribute could also be powered to some order to make it more or less relevant (order may assume real values less than 1). All those comments do not come without considering PCA or any other feature selection approach (Sect. 6.2). Note that this handcraft process does not explicitly use a typical kernel, but in fact this can be seen as transforming the input space and adding new dimensions, which also happens when applying kernels. The next section brings a very important discussion that influences in the kernel design.

## 6.5   SVM Kernel Trick

The SVM dual optimization is used to solve a binary classification problem considering data is linearly separable:

$$\max \quad -\frac{1}{2} \sum_{i=1}^{m} \sum_{j=1}^{m} \alpha_i \alpha_j y_i y_j < \mathbf{x}_i, \mathbf{x}_j > + \sum_{i=1}^{m} \alpha_i.$$

In order to deal with input data that it is not linearly separable (see Sect. 6.1.1), we need some function $\phi(.)$ to transform the input data into a more convenient features space. Instead of explicitly applying such transformation, it is common to design a kernel function to provide the *dot product of vectors in the features space*. This is used to avoid additional computational costs as well as to implicitly construct the features space in a general purpose manner, e.g., such as for polynomial kernels that basically add up some order.

The dot product itself, i.e., $k(x_i, x_j) = \langle \mathbf{x}_i, \mathbf{x}_j \rangle$ can be seen as a way to obtain an angular similarity value between vectors, for example, it outputs 0 if they are orthogonal to each other. But notice that this can also be represented as the dot product of two functions $\phi(\mathbf{x}_i)$ and $\phi(\mathbf{x}_j)$, i.e.:

$$k(x_i, x_j) = \phi(\mathbf{x}_i) \cdot \phi(\mathbf{x}_j)$$

In this way, as long as there is some higher dimensional space in which $\phi(.)$ is just the dot product of that higher dimensional space, the kernel can be written as a dot product and used directly into the optimization problem in form:

$$\max \quad -\frac{1}{2} \sum_{i=1}^{m} \sum_{j=1}^{m} \alpha_i \alpha_j y_i y_j k(\mathbf{x}_i, \mathbf{x}_j) + \sum_{i=1}^{m} \alpha_i,$$

All places in which dot products occur are then replaced by the kernel function. According to the properties of dot products, we need to have a kernel function such that $k(x_i, x_j) = k(x_j, x_i)$ and $k(x_i, x_j) \geq 0$, what is the same as requiring its kernelized matrix form to be positive semi-definite.

To exemplify, let the SVM problem in its matrix dual form:

$$\underset{\alpha}{\text{minimize}} \quad \mathbf{e}^T \alpha - \frac{1}{2} \alpha^T \mathbf{Q} \alpha$$

$$\text{subject to} \quad \mathbf{0} \leq \alpha \leq \mathbf{C}$$

$$\mathbf{y}^T \alpha = 0.$$

The important term to study in this case is $-\alpha^T \mathbf{Q} \alpha$, which is expected to form a concave objective function for the maximization problem. Similarly, in the primal form $\alpha^T \mathbf{Q} \alpha$, this term must form a convex objective function to be minimized. When using kernels, we must know matrix $\mathbf{Q}$, subject to $\mathbf{Q}_i, j = k(x_i, x_j)$, to confirm if those optimization problems have solutions. Such a matrix is only valid if it is Gramian, which is positive semi-definite and symmetric.

As an example, consider matrix $\mathbf{Q}$ obtained after a $p$-order polynomial kernel:

$$\mathbf{Q} = (YY^T)(c + XX^T)^p,$$

in which $Y$ represents the vector of classes, $X$ corresponds to the training examples already in some vectorial space (vectors along rows), $c$ is the coefficient to control the kernel homogeneity (it is homogeneous when $c = 0$, otherwise it is nonhomogeneous), and $p$ is the polynomial order. To proceed with further analysis, let $c = 1$:

$$\mathbf{Q} = (1 + XX^T)^p.$$

In order to show there is a mapping from the input space to some higher dimensional space so that $k(x_i, x_j) = \langle \phi(x_i), \phi(x_j) \rangle$, assume that $x_{i,0} = x_{j,0} = 1$, and explicitly separate each dot product of the power, writing it as a sum of the product of monomials (a polynomial with a single term), so that:

$$(1 + XX^T)^p = (1 + XX^T) \cdots (1 + XX^T)$$

$$= \left( \sum_{k=0}^{n} x_{i,k} x_{j,k} \right) \cdots \left( \sum_{k=0}^{n} x_{i,k} x_{j,k} \right)$$

$$= \sum_{K \in \{0,1,\cdots,n\}^p} \prod_{l=1}^{p} x_{i,K_l} x_{j,K_l}$$

$$= \sum_{K \in \{0,1,\cdots,n\}^p} \prod_{l=1}^{p} x_{i,K_l} \prod_{l=1}^{p} x_{j,K_l}.$$

In this case, the map is $\phi : \mathbb{R}^n \to \mathbb{R}^{(n+1)^p}$. If there is $\phi(\mathbf{x}) = \prod_{l=1}^{p} x_{K_l}$, for $K \in \{0, 1, \cdots, n\}^p$, we can say that the separating hyperplane in the range of $\phi(.)$ corresponds to a polynomial curve of degree $p$ in the original space. This is because function $\phi(.)$ contains all the single term polynomials up to the $p$th degree. Therefore, it is true that:

$$k(x_i, x_j) = \langle \phi(x_i), \phi(x_j) \rangle.$$

This result shows a practical example of the kernel trick: by using a polynomial kernel in the SVM optimization problem, we are in fact learning a predictor of $p$th degree over the input space. However, as mentioned before, in order to guarantee that the optimization is possible even when using kernels, $\mathbf{Q}$ must be a positive semi-definite matrix, so that $\alpha^T \mathbf{Q} \alpha \geq 0$. Let us introduce two simple examples, first:

$$\mathbf{Q} = \begin{bmatrix} 1 & 0 \\ 0 & 1 \end{bmatrix},$$

which is evaluated for every possible $\alpha \in \mathbf{R}^2$:

$$\begin{bmatrix} \alpha_1 & \alpha_2 \end{bmatrix} \begin{bmatrix} 1 & 0 \\ 0 & 1 \end{bmatrix} \begin{bmatrix} \alpha_1 \\ \alpha_2 \end{bmatrix} = \begin{bmatrix} 1\alpha_1 + 0\alpha_2 & 0\alpha_1 + 1\alpha_2 \end{bmatrix} \begin{bmatrix} \alpha_1 \\ \alpha_2 \end{bmatrix}$$

$$= \begin{bmatrix} \alpha_1 & \alpha_2 \end{bmatrix} \begin{bmatrix} \alpha_1 \\ \alpha_2 \end{bmatrix} = \alpha_1^2 + \alpha_2^2.$$

By plotting this function, we notice the surface is convex. Thus, considering a second example:

$$\mathbf{Q} = \begin{bmatrix} -1 & 0 \\ 0 & 1 \end{bmatrix},$$

and solving for $\mathbf{Q}$:

$$\begin{bmatrix} \alpha_1 & \alpha_2 \end{bmatrix} \begin{bmatrix} -1 & 0 \\ 0 & 1 \end{bmatrix} \begin{bmatrix} \alpha_1 \\ \alpha_2 \end{bmatrix} = \begin{bmatrix} -\alpha_1 & \alpha_2 \end{bmatrix} \begin{bmatrix} \alpha_1 \\ \alpha_2 \end{bmatrix} = -\alpha_1^2 + \alpha_2^2.$$

A saddle is obtained by plotting this function, what makes it unfeasible to proceed with the SVM optimization (no stationary solution would be obtained— no maximum and no minimum). The first matrix is positive semi-definite, while the second is not. However, it is common to have more complex matrices in practical scenarios. For example, a small matrix could be given by:

$$Q = \begin{bmatrix} 2.3898 & 2.8320 & 2.6503 \\ 0.9126 & 3.4200 & 2.2544 \\ 3.1522 & 3.8803 & 1.8666 \end{bmatrix}.$$

One way to investigate if this is a positive semi-definite matrix is by computing its eigenvalues. They represent the direction and magnitude of the eigenvectors and when no eigenvalue is negative, there is no reflection performed by the linear transformation. This is a useful tool because when linear transformations do not include the reflection operation, they correspond to positive semi-definite matrices. For this matrix, the eigenvalues are 7.5990, 1.1593 and −1.0822, so that it is not positive semi-definite, and therefore cannot be used in the context of SVM optimization.

As a more practical example, let us define a synthetical dataset using the following R code:

**Listing 6.26**  Practical example using a synthetical dataset

```
1  x = cbind(rnorm(mean=0, sd=1, n=500), rnorm(mean=0, sd=1, n
       =500))
2  x = rbind(x, cbind(rnorm(mean=10, sd=1, n=500), rnorm(mean
       =10, sd=1, n=500)))
3  y = c(rep(0,500), rep(1,500))
4  Q = (y%*%t(y))*(x%*%t(x))
```

Our matrix is symmetric, what guarantees that the eigenvalues will be real numbers, otherwise those may assume complex values. By computing the eigenvalues of the matrix, we observe some slightly negative values due to precision error in the numerical computation. Assuming symmetry, we can confirm if this is a valid matrix via the Cholesky decomposition which produces: $A = LL^T$, for some matrix $A$ containing real values. Notice that the Cholesky decomposition can only be used for positive definite matrices. Give $Q$ is positive semi-definite, one trick to solve this is by summing up a small $\epsilon$ value to every term on its diagonal in order to allow the decomposition.

The following R code performs this addition using $\epsilon = 1^{-10}$, and then it computes the divergence between the original matrix $Q$ and the new one obtained after the Cholesky decomposition:

**Listing 6.27**  Adding a constant to the diagonal to proceed with the Cholesky decomposition

```
1  Q = (y%*%t(y))*(x%*%t(x))
2  Qeps = Q
```

```
3   epsilon = 1e-10
4   for (i in 1:nrow(Q)) { Qeps[i,i] = Qeps[i,i] + epsilon; }
5   ch = chol(Qeps)
6   total_divergence = sum((Q - t(ch)%*%ch)^2)
7   newQ = t(ch) %*% ch
```

Now we have indeed a positive definite matrix (defined in the code as newQ) that can be used in the SVM optimization problem. In fact, even if we try to use the original matrix, SVM would be capable of reaching some feasible solution once the negative eigenvalues are numerically close to zero, indicating our assumption can be relaxed to consider the matrix is semi-definite, which in fact it is. However, this trick should be used with caution because we are changing matrix $\mathbf{Q}$ and therefore the original problem to some approximation. The greater $\epsilon$ is, the more relaxed is the approximation, which might become too dissimilar to the original problem.

In summary, we have now a general method to assess convexity and confirm there is some global solution under theoretical learning guarantees. In addition, kernels provide a tool to represent data in another space, for which the SVM optimization problem holds.

## 6.6    A Quick Note on the Mercer's Theorem

The conditions for a valid matrix in the context of the SVM optimization problem is due to Mercer's theorem [1]. It is analogue to the study of the eigenvalues to ensure matrices that are positive semi-definite, but also evaluates kernels in an infinite-dimensional space.

In particular, it states that the following is a sufficient and necessary condition for a valid kernel function: for all $x_1, \cdots, x_n$, matrix $\mathbf{Q}_{i,j} = k(x_i, x_j)$ is a positive semi-definite matrix (a.k.a. a Gramian matrix), i.e., a symmetric function $K : X \times X \to \mathbb{R}$ implements a dot product on some Hilbert space if and only if it is positive semi-definite.

The proof in two parts basically demonstrates, first, that if a kernel implements a dot product in some Hilbert space then a Gram matrix is obtained. It is mandatory that function $\phi(.)$ maps examples to a Hilbert space in which the dot product is defined.[2] In practice, it is then possible to verify how similar a vector is to others, and make projections in order to build separating hyperplanes, both crucial operations for classification problems.

Let $M$ be a linear subspace of a Hilbert space, then every $\mathbf{x}$ in this subspace can be written in form $\mathbf{x} = \mathbf{u} + \mathbf{v}$, with $\mathbf{u} \in M$ and $\langle \mathbf{v}, \mathbf{w} \rangle = 0$, for all $\mathbf{w} \in M$. Therefore, this first part is trivial.

---

[2]Recall a Hilbert space is simply a vector space for which: a distance metric, a vector norm and a dot product are defined. Because of that, it supports projection of vectors into spaces, and therefore geometry.

The second part of the proof involves the definition of a space of functions over $X$ as $\mathbb{R}^X = \{f : X \to \mathbb{R}\}$. Then, let function $\phi(x) : x \to k(., x)$ for each $x \in X$, a valid dot product on this vector space is:

$$\left\langle \sum_i \alpha_i k(., x), \sum_j \beta_j k(., x_j) \right\rangle = \sum_{i,j} \alpha_i \beta_j k(x, x_j),$$

because it is symmetric (as $k(.)$ is also symmetric), linear and it is positive definite since $k(x, x_j) \geq 0$. It shows, to conclude the proof, that:

$$\langle \phi(\mathbf{x}), \phi(\mathbf{x}_j) \rangle = \langle k(., \mathbf{x}), k(\mathbf{x}_j, .) \rangle = k(x, x_j)$$

## 6.7   Concluding Remarks

This chapter defined and discussed about the most common SVM kernels employed in the literature. Afterwards, we introduce the Principal Component Analysis and the Kernel Principal Component Analysis, as well as an exploratory data analysis, including a discussion on kernels and features spaces. Finally, the SVM Kernel trick and the Mercer's theorem are presented.

## 6.8   List of Exercises

1. Considering the MNIST database,[3] employ different kernels on its input space and then use SVM as classification algorithm. Assess the final results using the tenfold cross validation strategy;
2. Evaluate different SVM kernels using the Columbia University Image Library available at http://www.cs.columbia.edu/CAVE/software/softlib/coil-100.php;
3. Evaluate different SVM kernels using the CIFAR-10 dataset available at http://www.cs.utoronto.ca/~kriz/cifar.html;
4. Study Deep Learning algorithms and use them on the same problems listed before. Compare the overall results;
5. All those datasets are commonly employed to justify and assess Deep Learning algorithms. What can you conclude about the effects of a kernel transformation versus Deep Learning techniques?

---

[3]Please, refer to the MNIST database http://yann.lecun.com/exdb/mnist.

# Reference

1. R. Courant, D. Hilbert, *Methods of Mathematical Physics*. Methods of Mathematical Physics, vol. 2 (Interscience Publishers, Geneva, 1962)